COMPOST UTILIZATION in HORTICULTURAL CROPPING SYSTEMS

COMPOST UTILIZATION in HORTICULTURAL CROPPING SYSTEMS

Edited by
Peter J. Stoffella
Brian A. Kahn

CRC Press
Taylor & Francis Group
Boca Raton London New York

CRC Press is an imprint of the
Taylor & Francis Group, an **informa** business

CRC Press
Taylor & Francis Group
6000 Broken Sound Parkway NW, Suite 300
Boca Raton, FL 33487-2742

First issued in paperback 2019

ISBN-13: 978-1-56670-460-1 (hbk)
ISBN-13: 978-0-367-39759-3 (pbk)

Visit the Taylor & Francis Web site at
http://www.taylorandfrancis.com

and the CRC Press Web site at
http://www.crcpress.com

Library of Congress Cataloging-in-Publication Data

Compost utilization in horticultural cropping systems / edited by Peter J. Stoffella and Brian A. Kahn

p. cm.
Includes bibliographical references (p).
ISBN 1-56670-460-X (alk. paper)
1. Compost. 2. Horticulture. I. Stoffella, Peter J. II. Kahn, Brian A.

S661.C66 2000
635'.04895—dc21 00-046350

Library of Congress Card Number 00-46350

Preface

Compost production is increasing in the U.S. and throughout the world. Production methods vary from simple, inexpensive, static piles to scientifically computerized in-vessel operations. Traditionally local and regional municipalities were the primary operators of compost facilities. However, with new federal, state, and local government regulations prohibiting disposal of certain biologically degradable materials into landfills, and with the increased commercial demands for composts, the number of private composting facilities has increased during the past decade. Feedstocks, such as yard wastes, food scraps, wood chips, and municipal solid waste (MSW), and combinations of feedstocks have varied between compost operational facilities, depending on the local availability of biodegradable waste material. Several compost facilities mix feedstocks with treated sewage sludge (biosolids) as an inexpensive method to combine biosolids disposal with production of a plant-nutrient-enhanced compost. Innovative compost production methods have resulted in an expansion of operational facilities, which have generated a greater quantity of agricultural grade compost at an economical cost to agricultural users.

With the increased interest in and demand for compost from commercial horticultural industries throughout the world, a significant body of scientific information has been published in professional and trade outlets. The intent of this book is to provide a compilation of knowledge on the utilization of compost in various commercial horticultural enterprises at the dawn of a new millenium.

The major emphasis of the book is to provide a comprehensive review on the utilization of compost in horticultural cropping systems. However, we also felt it was important to include reviews of commercial compost production systems; the biological, chemical, and physical processes that occur during composting; and the attributes and parameters associated with measuring compost quality. A compilation of scientific information on compost utilization in vegetable, fruit, ornamental, nursery, and turf crop production systems is provided, as well as information on compost use in landscape management and vegetable transplant production. Benefits of compost utilization, such as soil-borne plant pathogen suppression, biological weed control, and plant nutrient availability, are reviewed in separate chapters. The economic implications of compost utilization in horticultural cropping systems are also included.

Although there are many good reasons to utilize compost in horticultural cropping systems, potential hazards such as heavy metals, human pathogens, odors, and phytotoxicity exist. These are particularly of concern to the public when biosolids are blended with various feedstocks. The U.S. and other countries introduced regulations on compost production, testing, and transportation in an attempt to provide a safe product to the horticultural consumer. Therefore, chapters are included to cover potential hazards, precautions, and regulations governing the production and utilization of compost.

This book is intended to encourage compost utilization in commercial horticultural enterprises. We attempted to have highly qualified scientists compile current scientific and research information within their areas of expertise. We hope that the knowledge gained from this book will generate an abundance of interest in compost utilization in horticulture among students, scientists, compost producers, and horticultural practitioners.

Peter J. Stoffella

Brian A. Kahn

The Editors

Dr. Peter J. Stoffella is a Professor of Horticulture at the Indian River Research and Education Center, Institute of Food and Agricultural Science, University of Florida, Fort Pierce, Florida. He has been employed with the University of Florida since 1980. Dr. Stoffella received a B.S. degree in Horticulture from Delaware Valley College of Science and Agriculture (1976), a M.S. in Horticulture from Kansas State University (1977), and a Ph.D. degree in Vegetable Crops from Cornell University (1980). He is an active member of several horticultural societies. Among his horticultural research interests, he established a research program on developing optimum compost utilization practices in commercial horticultural cropping systems. Specifically, he has interests in composts as biological weed controls, composts as peat substitutes for media used in transplant production systems, and composts as partial inorganic nutrient substitutes in field grown vegetable crop production systems. Recently, he developed a cooperative research program on utilization of compost in a vegetable cropping system as a mechanism of reducing nutrient leaching into ground water.

Dr. Brian A. Kahn is a Professor of Horticulture in the Department of Horticulture and Landscape Architecture, Oklahoma State University, Stillwater, Oklahoma. He has been at Oklahoma State since 1982, with a 75% research–25% teaching appointment. Dr. Kahn received a B.S. degree in Horticulture from Delaware Valley College of Science and Agriculture (1976), and M.S. (1979) and Ph.D. (1982) degrees in Vegetable Crops from Cornell University. He conducts research focused on sustainable cultural and management practices for improved yields and quality of vegetables. Dr. Kahn has served the American Society for Horticultural Science as an Associate Editor and as a member of the Publications Committee. His previous collaborations with Dr. Stoffella included a national symposium on root systems of vegetable crops, and 18 professional publications.

Contributors

Ron Alexander
R. Alexander Associates, Inc.
1212 Eastham Drive
Apex, North Carolina 27502
USA

Thomas G. Allen
University of Maine
Department of Resource Economics and Policy
5782 Winslow Hall
Orono, Maine 04469
USA

J. Scott Angle
University of Maryland
Symons Hall
College Park, Maryland 20742
USA

Allen V. Barker
University of Massachusetts
Department of Plant and Soil Sciences
Amherst, Massachusetts 01003
USA

Sally L. Brown
University of Washington
School of Forest Sciences (AR-10)
Seattle, Washington 98195
USA

David V. Calvert
University of Florida, IFAS
Indian River Research and Education Center
2199 South Rock Road
Fort Pierce, Florida 34945
USA

Rufus L. Chaney
United States Department of Agriculture
Agriculture Research Service
Environmental Chemistry Laboratory
Building 007, BARC-West
Beltsville, Maryland 20705
USA

George Criner
University of Maine
Department of Resource Economics and Policy
5782 Winslow Hall
Orono, Maine 04469
USA

Michael Day
Institute for Chemical Process and Environmental Technology
National Research Council of Canada
1500 Montreal Road, Room 119
Ottawa, Ontario K1AOR6
Canada

Eliot Epstein
E&A Environmental Consultants, Inc.
95 Washington Street, Suite 218
Canton, Massachusetts 02021
USA

George E. Fitzpatrick
University of Florida, IFAS
Fort Lauderdale Research and Education Center
3205 College Avenue
Fort Lauderdale, Florida 33314
USA

Nora Goldstein
Executive Editor, *Biocycle* Magazine
419 State Avenue
Emmaus, Pennsylvania 18049
USA

David Y. Han
Auburn University
Department of Agronomy
and Soils
252 Funchess Hall
Auburn University
Alabama 36849
USA

Zhenli He
Department of Resource Science
Zhejiang University
Hangzhou
China

Harry A. J. Hoitink
The Ohio State University
Ohio Agricultural Research
and Development Center
Department of Plant Pathology
Wooster, Ohio 44691
USA

Brian A. Kahn
Oklahoma State University
Department of Horticulture
and Landscape Architecture
360 Agricultural Hall
Stillwater, Oklahoma 74078
USA

Matthew S. Krause
The Ohio State University
Ohio Agricultural Research
and Development Center
Department of Plant Pathology
Wooster, Ohio 44691
USA

Urszula Kukier
Institute for Soil Science
and Plant Cultivation
24-100 Pulawy
Poland

Minnie Malik
University of Maryland, Symons Hall
College Park, Maryland 20742
USA

Robert O. Miller
Colorado State University
Soil and Crop Science Department
Fort Collins, Colorado 80523
USA

Thomas A. Obreza
University of Florida, IFAS
Southwest Florida Research
and Education Center
2686 State Road 29 North
Immokalee, Florida 34142
USA

Monica Ozores-Hampton
University of Florida, IFAS
Southwest Florida Research
and Education Center
2686 State Road 29 North
Immokalee, Florida 34142
USA

Flavio Pinamonti
Istituto Agrario di S. Michele all' Adige
Via E. Mach 1
S. Michele all' Adige 38010
Trento
Italy

Nancy E. Roe
Texas A&M University
Research and Extension Center
Route 2 Box 1
Stephenville, Texas 76401
USA
Current address:
Farming Systems Research, Inc.
5609 Lakeview Mews Drive
Boynton Beach, Florida 33437
USA

Thomas L. Richard
Iowa State University
Department of Agricultural
and Biosystems Engineering
214B Danutson Hall
Ames, Iowa 50011
USA

James A. Ryan
United States Environmental
Protection Agency
National Risk Reduction Laboratory
5995 Center Hill Road
Cincinnati, Ohio 45224
USA.

Robert Rynk
JG Press, Inc.
419 State Avenue
Emmaus, Pennsylvania 18049
USA

Raymond Joe Schatzer
Oklahoma State University
Department of Agricultural
Economics
420 Agricultural Hall
Stillwater, Oklahoma 74078
USA

Kathleen Shaw
Institute for Chemical Process
and Environmental Technology
National Research Council
of Canada
1500 Montreal Road,
Room G-3
Ottawa, Ontario K1AOR6
Canada

Luciano Sicher
Istituto Agrario di S. Michele all' Adige
Via E. Mach 1
S. Michele all' Adige 38010
Trento
Italy

Grzegorz Siebielec
Institute for Soil Science and Plant
Cultivation
24-100 Pulawy
Poland

Lawrence J. Sikora
United States Department of Agriculture
Agriculture Research Service
Soil Microbial Systems Laboratory
Building 001: BARC-West
10300 Baltimore Avenue
Beltsville, Maryland 20705
USA

Susan B. Sterrett
Virginia Polytechnic Institute
and State University
Eastern Shore Agriculture
Experiment Station
33446 Research Drive
Painter, Virginia 23420
USA

Peter J. Stoffella
University of Florida, IFAS
Indian River Research
and Education Center
2199 South Rock Road
Fort Pierce, Florida 34945
USA

Dan M. Sullivan
Oregon State University
Department of Crops and Soil Sciences
3017 Agricultural and Life
Sciences Building
Corvallis, Oregon 97331
USA

Robin A. K. Szmidt
Scottish Agricultural College
Center for Horticulture
Auchincruive
Ayr, Scotland KA6 5HW
United Kingdom

John Walker
United States Environmental
Protection Agency (4204)
1200 Pennsylvania Avenue, N.W.
Washington, D.C. 20460
USA

Xiaoe Yang
Department of Resource Science
Zhejiang University
Hangzhou
China

Contents

SECTION I

Compost Production Methods, Chemical and Biological Processes, and Quality

CHAPTER 1

The Composting Industry in the United States: Past, Present, and Future

Nora Goldstein

CONTENTS

I. INTRODUCTION

The horticulture industry is one of the primary consumers of organic amendments for use as its growing media. Consider these statistics for just nurseries and greenhouses (Gouin, 1995):

- Nearly 80% of all ornamental plants are marketed in containers and 75 to 80% of the ingredients in potting media consist of organic materials.

1-56670-460-X/01/$0.00+$.50

- When nurseries harvest balled and burlapped trees and shrubs, they also remove between 448 and 560 $Mg \cdot ha^{-1}$ (200 and 250 tons per acre) of topsoil with every crop.

The horticulture industry has used compost for many years, but not in the same quantities as other products such as peat. More recently, however, several factors have combined to make compost a competitive alternative in the horticulture industry. These include:

- Increased pressure on harvesting peat
- Proven benefits from compost use, including plant disease suppression, better moisture retention, and building soil organic matter
- Wider availability of quality compost products
- Creation of composting enterprises by the horticulture industry, in response to its own need for the end product; rising disposal fees for green waste; and consumer demand for compost at retail centers

Although landscapers, nurseries, and other entities in the horticulture industry can produce some of the compost to meet their own needs, demand exceeds what they can supply. Furthermore, certain composts that can better meet the needs of some crops may not be produced by the horticulture industry in adequate quantities.

Because of these factors, there is an excellent synergy between the horticulture industry and the composting industry. Currently, the largest dollar and volume markets for high quality compost producers are in the horticulture industry. This chapter provides an overview of where the composting industry in the U.S. is today, how it evolved, and where it is going.

II. COMPOSTING INDUSTRY OVERVIEW

Composting in the U.S. has come a long way in the past 30 years. A full range of organic residuals — from municipal wastewater biosolids and yard trimmings to manures and brewery sludges — are composted. Technologies and methods have grown in sophistication. The knowledge about what it takes to operate a facility without creating a nuisance and to generate a high-quality product has also expanded.

About 67% of the municipal waste stream in the U.S. (excluding biosolids) consists of organic materials. However, a considerable portion of the newspaper, office paper, and corrugated fiberboard is already recovered for recycling and thus is unavailable for composting. This leaves about 68 million Mg (75 million tons), or 36%, of the waste stream available for composting, including items such as yard trimmings, food residuals, and soiled or unrecyclable paper (U.S. EPA, 1999). However, in the general scheme of waste management alternatives, only a small percentage of residuals from the municipal, agricultural, commercial, industrial, and institutional sectors are composted at this time. Yet the significant level of composting experience in all those sectors lays the groundwork for growth in the future.

Although there is nothing new about the practice of composting, especially in agriculture, its application in the U.S. on a municipal or commercial scale did not occur until the middle of the 20th century. At that time, composting was viewed as a business opportunity — a way to turn garbage into a commercial product. However, before the industry had a chance to get off the ground, landfills came into the picture, making it nearly impossible for composting to be cost-competitive.

It was not until the 1970s that the current composting industry began to develop. The Clean Water Act was passed early in the decade, making millions of dollars available to invest in municipal wastewater treatment plants. One consequence of improved wastewater treatment was a greater amount of solids coming out of the wastewater treatment process. The U.S. Department of Agriculture (USDA) launched a project at its Beltsville, MD research laboratory to test composting of municipal sewage sludge (referred to in this chapter as biosolids). The research resulted in what was known as the Beltsville method of aerated static pile composting — essentially pulling air through a trapezoidal shaped pile to stimulate and manage the composting process (Singley et al., 1982).

At about the same time, European companies were developing technologies to compost municipal solid waste (MSW). These countries did not have the luxury of abundant land available for garbage dumps. As a result, many of the MSW composting technologies eventually marketed in the U.S. in the 1980s originated in Europe. These systems used enclosed, mechanical technologies, such as silos with forced air.

American companies also developed some in-vessel technologies during this time. These included rotating drums and vessels or bays with mechanical turning devices.

Although a handful of municipalities started to implement composting in the 1970s to manage biosolids or leaves, it was not until the 1980s that public officials and private developers paid any significant attention to this methodology. The drivers contributing to these developments differed somewhat for the different waste streams, but the net result is a significant base of knowledge and technological advancements that made composting a competitive management option for residuals from all sectors — municipal to agricultural.

This chapter will look at several different residual streams — biosolids, yard trimmings, MSW, and food residuals — and analyze composting developments in terms of the number and types of projects, technologies, end markets, commercial developments, public policies, and regulations. Much of the data will be provided from surveys conducted by *BioCycle*, a journal of composting and recycling.

III. BIOSOLIDS COMPOSTING

The first survey of biosolids composting appeared in *BioCycle* in 1983 (Willson and Dalmat, 1983). The survey was conducted by USDA staff in Beltsville, MD. At that time, a total of 90 projects were identified. These included 61 in operation and 29 in development. *BioCycle* began conducting the nationwide survey of biosolids composting in 1985. A survey was completed for every year from 1985 to 1998.

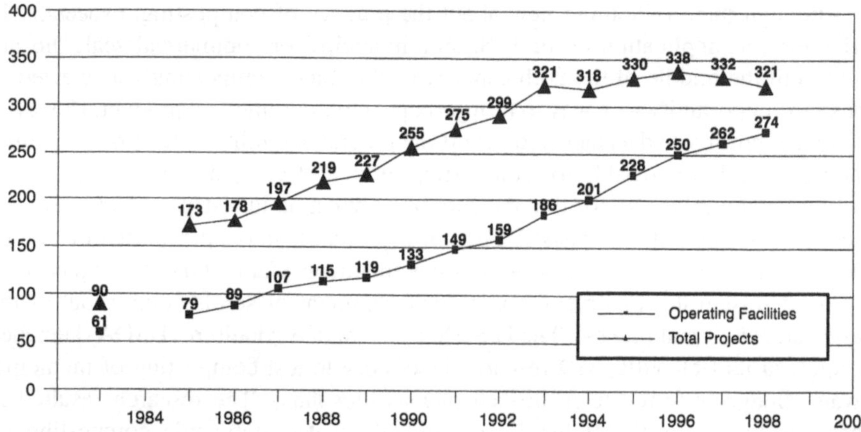

Figure 1.1 Biosolids composting project history in the U.S. (From *BioCycle* Annual Biosolids Composting Surveys: 1983–1998. With permission.)

Figure 1.1 provides a summary of the results of those surveys. Each year's report provides a state-by-state breakdown of biosolids composting projects, including the project's location, project status, composting methodology, and quantity composted. Projects that fall into the "in development" category include those in construction, permitting, planning, design, or active consideration.

A variety of configurations are used to compost biosolids. These include static piles, aerated static piles, actively and passively aerated windrows, enclosed versions of these methods, and in-vessel. The method chosen is dependent on a variety of factors, including climate, site location and proximity to neighbors, degree of process control desired (including the rate at which composting needs to proceed), and regulations. For example, a fairly isolated site in the Southwest can compost effectively in open air windrows. A facility in New England, with neighbors within view, might opt for an enclosed system — to better deal with the weather and with possible nuisance factors.

Biosolids are mixed with a bulking agent prior to composting. The bulking agent provides both a carbon source and pile structure. *BioCycle* survey data finds that the most common amendments for aerated static pile composting are wood chips, followed by leaves, grass, and brush. In-vessel systems without built-in agitation typically use sawdust and wood chips for amendments, while the agitated bay systems may utilize those materials and/or ground yard trimmings. The most common amendment at windrow facilities is yard trimmings, followed by wood chips. Other amendments utilized in biosolids composting include wood ash (which also helps with controlling odors), newsprint, manure, and peanut (*Arachis hypogaea* L.) and rice (*Oryza sativa* L.) hulls. Many facilities also use recycled compost.

Most biosolids composting facilities are fairly small to medium in size. According to *BioCycle*'s 1998 biosolids composting survey (Goldstein and Gray, 1999), three of the four largest sites are windrow operations composting between 82 and 91 dry Mg (90 and 100 dry tons) per day of biosolids (two in California and one in Kentucky); the fourth, in West Virginia, is an aerated static pile operation. Other larger scale facilities include a 54 dry Mg (60 dry ton) per day in-vessel plant in Ohio and a 36 dry Mg (40 dry ton) per day aerated static pile operation in Pennsylvania.

Overall, biosolids composting is fairly well represented across the country. The only states currently without any projects are Minnesota, Mississippi, North and South Dakota, Wisconsin, and Wyoming. In terms of the actual number of projects, New York State leads with 35, followed by Washington (19), California (18), Massachusetts (18), and 15 each in Colorado, Maine, and Utah.

Biosolids composting facilities typically are successful in marketing or distributing the compost produced. The top paying markets for biosolids compost are nurseries, landscapers, and soil blenders. Other end uses include public works projects (e.g., roadway stabilization, landfill cover), application on park land and athletic fields, and agriculture. Many composting plants distribute compost directly to homeowners.

A. Biosolids Composting Drivers

A number of "drivers" have contributed to the development of biosolids composting projects in the U.S. They revolve around potential difficulties in continuing current practices — such as landfilling, incineration, or in some cases, land application — to a confidence level to undertake the effort because of the success of other projects.

Although smaller plants may use composting as their primary management option, a number of facilities start a composting project in conjunction with a land application program. Composting provides a backup when fields are not accessible. For treatment plants in areas where agricultural land within a reasonable hauling distance is being developed, composting is a backup and is likely to become the primary management method in the future. In other areas, treatment plants that dispose of biosolids in landfills may start a composting facility because of the uncertainty of continuing landfill disposal in the future.

In the 1980s, landfill bans on yard trimmings forced many local governments to initiate composting projects to process leaves, brush, and grass clippings. In some cases, public works officials joined forces with wastewater treatment plant operators in their towns to create co-composting projects — using the yard trimmings as a bulking agent for the biosolids. This contributed to the growth of biosolids composting in the late 1980s and early 1990s.

Two other drivers — not just for biosolids composting but for other residuals — have been the evolution of the knowledge base and technologies to handle these materials and demand for compost products. In some municipalities, there is a higher comfort level with composting in a contained vessel or a bay-type system that is in a completely enclosed structure. The availability of these technologies, and the accompanying refinement in controlling odors from these types of systems, helped to fuel the growth in projects.

Research on compost utilization helped stimulate markets for biosolids compost, especially in the horticultural and landscaping fields. It is anticipated that demand for these kinds of products will grow in the future. For example, research in Massachusetts with utilization of biosolids compost in a manufactured topsoil showed significant potential for application in landscape architecture projects, an end use that can require vast amounts of finished product (Craul and Switzenbaum, 1996). In another case, landscape architects specified that biosolids compost be used in the soil mix for a recently completed riverside park in Pittsburgh, PA (Block, 1999).

A nursery in Ohio has used composted municipal biosolids for bed and container production for over 10 years (Farrell, 1998). It uses about 765 m^3 (1000 yd^3) per year of the compost, which it obtains from two sources. The nursery owner notes that the composted biosolids contributed to increased plant growth and plant disease suppression, and are a good source of mycorrhizal inoculum, organic material, and plant mineral nutrients. He adds that the compost made a tremendous difference in the quality and vigor of boxwoods (*Buxus* spp.) and reduced the cycle of growth so that more can be grown.

In the future, growth in the number of biosolids composting projects is expected to continue. At least four factors contribute to the increase. First, a high quality biosolids compost can meet the U.S. Environmental Protection Agency's Class A standards, which give a wastewater treatment plant more flexibility in product distribution and regulatory compliance. Second, increasing pressure on land application programs due to land development and public acceptance issues is forcing wastewater treatment plants to seek alternatives such as composting. Third, there is a growing demand for high-quality composts. Finally, continual technology and operational improvements result in more project successes, thus building confidence in composting as a viable management option.

There are some caveats that hamper the development of biosolids composting projects. The economics are such that composting can be more costly than other management alternatives, such as land application and landfilling. Also, there is adequate landfill capacity available in many regions, and some treatment plants are taking advantage of that option at this time. As a result, there is likely to be continued steady but not rapid growth in the number of biosolids composting projects in the U.S.

IV. YARD TRIMMINGS COMPOSTING

BioCycle began tracking the number of yard trimmings composting sites in the U.S. in 1989, as part of its annual "State of Garbage in America" survey. That first year, the survey found 650 projects. In the 1999 State of Garbage survey (which provides data for 1998), there were 3807 yard trimmings composting sites (Glenn, 1999).

A majority of the 3800-plus sites are fairly low technology, smaller operations that are municipally owned and operated. Typically, yard trimmings are composted in windrows. Some of these smaller sites utilize compost turning equipment. Most, however, turn piles with front-end loaders. Many operators simply build windrows, turn them occasionally in the beginning, and then let the piles sit for a number of months, moving material out only when there is a need for more space at the site.

There are some sizable municipal operations that utilize up-front grinding equipment, turners, and screens. These sites tend to be managed more intensively because of the higher throughput and thus the need to move finished compost off the site more quickly. There also is a healthy private sector that owns and operates yard trimmings composting facilities. These sites also tend to be managed more aggressively because the owners rely on income from tipping fees and from product sales. Although most of the larger sites also compost in windrows, some experienced odor problems (particularly from grass clippings) and started using aerated static piles in order to treat process air and not disturb the piles during active composting (Croteau et al., 1996).

Markets for yard trimmings compost include landscapers and nurseries (both wholesale and retail), soil blenders, other retail outlets, highway reclamation and erosion control projects, and agriculture. Many municipal projects provide free finished compost and mulch to residents.

A. Yard Trimmings Composting Drivers

State bans on the disposal of yard trimmings at landfills and incinerators were the primary driver in the development of yard trimmings composting projects. Currently, there are 23 states with disposal bans; several bans only apply to leaves, or leaves and brush. No state has passed a landfill ban on yard trimmings in recent years, but New York State was expected to consider such legislation in 2000. Growth of yard trimmings composting projects in the future will be driven primarily by localities trying to divert more green materials from landfills in order to save capacity or meet a state or locally mandated diversion goal (such as California's mandated 50% goal by 2000), or by market demand for composted soil products (and thus the need for more feedstocks).

Other possible drivers are the fact that yard trimmings are easy to source separate and thus are accessible for diversion; they are a good fit with biosolids composting; and most states' regulations make it fairly simple to compost yard trimmings, thus there are few entry barriers.

In the future, there likely will be some consolidation of yard trimmings projects. Smaller municipalities may opt to close their sites and send material to a private facility or a larger municipal site in their region. Private sector processors also offer mobile grinding, composting, and screening services, which eliminate the need to haul unprocessed feedstocks (a significant expense).

Municipal and privately owned yard trimmings sites also are starting to accept other source separated feedstocks, such as preconsumer vegetative food residuals (such as produce trimmings), manure, and papermill sludge. In some states, as long as the site is equipped to handle these other materials, getting a permit to take additional feedstocks is fairly straightforward. For example, a municipal yard trimmings composting site in Cedar Rapids, IA, takes papermill sludge and a pharmaceutical residual. A large-scale private site in Seattle, WA services commercial generators in its region.

V. MSW COMPOSTING

Historically, MSW generation grew steadily from 80 million Mg (88 million tons) in 1960 to a peak of 194 million Mg (214 million tons) in 1994. Since then, there has been a slight decline in MSW generation. Recovery of materials for recycling also increased steadily during this period. In 1996, about 56% of the MSW in the U.S. was landfilled; 17% was combusted, primarily in trash-to-energy plants; and 27% was recycled. Within the 27% of MSW that was recycled, about 10.2 million Mg (11.3 million tons) was composted, representing 5.4% of the total weight of MSW generated in 1996 (U.S. EPA, 1998).

MSW composting has been around in the U.S. for decades. Projects were started around 40 years ago, but closed with the advent of inexpensive landfill space. There was a resurgence in MSW composting in the 1980s due to a number of factors, including closure of substandard landfills in rural areas; rising tipping fees in some regions as well as perceived decreases in landfill capacity; minimal development of waste to energy facilities (due to cost and performance issues); a perceived natural "fit" with the growing interest in recycling; the existence of technologies, primarily European, so that projects did not have to start from scratch; flow control restrictions that could enable projects to direct MSW to their facilities; and a potential revenue stream from tip fees and product sales.

Solid waste composting in the U.S. emerged on two tracks during the 1980s. The first, the mixed waste approach, involves bringing unsegregated loads of trash (in some cases this includes the recyclables) and doing all separation at the facility, both through upfront processing and/or back end product finishing. The second track, the source separated approach, relies on residents and other generators to separate out recyclables, compostables, and trash.

BioCycle also conducts annual surveys of solid waste composting projects. Interest in MSW composting grew rapidly in the late 1980s and early 1990s, but the number of operating projects never grew very much (Table 1.1). At the peak in 1992, there were 21 operating MSW composting projects. As of November 1999, there were 19 operating facilities in 12 states, and 6 projects in various stages of development (Glenn and Block, 1999). The two most recent facilities to open are in Massachusetts. Operating projects range in size from 4.5 to 272 Mg (5 to 300 tons) per day of MSW.

Of the current operating projects, seven use rotating drums and either windrows, aerated windrows or aerated static piles for active composting and curing. Seven projects use windrows, two use aerated static piles (one contained in a tube-shaped plastic bag), two compost in vessels, and one uses aerated windrows. Fifteen projects receive a mixed waste stream; four take in source separated MSW. Currently, there are very few vendors in the U.S. selling solid waste composting systems.

Not all of the operating MSW composting facilities have paying markets for the finished compost. Some use the material as landfill cover, while others donate it to farmers. A few facilities market compost to the horticulture industry. These include Pinetop–Lakeside, AZ; Fillmore County, MN; and Sevierville, TN (Glenn and Block, 1999).

Table 1.1 Solid Waste Composting Project History in the U.S.

Year	Operational	Total
1985	1	1
1986	1	6
1987	3	18
1988	6	42
1989	7	75
1990	9	89
1991	18	—
1992	21	82
1993	17	—
1994	17	51
1995	17	44
1996	15	41
1997	14	39
1998	18	33
1999	19	25

From *BioCycle* Annual MSW Composting Surveys: 1985–1999. With permission.

A. MSW Composting Drivers

In the late 1980s, many in the solid waste field felt there would be a landfill crisis in some regions of the country, prompting a surge of interest in alternative management options. In addition, the federal regulations under Subtitle D of the Resource Conservation and Recovery Act (U.S. EPA, 1997) — which went into effect in 1994 — were expected to force the closure of many substandard landfills, again putting pressure on existing disposal capacity.

The expected landfill crisis never really materialized, at least on a national basis. Landfills definitely closed — from almost 8000 in 1988 to about 2300 in 1999 (Glenn, 1999). At the same time, however, new state of the art mega-landfills opened, serving disposal needs on a regional (vs. a local) basis. When landfills closed in small towns, instead of building small composting facilities, many communities opted to build solid waste transfer stations and to haul waste long distances for disposal. Today, there are more transfer stations than landfills in the U.S.

Tipping fees, which did start to rise in many places, never stayed high in most regions. In fact, tipping fees have dropped in the U.S., and it is not anticipated they will go up significantly any time in the near future.

Solid waste composting projects also were negatively impacted by a 1994 U.S. Supreme Court decision that struck down flow control laws that gave government agencies the ability to direct the waste stream to specific facilities (Goldstein and Steuteville, 1994). MSW flow into some composting plants dropped considerably as haulers opted to transport garbage further distances to landfills with lower tipping fees.

Other factors that have stymied the development of MSW composting in the U.S. include generation of odors at some of the larger, higher visibility projects, leading to their failures; inadequate capitalization to fix problems that caused odors and/or to install odor control systems; production of a marginal compost product; and significant skepticism about the technology due to the project failures.

In the future, there will be some development of MSW composting projects, perhaps in areas where it is difficult to implement recycling programs (e.g., major tourist areas). The application of the technology, however, will be very site specific. For example, there may be a few communities that decide to increase diversion by getting households to separate other organics beyond yard trimmings. Many towns, however, have opted to push backyard composting of household organics instead of getting involved in centralized collection.

Experience has shown that composting solid waste on a larger scale requires a significant amount of capital, as well as deep financial pockets to address problems that arise once the facility starts operating. Projects also need to be able to set tipping fees that are competitive with landfills, which can be difficult when a project needs to make a sizable capital investment in processing (upfront and product finishing) equipment.

VI. FOOD RESIDUALS COMPOSTING

Perhaps the fastest growing segment of the U.S. composting industry is diversion of institutional/commercial/industrial (ICI) organics, primarily food and food processing residuals, including seafood. *BioCycle* began tracking data on this sector in 1995, when there was a total of 58 projects (Kunzler and Roe, 1995). In 1998, the last time *BioCycle* surveyed projects in all ICI sectors individually, there were 250 total projects, with 187 in operation, 37 pilots, and 26 in development (Goldstein et al., 1998). The 1999 *BioCycle* survey excluded institutional projects (which in 1998 numbered 116) that only handle residuals generated at that institution (Glenn and Goldstein, 1999). Instead, the survey focused on projects that handle food residuals from a combination of ICI sources — or commercial only — and those handling food processing residuals from only industrial generators. A significant difference between the projects traced in 1999 and the on-site institutional ones is scale. Typically, the on-site projects have throughputs of 4.5 to 91 Mg (5 to 100 tons) per year. Those tallied in the 1999 food residuals composting survey can easily reach upwards of 90,720 Mg (100,000 tons) per year (though not all do).

The 1999 survey found a total of 118 projects in the U.S. Of those, 95 are full-scale facilities, and 9 are pilot projects, primarily at existing composting sites (including nurseries). Another 14 projects are in various stages of development. Geographically, there is a very sharp division in the distribution of food residuals composting projects, with the Northeast and West Coast containing the majority of the facilities. Most of the sites compost feedstocks in windrows; many use yard trimmings as a bulking agent. Feedstocks include pre- and post-consumer food residuals (e.g., vegetative trimmings, kitchen preparation wastes, plate scrapings, baked goods, meats), out-of-date or off-specification food products, and industrial organics such as crab and mussel residuals and brewery sludge. The economics of food residuals composting projects have to be competitive with disposal options because the generators typically deal with private haulers (and thus know current disposal costs) (Glenn and Goldstein, 1999).

As with biosolids compost, nurseries, landscapers, and soil blenders represent the highest volume and dollar markets. Agricultural markets also were cited by survey respondents (Glenn and Goldstein, 1999).

A. Food Residuals Composting Drivers

Several different factors combined to promote the initial diversion of food residuals to composting. On the institutional side, it was a combination of cost savings, legislated recycling goals, regulatory exemption, and a finished compost that could be used on site for landscaping or gardens. In most cases, these institutions had yard trimmings available to compost with the food residuals (or started composting yard trimmings and recognized that food residuals — generated in a fairly clean stream — could be co-composted with the yard trimmings).

On the commercial and industrial sides, which have been slower to develop, cost savings are a significant factor — again the ability to divert an already segregated stream to composting instead of disposal. Another benefit is that most food residuals composting sites also accept wet or recyclable waxed corrugated fiberboard, which otherwise would have to be disposed. This was and still remains a significant benefit to generators.

In terms of the composting process, food residuals provide additional moisture and nitrogen to the composting process, especially when the yard trimmings being composted are fairly high in woody materials (a carbon source). In addition, some states' regulations are designed to encourage diversion of source separated, preconsumer feedstocks such as vegetative food residuals. This made entry into food residuals composting more realistic on a permitting level.

With landfill prices holding fairly steady in the \$33 per Mg (\$30 per ton) range on a national basis, it is difficult for haulers and processors to convince generators to divert feedstocks to composting. Nonetheless, a growing number of commercial and municipal sites are finding the right combination of tools to encourage generators to sign on to a composting program.

VII. REGULATIONS

No discussion of composting is complete without a look at regulations. Because composting falls in the waste management spectrum, it is typically regulated under solid waste rules. Biosolids composting is an exception, as many states regulate it under their water divisions.

The federal government does not have specific regulations for composting, except for EPA's Part 503 rules for biosolids (U.S. EPA, 1994), which include stipulations for biosolids composting, particularly regarding pathogens and vectors. The Part 503 rules also set pollutant limits, which each state has to use as a minimum. These limits apply to biosolids compost.

Aside from applicability of the Part 503 rule at the state level, state composting regulations vary significantly. Some states, like California, Ohio, New York, Maine, and Oregon, have very specific composting regulations. In most of these cases, the regulations are "tiered," meaning the degree of permit restrictions changes with the feedstocks being composted. Typically, facilities composting yard trimmings have fairly minimal requirements (primarily addressing setback distances from ground and surface water and quantities processed). Wood processing operations also tend to have few regulatory requirements, as do those projects handling manure.

Regulatory requirements increase with source-separated food residuals (preconsumer) and then get more stringent with regard to postconsumer food residuals, biosolids, and MSW. Some states, like Maine, have few restrictions for sites which compost less than a certain quantity of feedstocks per year (e.g., 382 m^3 [500 yd^3] per year of preconsumer food residuals).

VIII. CONCLUSIONS

Composting serves as both a waste management method and a product manufacturer. As such, a project can generate revenue streams on both the front end (tipping fees) and the back end (product sales). Many companies got into composting mostly based on the upfront revenue from tipping fees, and did not focus a lot of attention on producing a high-quality product to maximize sales. But with steady or dropping tipping fees, projects are having to become more market driven and not tip fee driven. Successful companies and operations are those with excellent marketing programs. They have invested in equipment to service their markets, e.g., screens with various sizes to meet different end uses. In short, they know their markets and know how to service them.

There also are exciting developments on the end use side. Composts are used increasingly for their nutrient value and ability to build soil organic matter and also because of their ability to suppress plant diseases. There is an increase in agricultural utilization of compost, and many states are developing procurement programs for compost use on highways and for erosion control. Interesting projects also are developing in the use of compost for bioremediation. In short, although composting will always be available as a waste management option, it is becoming equally (and in some cases more) valuable as a producer of organic soil amendments.

For the most part, major solid waste initiatives that might have a positive impact on the development of composting projects are not expected. There may be some indirect impacts, e.g., from increasing regulation of manure management, which may lead to more composting on farms. But for the foreseeable future, growth in composting may be primarily due to market demand for compost.

In the final analysis, the composting industry knows how to make compost products that meet the needs of the horticulture industry. The combination of research and practical experience demonstrates the benefits, cost savings, and sustainability of compost use in horticulture. Furthermore, composting is an economically viable management tool for nurseries and other sectors of the horticulture industry that generate organic residuals.

If compost is going to play a more significant role in horticulture, it is critical that the composting industry has the capability to reliably (1) produce compost that is of a consistent quality, and (2) produce the volume of quality compost needed to match the demands of the horticulture industry.

Today's composting industry has the knowledge and technical ability to produce a compost product that consistently meets the needs of the end user. Adequate volumes are and can be produced. However, composters face a dilemma in that they need to secure long-term market contracts so that they can secure long-term sources of feedstocks and have adequate financing available for site expansion. A number of composters have found that balance; in fact, some actually pay for feedstocks in order to guarantee an adequate supply and to have the quality input desired.

In summary, the U.S. has a healthy and growing composting infrastructure. Around the country, private sector composters are running successful businesses, serving as models for other entrepreneurs and investors. Some individuals start composting companies from "scratch," while others add composting on to an existing business — such as a mining or excavation company, nursery, wood grinder, soil blender, or farmer. Many municipal projects are thriving as well, giving generators an excellent outlet for their residuals and providing end users with a steady supply of quality compost.

REFERENCES

Block, D. 1999. Compost plays role in riverfront restoration. *BioCycle* 40(8):26–29.

Craul, P.J. and M.S. Switzenbaum. 1996. Developing biosolids compost specifications. *BioCycle* 37(12):44–47.

Croteau, G., J. Allen, and S. Banchero. 1996. Overcoming the challenges of expanding operations. *BioCycle* 37(3):58–63.

Farrell, M. 1998. Composted biosolids are big plus to Ohio nursery. *BioCycle* 39(8):69–71.

Glenn, J. 1999. The state of garbage in America. *BioCycle* 40(4):60–71.

Glenn, J. and D. Block. 1999. MSW composting in the United States. *BioCycle* 40(11):42–48.

Glenn, J. and N. Goldstein. 1999. Food residuals composting in the U.S. *BioCycle* 40(8):30–36.

Goldstein, N., J. Glenn, and K. Gray. 1998. Nationwide overview of food residuals composting. *BioCycle* 39(8):50–60.

Goldstein, N. and K. Gray. 1999. Biosolids composting in the United States. *BioCycle* 40(12): 63–75.

Goldstein, N. and R. Steuteville. 1994. Solid waste composting seeks its niche. *BioCycle* 35(11):30–35.

Gouin, F. 1995. *Compost Use in the Horticultural Industries*. Green Industry Composting. *BioCycle* Special Report. The JG Press, Emmaus, Pennsylvania.

Kunzler, C. and R. Roe. 1995. Food service composting projects on the rise. *BioCycle* 36(4): 64–71.

Singley, M., A. Higgins, and M. Frumkin-Rosengaus. 1982. *Sludge Composting and Utilization: A Design and Operating Manual*. Cook College, Rutgers - The State University of New Jersey, New Brunswick, New Jersey.

United States Environmental Protection Agency (U.S. EPA). 1994. *A Plain English Guide to the EPA Part 503 Biosolids Rule*. Report No. EPA832-R-93-003. Office of Wastewater Management, Washington, DC.

United States Environmental Protection Agency (U.S. EPA). 1997. *RCRA: Reducing Risk from Waste*. Report No. EPA530-K-97-004. Office of Solid Waste, Washington, DC.

United States Environmental Protection Agency (U.S. EPA). 1998. *Characterization of Municipal Solid Waste in the United States: 1997 Update*. Report No. EPA530-R-98-007. Office of Solid Waste, Washington, DC.

United States Environmental Protection Agency (U.S. EPA). 1999. *Organic Materials Management Strategies*. Report No. EPA530-R-99-016. Office of Solid Waste and Emergency Response, Washington, DC.

Willson, G. and D. Dalmat. 1983. Sewage sludge composting in the U.S.A. *BioCycle* 24(5): 20–23.

CHAPTER 2

Biological, Chemical, and Physical Processes of Composting

Michael Day and Kathleen Shaw

CONTENTS

1-56670-460-X/01/$0.00+$.50

I. INTRODUCTION

The composting process was known and used by man since he changed from being a hunter to a gatherer. As our ancestors started to grow crops they observed that they grew better near rotting piles of vegetation and manure than elsewhere. This finding alone, although a casual observation, was scientific in nature and the discovery was not overlooked but passed on from generation to generation. Clay tablets unearthed in the Mesopotamian Valley dating back to the Akkadian Empire, 1000 years before Moses, attest to the use of compost in agriculture. However, it has only been since the Second World War that any major efforts have been made to focus on the scientific processes occurring during the actual composting period. Prior to the last few decades composting was mostly left to chance. However, today it is a big business and large private and public composting operations are now being accepted as the most environmentally acceptable way to divert about 50% of the waste destined for landfills. The development of these large composting operations has been stimulated by local and federal regulations prohibiting the disposal of yard wastes or other biodegradable materials in landfills.

The number of composting facilities, both aerobic and anaerobic, grows every year. Since 1985, the journal *Biocycle* has listed annually the number and type of composting facilities in the U.S. In 1998 there was a total of 250 food waste composting projects with 187 in operation, 37 pilots, and 26 in development in the U.S. (Goldstein et al., 1998). Biosolids composting facilities have decreased from a high of 338 in 1996 to 321 in 1998, with 274 operational (Goldstein and Block, 1999). Solid waste composting got a boost in 1998 with 18 municipal solid waste (MSW) composting facilities operating and 2 more scheduled to open in 1999 (Glenn, 1998). Anaerobic facilities are closed systems and so have the added advantage over the aerobic systems of controlling odors and capturing the gaseous methane that can bc used for fuel, but they can be more expensive.

Naylor (1996) observed that without the natural decomposition of organic wastes that has been going on for eons we would be miles deep in dead organic matter. Dindal's Food Web of the Compost Pile (Dindal, 1978) can be applied to the first stage of the natural decomposition of all types of organic wastes (Figure 2.1).

First level consumers at the compost restaurant are the microorganisms such as bacteria, actinomycetes, and fungi. These species are the true decomposers. They attack, feed on, and digest the organic wastes before they themselves are consumed by the second level organisms, such as the protozoa and beetle mites. The third level

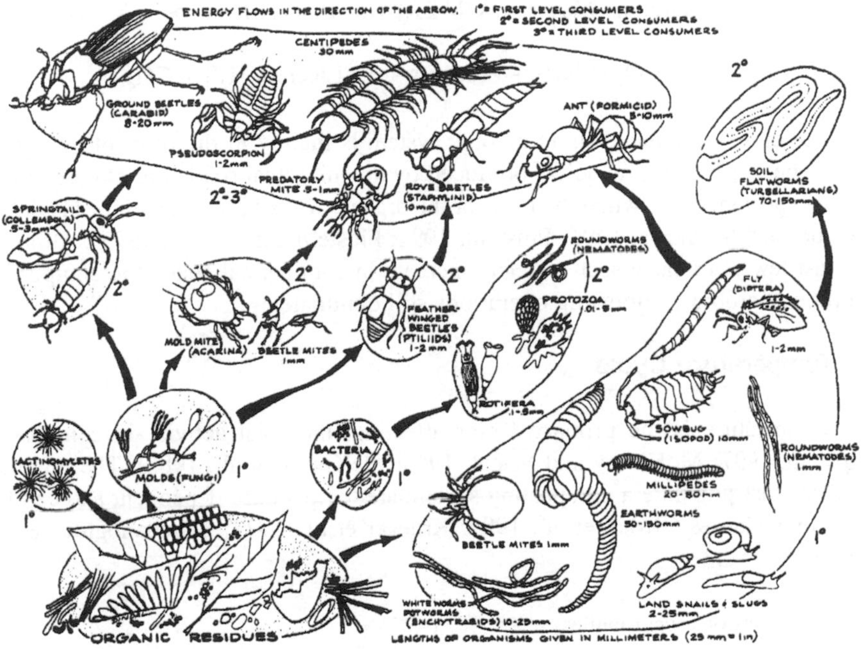

Figure 2.1 Food web of the compost pile. (From Dindal, D. L. 1978. Soil organisms and stabilizing wastes. *Compost Science/Land Utilization* 19(8): 8–11. WIth permission. www.jgpress.com)

consumers, e.g., centipedes and ground beetles, then prey on the second level consumers and on themselves. It is a very efficient system with the various levels of microflora being essential to the successful functioning of the composting process. The microflora dominate in most commercial (large-scale) operations. This chapter reviews the biological, chemical, and physical changes that occur during the actual composting process.

II. SPECIFIC BIOPROCESSES IN COMPOSTING

Composting is a mass of interdependent biological processes carried out by a myriad of microorganisms essential for the decomposition of organic matter. Most systems are aerobic, meaning the microorganisms require oxygen (O_2). The overall biochemical equation can be written:

$$\text{Organic Matter} + O_2 + \text{AEROBIC BACTERIA} \Rightarrow$$

$$CO_2 + NH_3 + \text{Products} + \text{ENERGY}$$

For anaerobic systems, oxygen is absent and the overall biochemical equation takes a different form:

$$\text{Organic Matter} + \text{ANAEROBIC BACTERIA} \Rightarrow$$

$$CO_2 + NH_3 + \text{Products} + \text{ENERGY} + H_2S + CH_4$$

The energy produced in an aerobic system is mainly in the form of low-grade heat. The self-heating, which is produced by the microbial oxidation of carbon (C), occurs spontaneously when the mass of the organic wastes is sufficient for insulation (Baader and Mathews, 1991; Finstein, 1992; Finstein and Morris, 1975). Although the last few years have seen a steady increase in commercial anaerobic composting facilities, aerobic composting operations still dominate.

A. Temperature Cycle

Temperature is the primary factor affecting microbial activity in composting (Epstein, 1997; McKinley and Vestal, 1985; McKinley et al., 1985). The microorganisms that populate a composting system are temperature dependent and can fall into three classes (Brock et al., 1984; Krueger et al., 1973; Tchobanoglous et al., 1993):

Cryophiles or psychrophiles	0–25°C
Mesophiles	25–45°C
Thermophiles	>45°C

Cryophiles are rarely found in composting, but winter composting does take place successfully in Canada and the northern U.S., where ambient temperatures range from –27 to 15°C (Brouillette et al., 1996; Fernandes and Sartaj, 1997; Lynch and Cherry, 1996). The organisms that predominate in commercial composting systems are mainly mesophiles and thermophiles each contributing at different times during the composting cycle. Temperature is also a good indicator of the various stages of the composting process. Frequently, the temperature profile of the composting process is shown as a simple curve such as Figure 2.2 (Burford, 1994; Polprasert, 1989). However, in many cases a more complex temperature profile is obtained as shown in Figure 2.3 (Day et al., 1998; Liao et al., 1996; Lynch and Cherry, 1996; Papadimitriou and Balis, 1996; Sikora et al., 1983; Wiley et al., 1955). In this case after the first increase in temperature, the temperature drops a few degrees before continuing to increase to 60°C or more. The temperature then plateaus briefly at 65 to 70°C and then starts to decrease slowly down through a second mesophilic phase to ambient temperature.

Based on microbial activity, the composting process can be divided into four different stages (Figures 2.2 and 2.3). The first stage is the mesophilic stage, where the predominant microbes are the mesophilic bacteria. The abundance of substrate at this time ensures that the microorganisms are very active, leading to the generation of large quantities of metabolic heat energy, which causes the temperature of the compost pile to increase. According to Burford (1994), Finstein (1992), and McKinley et al. (1985), the microbial activity in the 35 to 45°C range is prodigious (see Table 2.1). As the temperature rises past 45°C, conditions are less favorable for

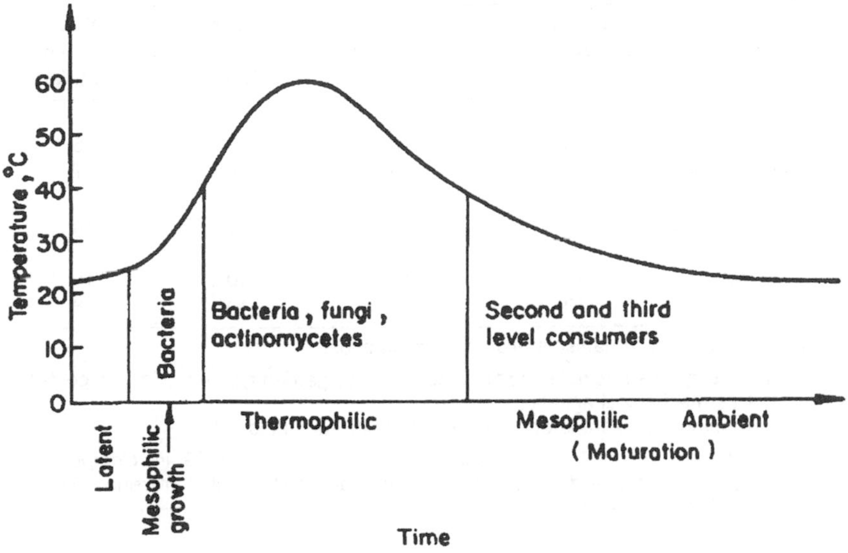

Figure 2.2 Patterns of temperature and microbial growth in compost piles. (From Polprasert, C. 1989. *Organic Waste Recycling*. John Wiley & Sons Ltd., Chichester, United Kingdom, p. 67. With permission.)

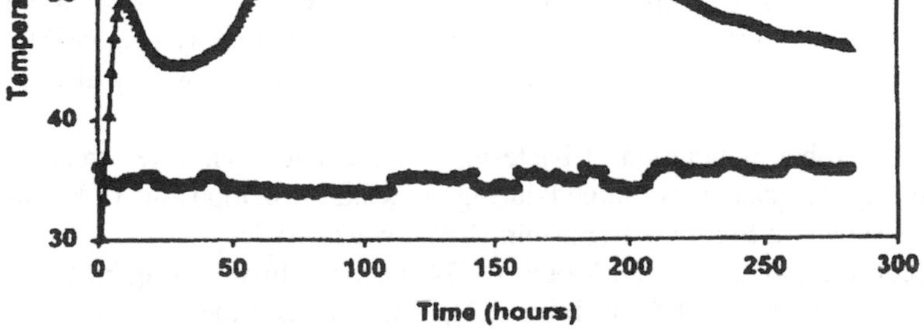

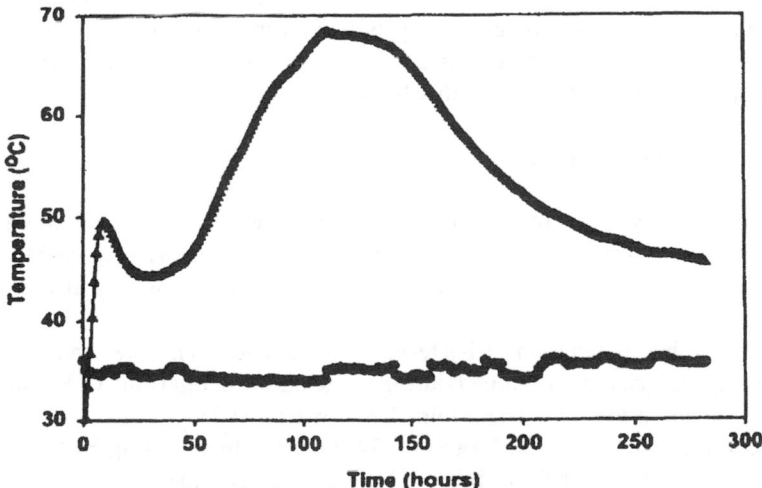

Figure 2.3 Temperatures recorded by the middle thermocouple (▲) in the laboratory composter as a function of time for a CORCAN test sample on day 0. Room temperature in the composting laboratory shown (•). (From Day, M., M. Krzymien, K. Shaw, L. Zaremba, W.R. Wilson, C. Botden and B. Thomas. 1998. An investigation of the chemical and physical changes occuring during commercial composting. *Compost Science & Utilization* 6(2):44-66. With permission. www.jgpress.com)

Table 2.1 Microfloral Population During Aerobic Composting[z]

Microbe	Mesophilic Initial Temp. <40°C	Thermophilic 40-70°C	Mesophilic 70°C to Cooler	Number of Species Identified
Bacteria				
Mesophilic	10^8	10^6	10^{11}	6
Thermophilic	10^4	10^9	10^7	1
Actinomyces				
Thermophilic	10^4	10^8	10^5	14
Fungi[y]				
Mesophilic	10^6	10^3	10^5	18
Thermophilic	10^3	10^7	10^6	16

Note: Number of organisms are per g of compost.

[z] Composting substrate not stated but thought to be garden-type material composted with little mechanical agitation.

[y] Actual number present is equal to or less than the stated value.

From Poincelet, R.P. 1977. The biochemistry of composting, p.39. in: *Composting of Municipal Sludges and Wastes.* Proceedings of the National Conference, Rockville, MD. With permission.

the mesophilic bacteria and instead begin to favor the thermophilic bacteria. The resulting increased microbial activity of the thermophiles causes the temperature in the compost pile to rise to 65 to 70°C. Eventually, with the depletion of the food sources, overall microbial activity decreases and the temperature falls resulting in a second mesophilic phase during the cooling stage. As the readily available microbial food supply is consumed, the temperature falls to ambient and the material enters the maturation stage. Microbial activity is low during this stage, which can last a few months. Methods of determining compost maturity for horticultural applications are discussed in other chapters in this book.

B. Microbial Population

Composting is a complex process involving a wide array of microorganisms attacking organic wastes. The microorganisms that are mainly responsible for the composting process are fungi, actinomycetes, and bacteria, possibly also protozoas and algae.

The microbial population of bacteria, fungi, and actinomycetes changes during composting. The changes obtained during the windrow composting of biosolids and bark are shown in Figure 2.4 (Epstein, 1997; Walke, 1975).

According to Finstein and Morris (1974) bacteria thrive during all the stages of composting. Poincelet (1977) (Table 2.1), who analyzed the microbial population as a function of temperature, found that bacteria are usually present in large numbers throughout the whole composting period and are the major microbial species responsible for the degradation processes.

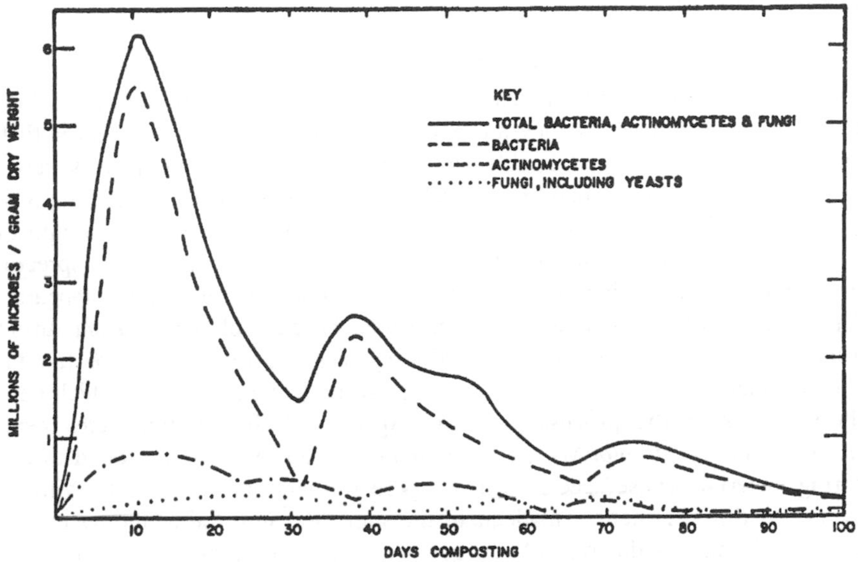

Figure 2.4 Fluctation of microbial population within windrow during composting. (From Walke, R. 1975. The preparation, characterization and agricultural use of bark-sewage compost, p.47. PhD Thesis, The University of New Hampshire, Durham, New Hampshire).

1. Bacteria

In most cases, bacteria are about 100 times more prevalent than fungi (Table 2.1; Poincelet, 1977). Golueke (1977) estimated that at least 80 to 90% of the microbial activity in composting is due to bacteria (see Figure 2.4). Actual bacteria populations are dependent upon the feedstock, local conditions, and amendments used. Burford (1994) observed that at the start of the composting process a large number of species are present including *Streptococcus* sp., *Vibrio* sp., and *Bacillus* sp. with at least 2000 strains. Corominas et al. (1987), in his study of microorganisms in the composting of agricultural wastes, identified species belonging to the genera *Bacillus, Psuedomonas, Arthrobacter,* and *Alicaligenes,* all in the mesophilic stage. In the thermophilic stage, Strom (1985b) identified 87% of the thermophilic bacteria to be of the *Bacillus* sp. such as *B. subtilis, B. stearothermophilic*, and *B. licheniformis*. However, colony variety has been found to decrease as the temperature increases (Carlyle and Norman, 1941; Finstein and Morris, 1974). This observation is consistent with that noted by Webley (1947) who reported the variation in the numbers of aerobic mesophilic bacteria in a study of three separate composts. During the high-temperature stage of composting the mesophilic bacteria are at their lowest level while the thermophilic bacteria are prevalent. However, as temperatures decrease to below 40°C there is a striking repopulation by the mesophilic bacteria, which have been inactive during the thermophilic stage (Webley, 1947).

2. Actinomycetes

Actinomycetes belong to the order Actinomycetales. Although they are similar to fungi, in that they form branched mycelium (colonies), they are more closely related to bacteria. Usually they are not present in appreciable numbers until the composting process is well established. Visual growth of actinomycetes may be observed under favorable conditions, usually between 5 to 7 days into the composting process (Finstein and Morris, 1974; Golueke, 1977). When present in a composting process they can be readily detected due to their greyish appearance spreading throughout the composting pile. With in-vessel composting this greyish appearance of the actinomycetes may not be as prevalent because of the constant turning. Golueke (1977) also suggests that actinomycetes are responsible for the faint "earthy" smell that the compost emits under favorable conditions and which generally increases as the process proceeds. Species of the actinomycetes genera *Micromonospora*, *Streptomyces*, and *Actinomyces* can regularly be found in composting material. These species can be spore formers and are able to withstand adverse conditions, such as inadequate moisture. Because the actinomycetes can utilize a relatively wide array of compounds as substrates, they play an important role in the degradation of the cellulosic component. To some extent they can also decompose the lignin component of wood (Golueke, 1977).

3. Fungi

Fungi appear within the composting process about the same time as the actinomycetes. More types of fungi have been identified in the composting process than either the bacteria or the actinomycetes. Kane and Mullins (1973a) identified 304 unifungal isolates in one batch of compost in a solid waste reactor composting system in Florida. Two general growth forms in fungi exist — molds and yeasts. The most commonly observed species of cellulolytic fungi (Bhardwaj, 1995) in composting materials are *Aspergillus, Penicillin, Fusarium, Trichoderma,* and *Chaetomoniun*. Although some fungi are very small, most are visible in the form of fruiting bodies — mushrooms — throughout the compost pile. While cellulose and hemicellulose (as in paper products) are slower to degrade than either sugars or starches, lignin is the most resistant organic waste and as such is usually the last in the food chain to be degraded (Epstein, 1997). However, the *Basidiomycetes*, or white rot fungi, play a very important role in the degradation of lignin.

The upper limit for fungal activity seems to be around 60°C. This inactivity of the mesophilic and thermophilic fungi above 60°C has been reported by Chang and Hudson (1967), Finstein and Morris (1974), Gray (1970), and Kane and Mullins (1973b). However, at temperatures below 60°C, the thermophilic fungi can recolonize the compost pile. At temperatures below 45°C, the mesophilic fungi reappear. One of the few thermophilic fungi that survive above 60°C is the thermotolerant species *Aspergillus fumigatus* (Haines, 1995). The spores of this species readily withstand temperatures above 60°C and this species becomes the dominant fungus in the compost pile at those temperatures. *Aspergilllus fumigatus* is a mold and has a special significance as a cellulose and hemicellulose degrader (Fischer et al., 1998). However, the air borne spores can be a health hazard at the composting facility, to site workers who have a history of respiratory illnesses (Olver, 1994). Human health issues are discussed in more detail in other chapters in this book.

4. Pathogens

One of the requirements of a commercial operation is to maximize the destruction of pathogens that may be present in the composting feedstock. Theoretically, if the feedstock does not contain manures or biosolids there should be few enteric pathogens. However, where composting operations allow disposable diapers and pet feces to be a part of their waste collection, this may not be the case. Other nonenteric pathogens can be found in meat scraps (*Trichinella spiralis*) and viruses of human origin (poliovirus) have also been found in refuse (Golueke, 1977). As the temperature rises in the composting process the pathogens are usually destroyed as they reach their thermal death points (Table 2.2). Viruses are killed in about 25 min at 70°C (Roediger, 1964). There is a relationship between temperature and time for pathogen kill. A high temperature for a short period of time may be just as effective as a lower temperature for longer duration (Haug, 1993).

Table 2.2 Thermal Death Points for Some Common Pathogens and Parasites

Organism	50°C	55°C	60°C
Salmonella thyphosa	—	30 min	20 min
Salmonella sp.	—	60 min	15–20 min
Shigella sp.	—	60 min	—
Escherichia coli	—	60 min	15–20 min
Streptococcus pyogens	—	10 min	
Mycobacterium diptheriae	—	45 min	—
Brucellus abortus or suis	—	60 min	3 min
Endamoeba histolytica (cysts)	—	1 sec	—
Trichinella spiralis	—	—	1 sec
Necator americanus	50 min	—	—
Ascaris lumbrigoides (ova)	—	60 min	—

Note: Data based on Burford (1994), Finstein and Morris (1974), Gotass (1956), Haug (1993), and Polprasert (1989).

The U.S. EPA in "Process to Further Reduce Pathogens" (Composting Council, 1993) established criteria for composts made with biosolids. According to the Federal Biosolids Technical Regulations, a windrow operation must reach a minimum temperature of 55°C for 15 days, with a minimum of five turnings. For an in-vessel or static pile system a minimum temperature of 55°C for 3 days is required. However, Hay (1996) suggested that bacterial regrowth may be possible under certain conditions following composting. Haug (1993) also indicated that a properly operated compost process should maintain an active population of nonpathogenic bacteria so as to prevent explosive regrowth of the pathogenic bacteria.

C. Recyclate

Several composting operators add amendments to their incoming feedstock to achieve desired properties. These amendments can include screened material, such as oversize wood chips, from previous runs. In studies conducted at our facilities (IPS at Joyceville, Ontario and Wright Environmental at Ste Anne des Plaines, Quebec), we found that the screened immature compost used as recyclate has attached microbial flora. This material, when mixed with the fresh feedstock, reintroduces microbial flora back into the composting process, facilitating the initiation of the compost process (Day et al., 1998).

III. CHEMICAL PROCESSES IN COMPOSTING

The fundamental elemental composition of compost is easy to determine using modern analytical equipment. Unfortunately the analytical precision usually far exceeds the sample homogeneity. Consequently, in the analysis of elemental composition, the question is not how accurate and reproducible are the analytical data, but how accurate and reproducible is the sample and how truly representative it is of the material being analyzed.

A. Elemental Composition: Carbon (C), Nitrogen (N), and the C:N Ratio

The elemental composition of the material processed at a composting operation is very much dependent upon the types of feed materials being processed. However, both C and N are essential to the composting process. Carbon provides the primary energy source, and N is critical for microbial population growth. For effective, efficient composting the correct C:N ratio is essential. Although various organic feedstocks have been successfully composted with C:N ratios varying from about 17 to 78 (McGaughey and Gotass, 1953; Nakasaki et al., 1992b), a much narrower range of between 25 to 35 is considered desirable (Hamoda et al., 1998; Keller, 1961; Schulze, 1962b). The concern at low C:N ratios is the loss of ammonia (NH_3) (Morisaki et al., 1989), but at higher levels slow rates of decomposition can be anticipated (Finstein and Morris, 1974).

Table 2.3 provides data for the C and N composition of a wide variety of possible compost feedstocks derived from a variety of reference sources. Clearly, organic feedstocks that can be processed by commercial composting operations can have a wide variety of C:N ratios. This requires that compost operators have a knowledge of their feedstocks to ensure that the desired mix for optimum composting is achieved. However, the C:N ratio is only one of a large number of variables that have to be controlled. Thus, computer programs have been developed to assist compost operators to achieve the desired mix for optimum composting (CRIQ, 1998; Naylor, 1996).

Although it is customary to express the C:N ratio as a function of the total concentration of C and N, this approach may not be appropriate for all materials (Kayhanian and Tchobanoglous, 1992) due to differences in the biodegradability

Table 2.3 Carbon and Nitrogen Composition of Some Compost Feedstocks (Based on Dry Wt. of Feedstocks)

Feedstock	C (%)	N (%)	C/N ratio	Reference
Urine	12.1	15.1	0.8	Polprasert, 1989
Fish scraps	32.8	8.2	4.0	Mathur, 1991
Activated sludge	35.3	5.6	6.3	Poincelet, 1977
Grass	41.6	2.46	17.0	Michel et al., 1993
Cow manure	30.6	1.7	18.0	Polprasert, 1989
Food waste	50	3.2	15.6	Kayhanian and Tchobanoglous, 1992
Yard waste	44.5	1.95	22.8	Kayhanian and Tchobanoglous, 1992
Leaves	44.5	0.93	48.0	Michel et al., 1993
Paper	43.3	0.25	173	Savage, 1996
Cardboard	48.2	0.20	254	Day et al., 1998
Sawdust	56.2	0.11	511	Willson, 1993

and bioavailability of different organic materials (Naylor, 1996). For example, Jeris and Regan (1973a) evaluated the compostability of a wide range of feedstocks and demonstrated the effect of different C sources. In the case of wood chips, which are frequently used as a bulking agent, not all woods have equal biodegradability (Allison, 1965); hardwoods are more biodegradable than softwoods. According to Chandler et al. (1980) these differences can, in part, be explained in terms of lignin content. More recently He et al. (1995) characterized the C content of compost into three classes — total extractable organic C, carbonate C, and residual C — and found the distribution on average to be 20%, 8%, and 72%, respectively.

Although the analysis for N content is usually more straightforward than for C, measurement of total Kjeldhal nitrogen (TKN) does not include all the nitrates and nitrites in the sample (Naylor, 1996). Fortunately, while TKN values range from 5000 to 60,000 $mg{\cdot}kg^{-1}$, the concentrations of the nitrates and nitrites together are generally less than 100 $mg{\cdot}kg^{-1}$.

Although the starting C:N ratio is important for effective and efficient composting, the final value is also important to determine the value of the finished compost as a soil amendment for growing crops. In general, a final C:N ratio of 15 to 20 is usually the range aimed for (Kayhanian and Tchobanoglous, 1993), although a value of 10 (Mathur, 1991) has been suggested as ideal. A final compost with a C:N ratio greater than 20 should be avoided since it could have a negative impact on plant growth and seed germination (Golueke, 1977). However, it is the availability of the C that is important, not the total measured C, so composts with C:N ratios higher than 20 can be acceptable when the C is not readily available (McGaughey and Gotass, 1953).

The composting process is essentially the bioconversion of biodegradable materials into carbon dioxide (CO_2) and H_2O. Consequently, it would be expected that the concentration of C in the compost material is reduced as composting proceeds, resulting in a corresponding reduction in the C:N ratio. In studies performed in our laboratory (Day et al., 1998), indeed, the concentration of C decreased during the composting process while that for N increased. As a result the C:N ratio decreased from 24.6 to 13.5 during 49 days of commercial composting. This was attributed to the loss in total dry mass due to losses of C as CO_2. These results are in keeping with those reported by others for commercial composting processes (Grebus et al., 1994; Liao et al., 1995; Lynch and Cherry, 1996; Mato et al., 1994; McGaughey and Gotass, 1953; Sesay et al., 1998) or for laboratory simulated systems (Hamoda et al., 1998; Iannotti et al., 1993; Michel et al., 1993; Morisaki et al., 1989; Wiley et al., 1955; Witter and Lopez-Real, 1987). However, some studies have shown a decrease rather than an increase in the concentration of N (Liao et al., 1996; Poincelet, 1977; Snell, 1957). Despite the generally accepted decline in the C:N ratio with composting, ammonium-N (NH_4-N) and nitrate-N (NO_3-N) concentrations can also undergo changes. One study showed increases in these species (Grebus et al., 1994), but another study showed decreases (Canet and Pomares, 1995). Alternatively, several reports indicate increases in NH_3 levels during the initial stages of composting before the values level off and ultimately decline (Liao et al., 1995; Nakaski et al., 1992b; Palmisano et al., 1993; Shin and Jeong, 1996; Snell, 1957). By contrast, NO_3 concentrations typically show a decrease at the beginning of the composting process followed by a progressive increase towards the end (Neto et al., 1987). However, still other studies have shown that NO_3-N remains relatively constant (Palmisano et al., 1993). It is the possible formation of NH_3 that has to be controlled if odor complaints are to be avoided and N losses from the compost are to be minimized.

B. Other Elements

1. Phosphorus (P)

Other chemical elements present in compost feedstocks can influence the composting process, the quality of the compost produced, and the general acceptance of the composting process. Although compost feedstocks must have C and N to provide the fundamental nutrients to the living organisms for the composting process, phosphorus (P) is also an essential element especially in composting MSW (Brown et al., 1998). Although feedstocks such as biosolids, yard debris, and agriculture wastes may have sufficient P, MSW (because it is high in cellulose) may not have sufficient P for effective composting. The quantities of P along with N and potassium (K) present in the final material also are important in determining the quality of the compost product because they are essential nutrients for plant growth. Although not as critical as the C:N ratio, a C:P ratio of 100 to 200 seems to be desirable (Howe and Coker, 1992; Mathur, 1991). Phosphorus composition and the C:P ratio can vary widely depending upon the source of the feedstocks (Table 2.4).

Based upon the assumption that loss of C occurs during composting while P is not lost by volatilization or lixiviation, the percentage P in the compost would be expected to increase as composting proceeds. These effects have indeed been noted (Chandler et al., 1980; Cooperband and Middleton, 1996; Grebus et al., 1994; Mato et al., 1994) resulting in compost containing 0.2 to 0.7% P (Canet and Pomares, 1995; Fricke and Vogtmann, 1994; He et al., 1995; Warman and Termeer, 1996).

Table 2.4 Carbon and Phosphorus Composition of Various Feedstocks

Feedstock	C (%)	P (%)	C/P Ratio	Reference
Grass	41.6	0.26	160	Michel et al., 1993
Leaves	44.5	0.05	890	Michel et al., 1993
Leaves	49.9	0.2	250	Polprasert, 1989
Mixed paper	48.9	0.05	978	Kayhanian and Tchobanoglous, 1993
Yard waste	43.1	0.07	700	Kayhanian and Tchobanoglous, 1993
Food waste	44.6	0.08	557	Kayhanian and Tchobanoglous, 1993
Liquid sludge	41.4	0.17	244	Neto et al., 1987
Poultry manure/peat	42.7	0.90	47	Fernandes and Sartaj, 1997

2. *Sulfur (S)*

Sulfur concentrations are not usually measured in most scientific investigations of the composting process, but the presence of S in sufficient quantities can lead to the production of volatile, odorous compounds detectable at low level concentrations (Day et al., 1998; Toffey and Hentz, 1995). The major sources of S in compost materials are the two amino acids cysteine and methionine found in protein materials. Typical S levels for some composts and compost feedstocks are listed in Table 2.5. Under microbiological processing conditions (Stevenson, 1986) such as composting, both reduction and oxidative processes can occur. Under well-aerated conditions the sulfides are oxidized to the sulfates. However, under anaerobic conditions volatile organic sulfides and H_2S, which would otherwise be absorbed by the humic material and be oxidized, are just vaporized into the atmosphere. It is these compounds (specifically carbon disulfide, carbonyl sulfide, methyl mercaptan, diethyl sulfide, dimethyl sulfide, dimethyl disulfide, and hydrogen sulfide) that are responsible for many of the malodors associated with composting (Kissel et al., 1992; Kuroda et al., 1996; Miller et al., 1991; Toffey and Hentz, 1995). Both volatile S compounds and water soluble sulfate anions have been measured during MSW composting. Values ranged from a low of 0.05% to a high of 0.33% while a typical value appears to be about 0.16% (He et al., 1995).

Table 2.5 Sulfur Content of Various Feedstocks and Compost Samples

Material	S (%)	Reference
Mixed paper	0.079	Kayhanian and Tchobanoglous, 1993
Mixed paper	0.008	Kayhanian and Tchobanoglous, 1992
Yard waste	0.202	Kayhanian and Tchobanoglous, 1993
Yard waste	0.33	Kayhanian and Tchobanoglous, 1992
Food waste	0.219	Kayhanian and Tchobanoglous, 1993
Food waste	0.54	Duggan, 1991
Compost	0.25	Kayhanian and Tchobanoglous, 1993
Compost	0.37	Polprasert, 1989

3. *Chlorine (Cl)*

While S is an element of interest from an odor point of view, chlorine (Cl) attracts interest regarding concerns about chlorinated pesticides and polychlorinated biphenyls (PCB), as well as the polychlorinated dibenzodioxins (PCDD) and dibenzofurans (PCDF). Here the concern is the possible release of these materials from the compost to the soil and their subsequent uptake by plants, or possible leachate runoff. However, information on the fate of these and other chlorinated species during composting is limited (Brown et al., 1997; Fricke and Vogtmann, 1994; Hsu et al., 1993; Kim et al., 1995). Generally, the chlorinated pesticides typically found in MSW and destined for composting pose no environmental or health risks. In fact several of these compounds have been shown to be mineralized during the composting process (Brown et al., 1997; Hsu et al., 1993; Michel et al., 1996), suggesting that composting is a possible decontamination route. Although measurements of PCB levels in several organic waste streams in Germany (Fricke and Vogtmann, 1994) indicate no immediate concern, recommendations have been made for the introduction of efficient and effective ways to reduce possible source contamination, such as avoiding the use of pentachlorophenol-treated woods. Fricke and Vogtmann (1994) also reported that the levels of PCDDs and PCDFs found in composted materials were consistent with the ubiquitous levels found in the environment as a whole.

4. *Heavy Metals*

Heavy metals in compost are a concern to all commercial composting operators and play an important role in determining compost quality. In fact many countries have established, or are establishing, compost quality standards that limit the permissabie concentrations for the metals arsenic, cadmium, chromium, cobalt, copper, lead, mercury, molybdenum, nickel, and zinc (Amlinger, 1996; Bourque et al., 1994; Chabbey, 1993; Chwastowska and Skalmowski, 1997; Composting Council, 1993; Genevini et al., 1997; Gies, 1997; Walker, 1996; Zucconi and de Bertoldi, 1987). Because of these regulations, many MSW composting facilities had to develop acceptable new procedures to restrict the introduction of possible contaminants in the feedstocks, by placing restrictions on specific materials (Richard and Woodbury, 1994). Table 2.6 provides a listing of the range of acceptable heavy metal levels for a variety of European countries (Gies, 1997) along with some typical values reported in the literature (Genevini et al., 1997; Vogtmann et al., 1993). Actual values depend very much on the raw materials being processed (Chabbey, 1993; Genevini et al., 1997; Kayhanian and Tchobanoglous, 1993; Mathur, 1991; Reinhart et al., 1993; Warman and Termeer, 1996). Because mineralization results in a reduction in organic content, the actual amounts of these heavy metals in the finished compost usually increase during composting (Chabbey, 1993). This means that although the original feedstock may have acceptable heavy metal levels, the concentration in the final compost may exceed regulatory levels. However, studies have shown that it is the chemical form of a heavy metal, rather than its presence, that is important in

Table 2.6 Heavy Metal Limits in European Compost Regulations and Measured Values ($mg \cdot kg^{-1}$)

Heavy Metals	Regulation Values[z]	MSW Compost[y]	Source Separation MSW Compost[y]	Biological Waste Compost[x]
Cd	1.2–4.0	4.4	1.22	0.84
Cr	50–750	90.8	34.9	35.8
Cu	60–1200	298.1	72.4	46.8
Pb	120–1200	455.0	147.4	83.1
Hg	0.3–25	—	—	0.38
Ni	20–400	76.3	17.5	20.5
Zn	200–4000	919.8	326.6	249.6

[z] After Gies, 1997.
[y] After Genevini et al., 1997.
[x] After Vogtmann et al., 1993.

determining compost quality, because the chemical form determines the metal's availability for plant uptake or leachability into the groundwater (Bourque et al., 1994; Chwastowska and Skalmowski, 1997; McBride, 1989; Petruzzelli et al., 1989; Tisdell and Breslin, 1995). These investigations suggest that although some MSW composts may contain heavy metals that exceed regulatory limits, only a small percentage of these metals may actually be bioavailable and pose health risks.

C. Chemical Functionality

Limited scientific information is available concerning the chemical reactions that occur during the composting processes. During composting approximately 50% of the organic matter is fully mineralized, producing CO_2 and H_2O. This applies specifically to the easily degradable materials such as protein, cellulose, and hemicellulose. Some of the organic material produces organic residuals, referred to as humic matter. This material has not received a great deal of attention until recently. Most of the early research in this area focused on extraction procedures to characterize the humic-like substances (Aoyama, 1991; Ciavatta et al., 1993; Jimenez and Garcia, 1992). However, more recently several research studies have been undertaken using sophisticated analytical techniques such as ^{13}C-NMR (carbon-13 nuclear magnetic resonance spectroscopy) (Chefetz et al., 1998b; Inbar et al., 1989; Preston et al., 1998), FTIR (Fourier-transform infrared spectroscopy) (Chefetz et al., 1998a; Inbar et al., 1989; Niemeyer et al., 1992; Proyenzano et al., 1998), pyrolysis-field ionization mass spectrometry (Schnitzer et al., 1993; van Bochove et al., 1996), and fluorescence spectroscopy (Chen et al., 1989; Senesi et al., 1991). These studies have provided some valuable information on the nature of the humic materials produced from a variety of waste streams, as well as the material sampled at various stages of maturity. The amount of humic acid (expressed as a percent of the organic matter) increases during composting (Chefetz et al., 1996; Inbar et al., 1990; Jimenez and Garcia, 1992; Roletto et al., 1985; Saviozzi et al., 1988). In terms of composition, research suggests that the following changes are taking place during composting:

- Aromatic structures increase (Chefetz et al., 1996; Chefetz et al., 1998b; Preston et al., 1998; Schnitzer et al., 1993)
- Phenolic structures increase (Chefetz et al., 1996; Chefetz et al., 1998b; Preston et al., 1998; Schnitzer et al., 1993)
- The proportion of carboxylic structures increase (Chefetz et al., 1996; Chefetz et al., 1998b; Preston et al., 1998; van Bochove et al., 1996)
- Alkyl structures remain essentially unchanged (Chefetz et al., 1998b; Schnitzer et al., 1993) or decrease slightly (Chefetz et al., 1996)
- O-alkyl structures decrease (Chefetz et al., 1998b; Preston et al., 1998)
- The concentration of amino acids appears to decrease (Chefetz et al., 1996; Proyenzano et al., 1998)
- The content of polysaccharides also decreases (Proyenzano et al., 1998)
- Data with respect to carbohydrates appear to be less consistent with studies showing no change (Schnitzer et al., 1993), increases (van Bochove et al., 1996), and decreases (Proyenzano et al., 1998)

D. Hydrogen Ion Concentration (pH)

The measurement of the pH of a compost sample is not a simple and straightforward procedure as most operators perceive. The actual pH measured is quite sensitive to sample size and sample preparation. Considerable variations in pH readings can be obtained from comparable samples unless standardized sampling and dilution procedures are used (Carnes and Lossin, 1971). Although the composting process is relatively insensitive to pH, because of the wide range of organisms involved (Epstein et al., 1977), the optimum pH range appears to be 6.5 to 8.5 (Jeris and Regan, 1973c; Willson, 1993). However, because of the natural buffering capacity of compost material, a much wider range of initial pH values can be tolerated (Willson, 1993). This allows a wide range of organic feedstocks to be composted whose pH can vary from a low of 5.0 to 6.5 for raw sludges (Haug, 1993) to highs of 11.0 for digested sludges treated with lime and ferric chloride (Shell and Boyd, 1969).

The initial pH of a typical MSW-based compost feedstock is usually slightly acidic (pH 6). During the early stages of composting the pH usually falls, due to the production of organic acids. However, as composting proceeds the pH becomes neutral again as these acids are converted to methane and CO_2. The pH of the final material is usually slightly alkaline (pH 7.5 to 8.5) (Poincelet, 1977; Polprasert, 1989; Snell, 1957). Compost mixtures with high pHs should be avoided because this can lead to loss of N as NH_3, and its associated odor problems (Miller et al., 1991). Slight increases in pH with composting time, following an initial drop in the early mesophilic stage, are characteristics of many composting studies involving agricultural wastes (Corominas et al., 1987), source-separated food wastes (Day et al., 1998; Shin and Jeong, 1996), and MSW (Burford, 1994; Canet and Pomares, 1995; Nakasaki et al., 1992b; Wiley et al., 1955). However, other reports do not show this initial drop in pH, but only a gradual increase in pH with time. Studies that show this type of behavior include those on MSW (Canet and Pomares, 1995; Jeris and Regan, 1973; Palmisano et al., 1993; Sesay et al., 1998), food wastes (Liao et al., 1995; Strom, 1985b), biosolids (McKinley and Vestal, 1985; Neto et al., 1987), and yard wastes (Michel et al., 1993). Notably, only two studies have shown a decrease in pH with composting time (Lau et al., 1992; Mathur et al., 1990).

E. Respiratory Rates (O_2 Uptake/CO_2 Formation)

Composting is essentially an oxidation process where O_2 is consumed and CO_2 is produced. Consequently monitoring these two gases during the composting process can provide a reliable indication of composting activity. It is highly recommended that composting operators use O_2 and CO_2 meters to ensure that they have sufficient aeration (van der Werf and Ormseth, 1997) to supply the necessary O_2 and remove the CO_2. Studies generally show a 1:1 ratio between O_2 consumption and CO_2 generation (Harper et al., 1992; MacGregor et al., 1981; Wiley et al., 1955), but because CO_2 can be produced by anaerobic respiration and fermentation in addition to aerobic composting (Citterio et al., 1987) its measurement alone is not a good indication of compost activity. On the other hand, the measurement of O_2 consumption is a more suitable parameter for monitoring the compost process (Haug, 1993). In fact several studies have been conducted where O_2 levels have been used to control the composting process (Citterio et al., 1987; de Bertoldi et al., 1988). Although O_2 and CO_2 levels are usually measured in the gases exiting the compost pile, the in situ O_2 consumption and CO_2 accumulation are more important indicators of whether aerobic or anaerobic conditions prevail (Jackson and Line, 1998). Thus the adherence to recommended minimum O_2 levels of 5% (Schulze, 1962a) or 10% (Suler and Finstein, 1977) can be misleading, especially where O_2 diffusion rates are restricted.

Numerous studies have reported values of O_2 depletion and CO_2 evolution and related them to the composting process. Although most of the data have been obtained using laboratory scale reactors, several studies have been made using actual commercial compost piles. Because O_2 is required for composting, it is essential to ensure that adequate aeration is available. Several studies have actually calculated aeration requirements based on temperature (Wiley et al., 1955) and free air space (Snell, 1957). Regan and Jeris (1970) and Jeris and Regan (1973b) demonstrated the correlation between O_2 uptake and free air space, and also showed that O_2 uptake was highest at low moisture levels where more free air space was available.

Typically, during a composting run, the O_2 concentration in the exit gas from a compost reactor mirrors the changes in the CO_2 evolution and temperature curves (Day et al., 1998; Palmisano et al., 1993). The O_2 will decrease from its initial value of 21% to a value approaching 10% over the first few days of composting as the compost temperature increases and the CO_2 evolution increases. Subsequently, as the rate of composting decreases, the O_2 level should gradually increase, slowly returning to the 21% level as the temperatures start to approach ambient. Based upon several controlled tests it would appear that typical O_2 utilization rates for composting at 50 to 70°C are within the range of 1 to 10 mg $O_2 \cdot g^{-1} \cdot h^{-1}$ (Strom, 1985b).

Several researchers observed correlations between CO_2 production and O_2 uptake and also noted two regions of peak composting activity (Ashbolt and Line, 1982; Atkinson et al., 1996; Sikora et al., 1983; Sikora and Sowers, 1985; Wiley et al.,

1955). One peak is associated with the mesophilic phase and the other is associated with the thermophilic phase. Using CO_2 formation as a measure of composting activity, temperatures of 56 to 60°C were shown to be optimum for maximum compost activity (Jeris and Regan, 1973a; Kuter et al., 1985; Walke, 1975; Wiley et al., 1955). Higher temperatures can result in decreased activity (Schulze, 1962a), provided that O_2 levels are maintained between 10 and 18%. Measured rates for CO_2 evolution of about 5.9 $mg{\cdot}g^{-1}{\cdot}h^{-1}$ (Kuter et al., 1985) are also of the order of these measured from O_2 consumption. More recently, Tseng et al. (1995) developed a kinetic model to determine the O_2 consumption and CO_2 evolution under controlled temperature and moisture conditions, which provides an insight into factors responsible for variations in microbial respiration and biomass formation.

The link between CO_2 evolution and O_2 consumption is sometimes referred to as the "respiratory quotient" (RQ). Typical RQ values for the composting process are usually about 0.9 (Atkinson et al., 1996; Schulze, 1960; Singley et al., 1982; Wiley et al., 1955). CO_2 evolution rates have been used by Nakasaki et al. (1985) to distinguish between thermophilic bacteria and thermophilic actimoycetes at different stages of composting. Nakasaki and his colleagues have used CO_2 evolution rates to study the different effects of a wide variety of factors such as moisture (Nakasaki et al., 1994), C:N ratio (Nakasaki et al., 1992b), feedstocks supplementations (Nakasaki et al., 1998), and O_2 concentration (Nakasaki et al., 1992a) on the composting process.

Several field measurements of O_2 and CO_2 have also been reported using both forced aeration and windrow turning. During these studies O_2 measurements were taken at various locations within the piles. In aerated piles, O_2 concentrations were highest in the lower pile sections and decreased on moving upwards (Epstein et al., 1976; Fernandes and Sartaj, 1997; Lynch and Cherry, 1996). In turned piles without aeration, the interior became rapidly depleted of O_2 soon after turning (Lynch and Cherry, 1996; Wiley and Spillane, 1962). O_2 concentration decreased sharply from 21 to about 10% as distance from the surface increased (Mato et al., 1994; Miller et al., 1991). This decrease in O_2 was accompanied by a corresponding increase in CO_2, which reached values as high as 60% in the interior (Miller et al., 1991). Clearly there is a need in aerated systems for a feedback loop to ensure O_2 levels do not fall too far, and to ensure that the CO_2 produced is swept out of the system (de Bertoldi et al., 1988; Hogan et al., 1989; MacGregor et al., 1981). Although temperature feedback loops are frequently employed (Finstein and Morris, 1974; Finstein et al., 1986; Hogan et al., 1989; MacGregor et al., 1981), O_2 feedback systems that maintain a 15 to 20% O_2 level within the compost pile (de Bertoldi et al., 1988) have advantages.

IV. PHYSICAL PROCESSES IN COMPOSTING

Although the composting process is a biochemical process, it is greatly influenced by physical factors such as moisture content and particle size. Both of these parameters can change during the composting process and influence the quality of compost and the time required to achieve a mature saleable product.

A. Moisture Content

The moisture content of compost is a critical criterion for optimum composting (Wiley, 1957). Optimum moisture values for a wide range of organic wastes were summarized by Jeris and Regan (1973b) with values ranging from 25 to 80%. However, it appears that moisture contents between 50 and 60% are most desirable (Bhardwaj, 1995; Golueke, 1989; Hachicha et al., 1992; Hamoda et al., 1998; McGaughey and Gotass, 1953; Miller, 1989; Neto et al., 1987; Poincelet, 1977; Stentiford, 1996). Water is essential for bacterial activity in the composting process (the nutrients for the microorganisms must be dissolved in water before they can be assimilated) (Fricke and Vogtmann, 1993; Hamoda et al., 1998). A minimum moisture content of 12 to 15% is essential for bacterial activity (Miller, 1989). However, even at levels of 45% or below, the moisture level can be rate limiting (Golueke, 1989; Jeris and Regan, 1973a; McGaughey and Gotass, 1953; Poincelet, 1977; Richard, 1992; Stentiford, 1996) causing composting facility operators to prematurely assume that their compost process has stabilized (Richard, 1992; Stentiford, 1996). On the other hand, excessive moisture in compost will prevent O_2 diffusion to the organisms, resulting in the material going anaerobic with the potential for odor formation (Golueke, 1989; Hamoda et al., 1998; McGaughey and Gotass, 1953; Poincelet, 1977; Wiley, 1957). A compost with too high a moisture content can also result in loss of nutrients and pathogens to the leachate, in addition to causing blockage of air passageways in the pile (Polprasert, 1989). Although moisture levels between 50 and 60% are generally accepted as optimum, detailed experiments performed by Snell (1957) suggested that for domestic garbage the range for optimum composting could be narrowed to between 52 to 58%. Suler and Finstein (1977) observed 60% to be the ideal moisture value for composting of food waste. Moisture in compost comes from two sources: moisture in the initial feedstock, and metabolic water produced by microbial action. Theoretical calculations by Finstein et al. (1983), Haug (1993), and Naylor (1996) suggest that between 0.6 and 0.8 g of water can be produced per gram of decomposed organic matter during composting. Experimental results suggest that the value is closer to 0.55 to 0.65 g per gram of organic material (Griffin, 1977; Wiley et al., 1955). However, the aerobic decomposition of 1 g of organic matter releases approximately 25 kJ of heat energy, which is enough to vaporize 10.2 g water (Finstein et al., 1986). Thus there is a tenfold excess of energy for water vaporization, which when coupled with losses due to aeration (Naylor, 1996) accounts for the major loss of water during composting. Typically, a compost operator would aim for an initial moisture content of about 60%, which during composting will decrease to about 40% to facilitate downstream processing such as sieving, mixing, and bagging (Fricke and Vogtmann, 1993).

The changes in moisture during composting are very dependent upon the feedstock bulking agents and method of composting. When outdoor windrow composting is being considered, environmental effects such as precipitation (Canet and Pomares, 1995; Lynch and Cherry, 1996) or the lack thereof (Reinhart et al., 1993) need to be considered. However, when external climatic conditions are eliminated, moisture levels decrease due to evaporation, as already noted (Day et al., 1998; Kuter et al., 1985; Liao et al., 1995; Liao et al., 1996; Neto et al., 1987; Papadimitriou and Balis,

1996; Sesay et al., 1998; Tseng et al., 1995; van der Werf and Ormseth, 1997). Consequently, in most commercial composting operations water addition may be required to maintain the desired biological activity (Kuter et al., 1985; Liao et al., 1995; Neto et al., 1987; Sesay et al., 1998; Tseng et al., 1995).

B. Particle Size

Another physical property of importance to the compost process is particle size. This not only affects moisture retention (Jeris and Regan, 1973b) but the free air space (Jeris and Regan, 1973b; Schulze, 1961) and porosity of the compost mixture (Naylor, 1996). Large particle size materials result in increased free air space and high porosity, but smaller particles result in the reverse effect. However, because aerobic decomposition occurs on the surface of particles, increasing the surface to volume ratio of the particles by decreasing particle size increases composting activity (Gotass, 1956; U.S. EPA, 1971; Willson, 1993). Consequently a compromise in particle size is required, with good results reported with material ranging in size from 3 to 50 mm in diameter (Gray and Biddlestone, 1974; Hamoda et al., 1998; Haug, 1993; Snell, 1991; Willson, 1993). The ideal free air space for optimum composting has been estimated to be 32 to 36% (Epstein, 1997). Jeris and Regan (1973b) calculated this range from field studies using a variety of materials with different densities and particle sizes, where the relationship between free air space, moisture, and O_2 consumption was determined. Fermor (1993) determined a similar value of 30%.

Compaction can also influence the free air space, although free air space is related to particle size. Any form of compaction that will reduce the free air space will reduce air permeably and increase resistance to air flow (Singley et al., 1982). In view of the importance of particle size distribution, compost operators usually employ grinding and sieving equipment to achieve material of the desired size for easier handling and processing (McGaughey and Gotass, 1953; Poincelet, 1977; Richard, 1992; Savage, 1996) when dealing with oversize wastes. However, when dealing with sludges and animal manures that contain fine particulate matter, organic amendments and/or bulking agents such as wood chips, sawdust, rice (*Oryza sativa* L.), straw, peat, rice hulls, etc. may be required to increase the free air space of the feedstock materials (Polprasert, 1989).

Although several methods exist for increasing the particle size distribution and air voids in compost (Day and Funk, 1994; Gabriels and Verdonck, 1992; Kayhanian and Tchobanoglous, 1993), Jeris and Regan (1973b) proposed that free air space be calculated from the bulk density (BD) and specific gravity (SG) of the material using the equation:

$$\text{Free Air Space} = 100\ (1 - \text{BD/SG}) \times \text{dry mass}$$

Although methods to measure air volume are available (Toffey and Hentz, 1995), most studies just report the bulk density as this is the easiest to measure, and from the operator's point of view, the most meaningful (van der Werf and Ormseth, 1997). Using data for samples taken from different depths in a compost pile (Brouillette et al., 1996), it is possible to plot both the measured porosity and free air space as a function of bulk density. From these data (shown in Figure 2.5), it is possible to establish the following relationships among bulk density (BD), porosity (P), and free air space (FAS):

$$P = 100.3 - 0.0263\ BD$$

$$FAS = 99.5 - 0.0788\ BD$$

The bulk densities for a variety of compost feedstocks, which are presented in Table 2.7, merely represent typical values reported in the literature, and in many cases the moisture content and particle size distribution have not been provided. Similarly, bulk density values of initial and final composts reported in the literature show wide variations from a low of 178 $kg \cdot m^{-3}$ to a high of 740 $kg \cdot m^{-3}$ (Grebus et al., 1994; He et al., 1995; Howe and Coker, 1992; Kayhanian and Tchobanoglous, 1993; Marugg et al., 1993; Reinhart et al., 1993). During composting, the bulk density of compost would be expected to increase due to the breakdown in the particle size of the material. This results in a more compact compost, as confirmed by several studies (Jackson and Line, 1998; Kayhanian and Tchobanoglous 1993; Marugg et al., 1993; Reinhart et al., 1993; van der Werf and Ormseth, 1997). However, in some compost systems where substantial evaporation and loss of water is possible, the measured bulk density may decrease as the material dries out during the composting period (Day et al., 1998).

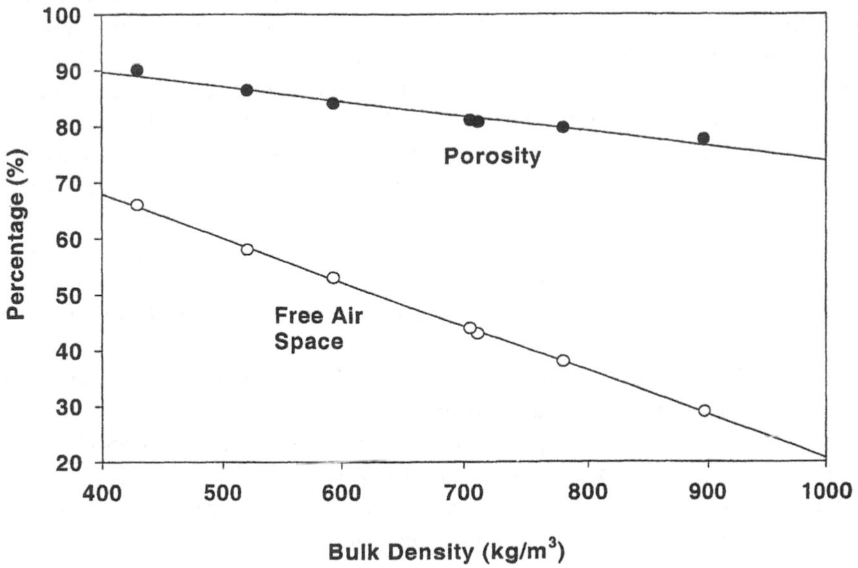

Figure 2.5 Relationship between porosity, free air space, and bulk density. (Using data from Brouillette, M., L. Trepanier, J. Gallichand, and C. Beauchamp.1996. Composting paper mill deinking sludge with forced aeration. *Canadian Agricultural Engineering* 38(2):115–122.)

Table 2.7 Typical Bulk Densities for Some Compost Feedstocks

Feedstock	Bulk Density (kg.m^{-3})	Reference
Mixed paper	80	Kayhanian and Tchobanoglous, 1993
Cardboard	130	Day et al., 1998
Yard waste	215	Reinhart et al., 1993
Yard waste	330	Day et al., 1998
Food waste	352	Kayhanian and Tchobanoglous, 1993
Leaves (shredded)	420	Howe and Coker, 1992
Restaurant waste	990	Day et al., 1998
Dewatered biosolids	1010	Glass, 1993

V. OVERALL CHANGES

A. Changes in Temperature

Temperature is a key factor affecting biological activity within a composting operation and is one factor that is maintained and controlled in any composting operation to ensure optimum growth and activity of the microbes. However, temperature is only a manifestation of the heat energy being released by the metabolic oxidation of the organic matter by microbes. A wide range of microorganisms exist in a composting environment and each has its own optimum temperature for growth. Mesophiles prefer temperatures around 15 to 45°C, while thermophiles prefer temperatures between 45 to 70°C (Burford, 1994; Finstein, 1992; Golueke, 1989; Poincelet, 1977). Although temperature is viewed by most compost operators as a key operating parameter and is used by many to control the process and optimize the degradation, it is only part of the whole thermodynamics of the process (Finstein et al., 1986; Harper et al., 1992; Haug, 1993; MacGregor et al., 1981; Naylor, 1996). However, when dealing with similar feedstocks of reproducible heat capacities, moisture contents, and porosities in piles of reproducible dimensions, temperature is an exceedingly useful tool for following and controlling the composting process. For the compost operator, the temperature of the compost is important for two reasons: (1) to maximize the decomposition rate and (2) to produce a "safe" product by maximizing pathogenic inactivation (Mathur, 1991; Polprasert, 1989; Stentiford, 1987).

Some debate exists concerning optimum temperature conditions for composting. These differences of opinion seem to originate because of the different feedstocks used in the different studies (Epstein, 1997). A temperature of about 55°C seems to be most commonly aimed for (Polparsert, 1989) with operating temperature ranges between 35 to 60°C considered normal. This temperature range also allows the operator to reconcile the trade-offs between pathogenic reduction and maximized biological activity.

Because of the simplicity of its measurement, most compost operators use temperature regulation as a means of controlling the compost operation. Operators typically link air ventilation with a temperature feedback control mechanism. In standard windrow operations this can be accomplished by monitoring the temperature with a thermometer and turning the pile when required (Atkinson et al., 1996). In more sophisticated operations this can involve negative pressure aeration or forced air ventilation (Stentiford, 1987), and a wide variety of systems have been developed and evaluated in both bench scale (Hogan et al., 1989; Sikora et al., 1983; Suler and Finstein, 1977; Tseng et al., 1995) and commercial operations (Finstein et al., 1987; Lau et al., 1992; MacGregor et al., 1981; Sesay et al., 1998).

In nearly all scientific studies of the composting process, temperature–time relationships are usually presented to represent the rate of microbiological activity as a function of time. In most of these cases, the data show the typical temperature–time response, illustrated in Figures 2.2 and 2.3. Initially the temperature of the compost usually increases rapidly to about 40°C within the first 24 hours, as the population of mesophilic microbes is established. At this point the temperature may show an actual decrease for approximately 24 hours (see Figure 2.3) (Canet and Pomares, 1995; Day et al., 1998; Liao et al., 1996; Papadimitriou and Balis, 1996; Sikora et al., 1983). The temperature usually then increases rapidly into the thermophilic range, reaching peak temperatures of about 65°C over the next 2 or 3 days. These temperatures can usually be maintained for about 7 days before decreasing. However, because optimum decomposition has been shown to occur around 55°C (Bach et al., 1984; Epstein, 1997; Jeris and Regan, 1973a; McKinley et al., 1985; Suler and Finstein, 1977; Wiley, 1957), turning and/or aeration may be applied to achieve maximum degradation rates, which can be maintained for a longer period of time. Although studies with MSW compost have shown the temperature to drop 5 to 10°C as a result of the turning process, temperatures within the center of these piles were rapidly reestablished (Canet and Pomares, 1995; Fischer et al., 1998; Kochtitzky et al., 1969; Papadimitriou and Balis, 1996; Wiley and Spillane, 1962). During normal composting operations the temperature of the compost then gradually cools down as the mineralizable organic material is consumed, with the temperature gradually approaching ambient. However, within static piles and aerated bed systems, the temperature distribution can vary widely from the center of the piles to the outer layers. This effect has been noted in controlled laboratory experiments (Finstein et al., 1986; Liao et al., 1996) as well as in full-scale systems using both passive aeration (Fischer et al., 1998; Lynch and Cherry, 1996; Sartaj et al., 1995) and forced aeration (Epstein et al., 1976; Fernandes and Sartaj, 1997; Kuter et al., 1985; Sesay et al., 1998). In all test cases the hottest temperatures are recorded near the middle of the piles, while the coolest temperatures are recorded near the surfaces. Because of the need for a minimum temperature of about 20°C to maintain mesophilic activity, the question of the effect of harsh winters on year-round composting needs to be addressed for those operations in northern climates such as Canada (Lynch and Cherry, 1996). While in-vessel composting is one solution to this dilemma, passively aerated windrow systems also can be used at temperatures ranging from –27° to 15°C (Brouillette et al., 1996; Lynch and Cherry, 1996). Under these conditions the metabolic activity is principally mesophilic and the use of insulating materials, such as peat or finished compost, may be desirable for heat retention (Fernandes and Sartaj, 1997; Lynch and Cherry, 1996).

B. The Mineralization Process

One of the major objectives of any aerobic composting process is the transformation of a purifactable organic waste stream into a stabilized soil amendment that will improve soil physical properties, increase soil buffer capacity, add plant nutrients to the soil, increase soil water-holding capacity, and support and enhance a microbial population (Epstein, 1997).

In simplistic terms, compost can be considered to be composed of water, organic matter, and inorganic matter. The amount of water in a sample is usually determined by appropriate drying methods, whereas the organic and inorganic fractions are determined by a combustion process. The organic fraction is burned and volatilized leaving an ash residue considered to be the inorganic faction. The combustible fraction, sometimes referred to as the volatile solids, is a good indication of the organic content (Naylor, 1996). While most compost feedstocks have high volatile solids contents, these values can vary from a low of 65% noted for dewatered biosolids (Glass, 1993) to a high of about 99% for newspaper (Jeris and Regan, 1973a; Tchobanoglous et al., 1993). Typical values for a variety of compost feedstocks are provided in Table 2.8. Values for volatile solids reported for commercial compost operations vary from 23.2 to 85.7% depending upon the type of feedstocks being processed (He et al., 1995), although values between 55 to 80% are more common (Canet and Pomares, 1995; Day et al., 1998; Glass, 1993; Sikora and Sowers, 1985; Witter and Lopez-Real, 1987).

Table 2.8 Typical Volatile Solids for Some Compost Feedstocks (Dry Mass Basis)

Feedstock	Volatile Solids (%)	Reference
Dewatered biosolids	65	Glass, 1993
Poultry manure	77	Sartaj et al., 1995
Biosolids	85	Kosaric and Velayudhan, 1991
Food waste	84	Tchobanoglous et al., 1993
Food waste	86.3	Shin and Jeong, 1996
Food waste	96.8	Kayhanian and Tchobanoglous, 1992
Grass	89	Michel et al., 1993
Yard wastes	93.2	Tchobanoglous et al., 1993
Office paper	85.7	Shin and Jeong, 1996
Office paper	94.0	Tchobanoglous et al., 1993
Food wastes	96.8	Kayhanian and Tchobanoglous, 1992
Newspaper	95.6	Shin and Jeong, 1996
Newspaper	98.5	Tchobanoglous et al., 1993
Newspaper	99.5	Jeris and Regan, 1973a

During the composting process the ash or inorganic component increases due to the loss of the organic fraction or volatile solids as CO_2. Consequently, the measurement of ash content is a crude indicator of extent of composting. However, the measurement of ash content alone tends to lack sensitivity due to its dependence upon sampling practices (Papadimitriou and Balis, 1996) and sample sizes taken (Atkinson et al., 1996). Wiley et al. (1955) performed a mass balance of compost and found that, in general, losses in volatile solids varied from 17 to 53% with an average of 30%. This suggests that approximately one third of the organic material is decomposed into water and CO_2. However, losses in volatile solids are very much dependent upon the feedstocks used. In the case of MSW composting studies, volatile solids loss values close to 30% have been recorded (Brown et al., 1998; Harper et al., 1992; Iannotti et al., 1993; Poincelet, 1977; Tseng et al., 1995), while other studies have shown losses of about 10 to 15% (Canet and Pomares, 1995; de Bertoldi et al., 1988; Kuter et al., 1985) or intermediate values close to 20% (Day et al., 1998; McGaughey and Gotass, 1953). As would be expected, the values can be influenced by aeration (Sesay et al., 1998) and temperature control (Tseng et al., 1995) as well as nutrient level (Brown et al., 1998). When grass or leaf mixtures were composted, the decreases in volatile solids were close to 30%, with the greatest losses being associated with mixtures containing the larger quantities of grass (Michel et al., 1993). For biosolids the losses in volatile solids are very much dependent upon the bulking agents used. With straw as a bulking agent, the losses in volatile solids were 24% (Witter and Lopez-Real, 1987). However, when fewer biodegradable bulking agents were employed the losses were less than 10% (Liao et al., 1996; McKinley and Vestal, 1985). Similar results have been noted for animal manure (Sartaj et al., 1995; Lynch and Cherry, 1996).

VI. SUMMARY

Many biological, chemical, and physical changes take place during the composting process. Under the influence of microbial attack, many of the organic compounds such as carbohydrates, sugars, and cellulose undergo chemical transformations producing heat, water, and CO_2 in addition to a wide variety of new and modified chemical species. The transformations not only provide valuable information on the actual composting process, but many can be used as control mechanisms to achieve optimum composting and a beneficial product. A knowledge of these fundamental changes is important if composting is to become a widely acceptable technology for the recovery of the organic fraction from our waste stream.

REFERENCES

Allison, L. 1965. Organic matter and crop management problems, p. 1367. In: C.A. Black (ed). *Methods of Soil Analysis*. American Society of Agronomy, Madison, Wisconsin.

Amlinger, F. 1996. Biowaste compost and heavy metals: a danger for soil and environment?, p. 314–328. In: M. de Bertoldi, P. Sequi, B. Lemmes, and T. Papi. (eds.) *The Science of Composting, Part 1*. Blackie Academic and Professional, Glasgow, United Kingdom.

Aoyama, M. 1991. Properties of fine and water soluble fractions of several composts. *Soil Science Plant Nutrition* 37:629–637.

Ashbolt, N.J. and M.A. Line. 1982. A bench-scale system to study the composting of organic wastes. *Journal of Environmental Quality* 11(3): 405–408.

Atkinson, C.F., D.D. Jones, and J.J. Gauthier. 1996. Biodegradabilities and microbial activities during composting of municipal solid waste in bench-scale reactors. *Compost Science & Utilization* 4(4):14–23.

Baader, W. and J. Mathews. 1991. Biological waste treatment, p. 305-327. In: W. Baader and J. Matthews (eds.). *Progress in Agricultural Physics and Engineering.* CAB International, Wallingford, United Kingdom.

Bach, P.D., M. Shoda, and M. Kubota. 1984. Rate of composting of dewatered sewage sludge in continually mixed isothermal reactor. *Journal of Fermentation Technology* 62(3):285–292.

Bhardwaj, K.K.R. 1995. Improvements in microbial compost technology: a special reference to microbiology of composting, p. 115–135. In: S. Khawna and K. Mohan (eds.). *Wealth from Waste.* Tata Energy Research Institute, New Delhi, India.

Bourque, C.L., D. LeBlanc, and M. Losier. 1994. Sequential extraction of heavy metals found in MSW-derived compost. *Compost Science & Utilization* 2(3):83–99.

Brock, T.D., D.W. Smith, and M.T. Madigan. 1984. *Biology of Microorganisms.* Prentice-Hall Inc., Englewood Cliffs, New Jersey, p. 240.

Brouillette, M., L. Trepanier, J. Gallichand, and C. Beauchamp. 1996. Composting paper mill deinking sludge with forced aeration. *Canadian Agricultural Engineering* 38(2):115–122.

Brown, K.H., J.C. Bouwkamp, and F.R. Gouin. 1998. The influence of C:P ratio on the biological degradation of municipal solid waste. *Compost Science and Utilization* 6(1):53–58.

Brown, K.W., J.C. Thomas, and F. Whitney. 1997. Fate of volatile organic compounds and pesticides in composted municipal solid waste. *Compost Science & Utilization* 5(4):6–14.

Burford, C. 1994. The microbiology of composting, p. 10–19. In: A. Lamont (ed.). *Down to Earth Composting.* Institute of Waste Management, Northampton, United Kingdom.

Canet, R. and F. Pomares. 1995. Changes in physical, chemical and physico-chemical parameters during the composting of municipal solid wastes in two plants in Valencia. *Bioresource Technology* 51:259–264.

Carlyle, R.E. and A.G. Norman. 1941. Microbial thermogenesis in the decomposition of plant materials . Part II. Factors involved. *Journal of Bacteriology* 41:699–724.

Carnes, R.A. and R.D. Lossin. 1971. An investigation of the pH characteristics of compost. *Compost Science* 5:18–21.

Chabbey, L. 1993. Heavy metals, maturity and cleanness of the compost produced on the experimental site of Chatillon, p. 62–68. In: *Proceedings of the ReC'93 International Recycling Congress*, Palexpo, Geneva, Switzerland.

Chandler, J.A., W.J. Jewell, J.M. Gassett, P.J. VanSoest, and J.B. Robertson. 1980. Predicting methane fermentation. *Biotechnology and Bioengineering Symposium No. 10.* John Wiley & Sons Inc., New York.

Chang, Y. and H.J. Hudson. 1967. The fungi of wheat straw compost. I. Ecological studies. *Transcripts of the British Mycologia Society* 50(4):649–666.

Chefetz, B., P.G. Hatcher, Y. Hadar, and Y. Chen. 1996. Chemical and biological characterization of organic matter during composting of municipal solid waste. *Journal of Environmental Quality* 25:776–785.

Chefetz, B., F. Adani, P. Genevini, F. Tambone, Y. Hadar, and Y. Chen. 1998a. Humic acid transformation during composting of municipal solid waste. *Journal of Environmental Quality* 27:794–800.

Chefetz, B., P.G. Hatcher, Y. Hadar, and Y. Chen. 1998b. Characterization of dissolved organic matter extracted from composted municipal solid waste. *Soil Science Society of America Journal* 62:326–332.

Chen, Y., Y. Inbar, Y. Hadar, and R.L. Malcom. 1989. Chemical properties and solid-state CPMAS ^{13}C-NMR of composted organic matter. *Science of the Total Environment* 81/82:201–208.

Chwastowska, J. and K. Skalmowski. 1997. Speciation of heavy metals in municipal composts. *International Journal of Environmental Analytical Chemistry* 68:13–24.

Ciavatta, C., M. Gavi, L. Pastotti, and P. Sequi. 1993. Changes in organic matter during stabilization of compost from municipal solid waste. *Bioresource Technology* 43:141–145.

Citterio, B., M. Civilini, A. Rutili, A. Pera, and. M. de Bertoldi. 1987. Control of a composting process in bioreactor by monitoring chemical and microbial parameters, p. 642. In: M. de Bertoldi, M.P. Ferranti, P. L'Hermite, and F. Zucconi (eds.). *Compost: Production, Quality and Use.* Elsevier Applied Science, London, United Kingdom.

Composting Council of the United States. 1993. EPA Guideline, 40 CFR Part-3. Composting Council Fact Sheet, Alexandria, Virginia.

Cooperband, L.R. and L.H. Middleton. 1996. Changes in chemical, physical and biological properties of passively-aerated co-composted poultry litter and municipal solid waste compost. *Compost Science & Utilization* 4(4):24–34.

Corominas, E., F. Perestelo, M.L. Perez, and M.A. Falcon. 1987. Microorganisms and environmental factors in composting of agricultural waste of the Canary Islands, p. 127–138. In: M. de Bertoldi, M. P. Ferranti, P. L'Hermite, and F. Zucconi (eds.). *Compost: Production, Quality and Use.* Elsevier Applied Science, London, United Kingdom.

CRIQ, 1998. Composting Formulation Software Force 3, Version 2. Centre de Reserche Industrielle du Quebec (CRIQ), Sainte-Foy, Quebec, Canada.

Day, D.L. and T.L Funk. 1994. Processing manure: physical, chemical and biological treatment, p. 244-282. In: J.L. Hatfield and B.A. Stewart (eds.) *Animal Waste Utilization: Effective Use of Manure as a Soil Resource.* Ann Arbor Press, Chelsea, Michigan.

Day, M., M. Krzymien, K. Shaw, L. Zaremba, W.R. Wilson, C. Botden, and B. Thomas. 1998. An investigation of the chemical and physical changes occuring during commercial composting. *Compost Science & Utilization* 6(2):44–66.

de Bertoldi, M., A. Rutiki, B. Citterio, and M. Civilini. 1988. Composting management: a new process control through O_2 feedback. *Waste Management & Research* 6:239–259.

Dindal, D.L. 1978. Soil organisms and stabilizing wastes. *Compost Science/Land Utilization* 19(8):8–11.

Duggan, J.J. 1991. The relationships between temperature, oxygen consumption and respiration in the passively aerated composting of solid manures. *Report for Agriculture Canada.* Land Resources Division, Ottawa, Ontario, Canada.

Epstein, E., G.B. Willson, W.D. Burge, D.C. Mullen, and N.K. Enkiri. 1976. A forced aeration system for composting wastewater sludge. *Water Pollution Control Federation* 48:688–694.

Epstein, E., G.B. Willson, and J.F. Parr. 1977. The Beltsville aerated pile method for composting sewage sludge, p. 201-213. In: *New Processes of Waste Water Treatment and Recovery.* Society of Chemical Industry, London, United Kingdom.

Epstein, E. 1997. *The Science of Composting*. Technomic Publishing Inc., Lancaster, Pennsylvania, p. 83.

Fermor, T.R. 1993. Applied aspects of composting and bioconversion of lignocellulosic materials: an overview. *International Biodeterioration & Biodegradation* 31:87–106.

Fernandes, L. and M. Sartaj. 1997. Comparative study of static pile composting using natural, forced and passive aeration methods. *Compost Science & Utilization* 5(4):65–77.

Finstein, M.S. 1992. Composting in the context of municipal solid waste management, p. 355–374. In: R. Mitchell (ed.). *Environmental Microbiology.* Wiley-Liss, Inc., New York.

Finstein, M.S. and M.L. Morris. 1974. Microbiology of municipal solid waste composting. *Advances in Applied Microbiology* 19:113–151.

Finstein, M.S., F.C. Miller, and P.F. Strom. 1986. Waste treatment composting as a controlled system, p. 363–398. In: W. Schenborn (ed.). *Biotechnology, Vol. 8-Microbial Degradations.* VCH Verlaqsgedellschaft [German Chemical Society]: Weinheim F.R.G.

Finstein, M.S., F.C. Miller, P.F. Strom, S.T. MacGregor, and K.M. Psarlanos. 1983. Composting ecosystems management for waste treatment. *Biotechnology* 1:347–353.

Finstein, M.S., F.C. Miller, J.A. Hogan, and P.F. Strom. 1987. Analysis of EPA guidance on composting sludge. *Biocycle* 28(2):42–47.

Fischer, J.L., T. Brello, P.F. Lyon, and M. Aragno. 1998. Aspergillus fumigatus in windrow composting: effect of turning frequency. *Waste Management and Research* 16(4):320–329.

Fricke, K. and H. Vogtmann. 1993. Quality of source separated compost. *Biocycle* 34(10):64–70.

Fricke, K. and H. Vogtmann. 1994. Compost quality: physical characteristics, nutrient content, heavy metals and organic chemicals. *Toxicological and Environmental Chemistry* 43:95–114.

Gabriels, R. and O. Verdonck. 1992. Reference methods for analysis of composts, p. 173–183. In: *Composting and Compost Quality Assurance Criteria.* Commission of the European Communities: Proceedings. Angers, France.

Genevini, P.L., F. Adani, D. Borio, and F. Tambone. 1997. Heavy metal content in selected European commercial composts. *Compost Science & Utilization* 5(4):31–39.

Gies, G. 1997. Developing compost standards in Europe. *Biocycle* 38(10):82–83.

Glass, J.S. 1993. Composting wastewater biosolids. *Biocycle* 34(1):68–72.

Glenn, J. 1998. Solid waste composting trending upward. *Biocycle* 39(11):65–72.

Goldstein, N. and D. Block. 1999. Biocycle nationwide survey — Biosolids composting in the states. *Biocycle* 40(1):63–76.

Goldstein, N., J. Glenn, and K. Gray. 1998. Nationwide overview of food residuals composting. *Biocycle* 39(1):50–60.

Golueke, C.G. 1977. *Biological Reclamation of Solid Wastes.* Rodale Press, Emmaus, Pennsylvania, p. 9.

Golueke, C.G. 1989. Putting principles into successful practice, p. 106–110. In: The staff of *Biocycle* (eds.). *The Biocycle Guide to Yard Waste Composting.* The JG Press, Inc., Emmaus, Pennsylvania.

Gotass, H.B. 1956. *Composting – Sanitary Disposal and Reclamation of Organic Wastes.* World Health Organisation Monograph Series No. 31.

Gray, K. 1970. Research on composting in British universities. *Compost Science* 5:12–15.

Gray, K.R. and A.J. Biddlestone. 1974. Decomposition of urban waste, p. 743–775. In: C.H. Dickenson and G.J.F. Pugh (eds.). *Biology of Plant Litter Decomposition.* Academic Press, London, United Kingdom.

Grebus, M.E., M.E. Watson, and H.A.J. Hoitink. 1994. Biological, chemical and physical properties of composted yard trimmings as indicators of maturity and plant disease suppression. *Compost Science & Utilization* 2(1):57–71.

Griffin, D.M. 1977. Water potential and wood-decay fungi. *Annual Review of Phytopathology* 15:319–329.

Hachicha, R., A. Hassen, N. Jedidi, and H. Kallali. 1992. Optimal conditions for MSW composting. *Biocycle* 33(6):76–77.

Haines, J. 1995. Aspergillus in compost: straw man or fatal flaw? *Biocycle* 36(4): 32–35.

Hamoda, M.F., H.A. Abu Qdais, and J. Newham. 1998. Evaluation of municipal solid waste composting kinetics. *Resources, Conservation and Recycling* 23:209–223.

Harper, E., F.C. Miller, and J. Macauley. 1992. Physical management and interpretation of an environmentally controlled composting ecosystem. *Australian Journal of Experimental Agriculture* 32:657–667.

Haug, R.T. 1993. *The Practical Handbook of Composting Engineering*. Lewis Publishers, Boca Raton, Florida.

Hay, J. 1996. Pathogen destruction and biosolids composting. *Biocycle* 37(6):67–76.

He, X.T., T.J. Logan, and S.J. Traina. 1995. Physical and chemical characteristics of selected U.S. municipal solid waste composts. *Journal of Environmental Quality* 3:543–552.

Hogan, J.A., F.C. Miller, and M.S. Finstein. 1989. Physical modeling of the composting ecosystem. *Applied and Environmental Microbiology* 55(5):1082–1092.

Howe, C.A. and C.S. Coker. 1992. Co-composting municipal sewage sludge with leaves, yard wastes and other recyclables a case study. In: *Air Waste Management Association*. 85th Annual Meeting and Exhibition, Kansas City, Missouri, 21–26 June 1992.

Hsu, S.M., J.L. Schnoor, L.A. Licht, M.A. St.Clair, and S.A. Fannin. 1993. Fate and transport of organic compounds in municipal solid waste compost. *Compost Science & Utilization* 1(4):36–48.

Iannotti, D.A., T. Pang, B.L. Toth, D.L. Elwell, H.M. Keener, and H.A.J. Hoitink. 1993. A quantitative respirometric method for monitoring compost stability. *Compost Science & Utilization* 1(3):52–65.

Inbar, Y., Y. Chen, and Y. Hadar. 1989. Solid-state carbon-13 nuclear magnetic resonance and infrared spectroscopy of composted organic matter. *Soil Science Society of America Journal* 53:1695–1701.

Inbar, Y., Y. Chen, and Y. Hadar. 1990. Humic substances formed during the composting of organic matter. *Soil Science Society of America Journal* 54:1316–1323.

Jackson, M.J. and M.A. Line. 1998. Assessment of periodic turning as an aeration mechanism for pulp and paper mill sludge composting. *Waste Management Research* 4:312–319.

Jeris, J.S. and R.W. Regan. 1973a. Controlling environmental parameters for optimum composting (Part I). *Compost Science* 14(1):10–15.

Jeris, J.S. and R.W. Regan. 1973b. Controlling environmental parameters for optimum composting (Part II). *Compost Science* 14(2):8–15.

Jeris, J.S. and R.W. Regan. 1973c. Controlling environmental parameters for optimum composting (Part III). *Compost Science* 14(3):16–22.

Jimenez, E.I. and V.P. Garcia. 1992. Determination of maturity indices for city refuse composts. *Agricultural Ecosystem Environment* 38:331–343.

Kane, B.E. and J.T. Mullins. 1973a. Thermophilic fungi and the compost environment in a high-rate municipal composting system. *Compost Science* 14(6):6–7.

Kane, B.E. and J.T. Mullins. 1973b. Thermophilic fungi in a municipal waste compost system. *Mycologia* 65:1087–1100.

Kayhanian, M. and G. Tchobanoglous. 1992. Computations of C/N ratio for various organic fractions. *Biocycle* 33(5):58–60.

Kayhanian, M. and G. Tchobanoglous. 1993. Characteristics of humus produced from the anaerobic composting of the biodegradable organic fraction of municipal solid waste. *Environmental Technology* 14:815–829.

Keller, P. 1961. Methods of evaluating maturity of compost. *Compost Science* 2:20–26.

Kim, J.Y., J.K. Park, B. Emmons, and D.E. Armstrong. 1995. Survey of volatile organic compounds at a municipal solid waste composting facility. *Waste Environment Research* 67(7):1044–1051.

Kissel, J.C., C.L. Henry, and R. B. Harrison. 1992. Potential emissions of volatile and odorous organic compounds from municipal solid waste composting facilities. *Biomass and Bioenergy* 3(3-4):181–194.

Kochtitzky, I.W., W.K. Seaman, and J.S. Wiley. 1969. Municipal composting research at Johnson City, Tennessee. *Compost Science* 9(4):5–16.

Kosaric, N. and R. Velayudhan. 1991. Biorecovery processes: fundamental and economic considerations, p. 3-37. In: A.M. Martin (ed.). *Bioconversion of Waste Materials to Industrial Products*. Elsevier Applied Science, New York.

Krueger, R.G., N.W. Gillam, and J.H. Coggin, Jr. 1973. *Introduction to Microbiology*. The Macmillan Co., New York.

Kuroda, K., T. Osada, M. Yonaga, A. Kanematu, T. Nitta, S. Mouri, and T. Kojima. 1996. Emissions of malodorous compounds and greenhouse gases from composting swine feces. *Bioresource Technology* 56:265–271.

Kuter, G.A., H.A.J. Hoitink, and L.A. Rossman.1985. Effects of aeration and temperature on composting of municipal sludge in a full-scale vessel system. *Journal of Water Pollution Control Federation* 57(4):309–315.

Lau, A.K., K.V. Lo, P.H Liao, and J.C. Yu. 1992. Aeration experiments for swine waste composting. *Bioresource Technology* 41:145–152.

Liao, P.H., A.C. May, and S.T. Chieng. 1995. Monitoring process efficiency of a full-scale in-vessel system for composting fisheries wastes. *Bioresource Technology* 54:159–163.

Liao, P.H., A.T. Vizcarra, A. Chen, and K.V. Lo. 1996. Composting of salmon farm mortalities with passive aeration. *Compost Science & Utilization* 2(4):58–66.

Lynch, N.J. and R.S. Cherry. 1996. Winter composting using the passively aerated windrow system. *Compost Science & Utilization* 4(3): 44–52.

MacGregor, S.T., F.C. Miller, K.M. Psarianos, and M.S. Finstein. 1981. Composting process control based on interaction between microbial heat output and temperature. *Applied and Environmental Microbiology* 41(6):1321–1330.

Marugg, C., M. Grebus, R.C. Hansen, H.M. Keener, and H.A.J. Hoitink. 1993. A kinetic model of the yard waste composting process. *Compost Science & Utilization* 1(1):38–51.

Mathur, S.P. 1991. Composting processes, p. 147–183. In: A.M. Martin (ed.). *Bioconversion of Waste Materials to Industrial Products*. Elsevier Applied Science, New York.

Mathur, S.P., N.K. Patri, and M.P. Levesque. 1990. Static pile, passive aeration composting of manure slurries, using peat as a bulking agent. *Biological Wastes* 34:323–334.

Mato, S., D. Otero, and M. Garcia. 1994. Composting of <100mm fraction of municipal solid waste. *Waste Management and Research* 12:315–325.

McBride, M.B. 1989. Reactions controlling heavy metal solubility in soils, p. 1–56. In: B.A. Stewart (ed.). *Advances in Soil Sciences*. Springer-Velag, New York.

McGaughey, P.H. and H.B Gotass. 1953. Stabilisation of municipal refuse by composting, p. 897–920. *American Society of Civil Engineers Transactions*. Proceedings-Separate No.302 Paper No.2767.

McKinley, V.L. and J.R. Vestal. 1985. Physical and chemical correlates of microbial activity and biomass in composting municipal sewage sludge. *Applied and Environmental Microbiology* 50(6):1395–1403.

McKinley, V.L., J.R. Vestal, and A.E. Eralp. 1985. Microbial activity in composting. *Biocycle* 26(10):47–50.

Michel, F.C. Jr., C.A. Reddy, and L.J. Forney. 1993. Yard waste composting: studies using different mixes of leaves and grass in a laboratory scale system. *Compost Science and Utilization* 1(3): 85–96.

Michel, F.C., Jr., D.Graeber, L.J. Forney, and C.A. Reddy. 1996. The fate of lawn care pesticides during composting. *Biocycle* 37(3):64–66.

Miller F.C. 1989. Matric water potential as an ecological determinant in compost, a substrate dense system. *Microbial Ecology* 18:59–71.

Miller, F.C., B.J. Macauley, and E.R. Harper. 1991. Investigation of various gases, pH and redox potential in mushroom composting phase I stacks. *Australian Journal of Experimental Agriculture* 31:415–425.

Morisaki, N., C.G. Phae, K. Nakasaki, M. Shoda, and H. Kubota. 1989. Nitrogen transformation during thermophilic composting. *Journal of Fermentation and Bioengineering* 1:57–61.

Nakasaki, K., M. Sasaki, M. Shoda, and H. Kubota. 1985. Change in microbial numbers during thermophilic composting of sewage sludge with reference to CO_2 evolution rate. *Applied and Environmental Microbiology* 49(1):37–41.

Nakasaki, N., A. Watanbe, and H. Kubota. 1992a. Effects of oxygen concentration on composting organics. *Biocycle* 33(6):52–54.

Nakasaki, K., H. Yaguchi, Y. Sasaki, and H. Kubota. 1992b. Effects of C/N ratio on thermophilic composting of garbage. *Journal of Fermentation and Bioengineering* 1:43–45.

Nakasaki, K., N. Aoki, and H. Kubota. 1994. Accelerated composting of grass clippings by controlling moisture level. *Waste Management and Research* 12:13–20.

Nakasaki, K., N Akakura, and K. Atsumi. 1998. Degradation patterns of organic material in batch and fed-batch composting operations. *Waste Management and Research* 16(5):484–489.

Naylor, L.M. 1996. Composting. *Environmental and Science and Pollution Series* 18(69):193–269.

Neto, J.T.P., E.I. Stentiford, and D.D. Mara. 1987. Comparative survival of pathogenic indicators in windrow and static pile, p. 276–295. In: M. de Bertoldi, M.P. Ferranti, P. L'Hermite, and F. Zucconi (eds.). *Compost: Production, Quality and Use.* Elsevier Applied Science, London, United Kingdom.

Niemeyer, J., Y. Chen, and J.M. Bollag. 1992. Characterization of humic acids, composts and peat by diffuse reflectance fourier-transform infrared spectroscopy. *Soil Science Society of America Journal* 56:135–140.

Olver, W.M. Jr. 1994. The Aspergillus fumigatus problem. *Compost Science & Utilization* 2(1):27–31.

Palmisano, A.C., D.A. Maruscik, C.J. Ritchie, B.S. Schwab, S.R. Harper, and R.A. Rapaport. 1993. A novel bioreactor simulating composting of municipal solid waste. *Journal of Microbiological Methods* 18:99–112.

Papadimitriou, E.K. and C. Balis. 1996. Comparative study of parameters to evaluate and monitor the rate of a composting process. *Compost Science & Utilization* 4(4):52–61.

Petruzzelli, G., I. Szymura, L. Lubrano, and B. Pezzarossa. 1989. Chemical separation of heavy metals in different size fractions of compost from solid urban wastes. *Environmental Technology Letter* 10:521–526.

Poincelet, R.P. 1977. The biochemistry of composting, p.33–39. In: *Composting of Municipal Sludges and Wastes*. Proceedings of the National Conference, Rockville, Maryland.

Polprasert, C. 1989. *Organic Waste Recycling*. John Wiley & Sons Ltd., Chichester, United Kingdom.

Preston, C.M., B.J. Cade-Benun, and B.G. Sayer. 1998. Characterization of Canadian backyard composts chemical & spectroscopy analysis. *Compost Science & Utilization* 6(6):53–66.

Proyenzano, M.R., N. Senesi, and G. Piccone. 1998. Thermal and spectroscopic characterization of composts from municipal solid waste. *Compost Science & Utilization* 1(6):67–73.

Regan, R.W. and J.S. Jeris. 1970. A review of the decomposition of cellulose and refuse. *Compost Science* 11(1):17–20.

Reinhart, D.R., A.R. deForest, S.J. Keely, and D.R. Vogt. 1993. Composting of yard waste and wastewater treatment plant sludge mixtures. *Compost Science & Utilization* 1(2):58–64.

Richard, T.L. 1992. Municipal solid waste composting: physical and biological processing. *Biomass and Bioenergy* 3(3–4):163–180.

Richard, T.L. and P.B. Woodbury. 1994. What materials should be composted. *BioCycle* 35(9):63–68.

Roediger, H.J. 1964. The technique of sewage-sludge pasteurization: actual results obtained in existing plants economy. *International Research Group on Refuse Disposal Information*, (now International Solid Wastes Association), Bulletin 21–31.

Roletto, E., R. Chiono, and E. Barberis. 1985. Investigation on humic mat from decomposing poplar bark. *Agricultural Wastes* 12:261–272.

Sartaj, M., L. Fernandes, and N.K. Patni. 1995. Influence zone of aeration pipes and temperature variations in passively aerated composting of manure slurries. *Transactions of the American Society of Agricultural Engineers* 38(6):1835–1841.

Savage, G.M. 1996. The importance of waste characteristics and processing in the production of quality compost, p. 784–791. In: M. de Bertoldi, P. Sequi, B. Lemmes, and T. Papi. (eds.). *The Science of Composting, Part 2*. Blackie Academic and Professional, Glasgow, United Kingdom.

Saviozzi, A., R. Levi-Minzi, and R. Riffaldi. 1988. Maturity evaluation of organic waste. *BioCycle* 29:54–56.

Schnitzer, M., H. Dinel, S.P. Mathur, H.R. Schulten, and G. Owen 1993. Determination of compost maturity. III. Evaluation of a calorimetric test by ^{13}C-NMR spectroscopy and pyrolysis field ionization mass spectrometry. *Biological Agriculture and Horticulture* 10:109–123.

Schulze, K.L. 1960. Rate of oxygen consumption and respiratory quotients during the aerobic decomposition of a synthetic garbage. *Compost Science* 1(1):36–40.

Schulze, K.L. 1961. Aerobic decomposition of organic waste materials. *Final Report, Project RG-4180 (C5R1)*. National Institutes of Health, Washington, DC.

Schulze, K.L. 1962a. Continuous thermophilic composting. *Applied Microbiology* 10:108–122.

Schulze, K.L. 1962b. Continuous thermophilic composting. *Compost Science* 3(1):22–34.

Senesi, N., T.M. Miano, M.R. Provenzano, and G. Brunetti. 1991. Characterization, differentiation and classification of humic substances by fluorescence spectroscopy. *Soil Science* 152(4):259–271.

Sesay, A.A., K.E. Lasaridi, and E.I. Stentiford. 1998. Aerated static pile of composting of municipal solid waste (MSW): a comparison of positive pressure aeration with hybrid positive and negative aeration. *Waste Management and Research* 3:264–272.

Shell, G.L. and J.L. Boyd. 1969. Composting dewatered sewage sludge. *Public Health Service Publication No. 1936*. U.S. Department of Health, Education and Welfare, Washington, DC.

Shin, H.S. and Y.K. Jeong. 1996. The degradation of cellulosic fraction in composting of source separated food waste and paper mixture with change of C/N ratio. *Environmental Technology* 17:433–438

Sikora, L.J. and M.A. Sowers. 1985. Effect of temperature control on the composting process. *Journal of Environmental Quality* 14(3):434–439.

Sikora, L.J., M.A. Ramirez, and T.A. Troeschel. 1983. Laboratory composter for simulation studies. *Journal of Environmental Quality* 12(2):219–224.

Singley, M.E., A.J. Higgins, and M. Frumkin-Rosengau. 1982. *Sludge Composting and Utilization - A Design and Operating Manual.* New Jersey Agricultural Experimental Station, Rutgers University, New Brunswick, New Jersey.

Snell, J.R. 1957. Some engineering aspects of high-rate composting. *Journal of the Sanitary Engineering Division of the American Society of Civil Engineers* 83, No. SA 1, paper no. 1178.

Snell, J.R. 1991. Proper grinding for efficient composting. *BioCycle* 32(4):54–55.

Stentiford, E.I. 1987. Recent developments in composting, p. 52–60. In: M. de Bertoldi, M.P. Ferranti, P. L'Hermite, and F. Zucconi (eds.). *Compost: Production, Quality and Use.* Elsevier Applied Science, London, United Kingdom.

Stentiford, E.I. 1996. Composting control: principles and practice, p.49–59. In: M. de Bertoldi (ed.). *The Science of Composting, Part 1.* Blackie Academic and Professional, Glasgow, United Kingdom.

Stevenson, F. 1986. *Cycles of Soil: Carbon, Nitrogen, Phosphorous, Sulfur, Micronutrients.* John Wiley & Sons, New York.

Strom, P.F. 1985a. Effect of temperature on bacterial species diversity in thermophilic solid-waste composting. *Applied and Environmental Microbiology* 50(4):899–905.

Strom, P.F. 1985b. Identification of thermophilic bacteria in solid-waste composting. *Applied and Environmental Microbiology* 50(4):906–913.

Suler, D.J. and M.S. Finstein. 1977. Effect of temperature, aeration and moisture on CO_2 formation in bench-scale, continuously thermophilic composting of solid waste. *Applied and Environmental Microbiology* 33(2):345–350.

Tchobanoglous, G., H. Theisen, and S. Vigil. 1993. Biological and chemical conversion technologies, p. 671–716. In: G. Tchobanoglous (ed.). *Integrated Solid Waste Management: Engineering Principles and Management Issues*, 2nd edition. McGraw-Hill Higher Education, New York.

Tisdell, S.E. and V.T. Breslin. 1995. Heavy metals in the environment. *Journal of Environmental Quality* 24:827–833.

Toffey, W.E. and L.H. Hentz. 1995. Measurement and control of odour and VOC emissions from the largest municipal aerated-static pile biosolids composting facility in the United States, p. 141–152. In: *The 68th Water Environment Federation Annual Conference: Residuals and Biosolids Management*, Miami, Florida, 21–25 October 1995.

Tseng, D.Y., J.J. Chalmers, O.H. Tuovinen, and H.A.J. Hoitink. 1995. Characterization of a bench-scale system for studying the biodegradation of organic wastes. *Biotechnology Progress* 11:443–451.

United States Environmental Protection Agency (U.S. EPA). 1971. *Composting of Municipal Solid Wastes in the United States.* Solid Waste Management Series SW–47r, p. 103.

van Bochove, E., D. Couillard, M. Schnitzer, and H.-R. Schulten. 1996. Pyrolysis – field ionization mass spectrometry of the four phases of cow manure composting. *Soil Science Society of America Journal* 60:1781–1786.

van der Werf, P. and J. Ormseth. 1997. Measuring process parameters at an enclosed composting facility. *Biocycle* 38(5):58–61.

Vogtmann, H., K. Fricke, and T. Turk. 1993. Quality, physical characteristics, nutrient content, heavy metals and organic chemicals in biogenic waste compost. *Compost Science & Utilization* 1(4):69–87.

Walke, R. 1975. *The Preparation, Characterization and Agricultural Use of Bark-Sewage Compost*, p. 47. Ph.D. Thesis, The University of New Hampshire, Durham, New Hampshire.

Walker, J.M. 1996. U.S.A. Environmental Protection Agency regulations for compost production and use, p. 357–369. In: M. de Bertoldi, P. Sequi, B. Lemmes, and T. Papi. (eds.) *The Science of Composting, Part I.* Blackie Academic and Professional, Glasgow, United Kingdom.

Warman, P.R. and W.C. Termeer. 1996. Composting and evaluation of racetrack manure, grass clippings and sewage sludge. *Bioresource Technology* 55:95–101.

Webley, D.M. 1947. The microbiology of composting. I. The behavior of the aerobic mesophilic bacteria flora of composts and its relation to other changes taking place during composting. *Proceedings of the Society of Applied Bacteriology* 2: 83–89.

Wiley, J.S., A.M. Asce, and G.W. Pearce. 1955. A preliminary study of high-rate composting, p. 1009–1034. In: *American Society of Civil Engineers Transactions*, Paper no. 2895.

Wiley, J.S. 1957. High rate composting. *American Society of Civil Engineering Transcripts* 2895.

Wiley, J.S. and J.T. Spillane. 1962. Refuse-sludge composting in windrows and bins. *Compost Science* 2(4):18–25.

Willson, G.B. 1993. Combining raw materials for composting, p. 102–105. In: J. Goldstein (ed.). *The Biocycle Guide to Yard Waste Composting.* The JG Press, Emmaus, Pennsylvania.

Witter, E. and J.M. Lopez-Real. 1987. Monitoring the composting process using parameters of compost stability, p. 351–358. In: M. de Bertoldi, M.P. Ferranti, P. L'Hermite, and F. Zucconi (eds.). *Compost: Production, Quality and Use.* Elsevier Applied Science, London, United Kingdom.

Zucconi, F. and M. de Bertoldi. 1987. Compost specifications for the production and characterization of compost from municipal solid waste, p. 30–50. In: M. de Bertoldi, M. P. Ferranti, P. L'Hermite, and F. Zucconi (eds.). *Compost: Production, Quality and Use.* Elsevier Applied Science, London, United Kingdom.

CHAPTER 3

Commercial Compost Production Systems

Robert Rynk and Thomas L. Richard

CONTENTS

1-56670-460-X/01/$0.00+$.50

I. INTRODUCTION

Composting is a treatment technique for organic materials, a manufacturing process for soil products, a method of recycling organic matter and nutrients, a means to pasteurize pathogen-infested media, and a disposal strategy for troublesome materials like animal mortalities. It is a treatment operation, a commercial venture, an agricultural practice, and a backyard hobby. Composting serves all of these purposes, and more, because it is simple, flexible, and applicable to a broad range of scales. In short, there are many reasons to produce compost and many ways to accomplish it.

The utility of composting stems from two basic functions: (1) it changes the qualities of difficult and often unwanted materials, yielding a product that is, at minimum, more easily handled and used; and (2) it creates compost, a product that has better uses and more value than the feedstocks from which it is made. Because of the first function, composting is a method for treating organic byproducts, or residuals, allowing them to be safely or economically recycled. Because of the second function, composting is a production process for an industry that manufactures and sells compost-based products for agricultural, horticultural, and environmental uses (e.g., soil amendments, mulch, potting blends, organic fertilizers, and media for controlling erosion and for remediating contaminated soils). Other applications of composting fall between its treatment and manufacturing roles. For example, livestock manure is composted without necessarily creating a product for sale but to improve its value for farm use (e.g., to reduce weight, volume, weed seeds, and pathogens).

The many applications of composting have given rise to many types of composting systems. Composting systems have evolved and proliferated to handle different raw materials or "feedstocks." They reflect various scales of operation, facility locations, end products, economic constraints, and differing philosophies and preferences among managers and operators. Therefore, composting methods range from simple freestanding piles to highly controlled reactors with automated materials handling and sophisticated process control. However, they share a common link to the underlying biological principles of composting. They all depend on the microorganisms that transform raw feedstocks into compost. Where the systems differ is in how, and how much, they attempt to control the biological processes at work. They also differ in how they interact with the system operators and the surrounding environment.

This chapter describes the process of composting and composting methods that are commonly used by municipalities, commercial companies, and farms to produce compost for horticultural uses. A distinction is made here between composting "systems" and composting "methods." Because the composting method is so important to the system, the terms, system and method, are frequently used interchangeably. A system encompasses all of the operations, equipment, structures, and procedures used to produce compost including grinding, screening, materials handling, odor treatment, process monitoring, and features of the composting site. Composting systems can range from a simple backyard pile and pitchfork to a complex industrial facility (Dickson et al., 1991; Haug, 1993; Richard, 1992; Rynk et al., 1992). The term system also applies to a particular set of procedures and equipment used at a given facility or offered by a company selling a commercial system. The composting method is the central part of the system. It is the primary operation within which the raw feedstocks become compost. This chapter largely covers composting methods. Other elements of the systems enter into the descriptions because such elements are often integrated into the composting method and because it is difficult to separate the method from the system it serves. For example, odor control and site management are critical elements in the success of any composting system. Because of their importance, these elements are discussed in some detail as well.

II. THE COMPOSTING PROCESS IN BRIEF

Composting is a biological process through which microorganisms convert organic materials into compost. It is predominantly an aerobic or oxygen (O_2)-requiring process. The microorganisms consume O_2 to extract energy and nutrients from organic matter. In doing so they produce carbon dioxide (CO_2), water, heat, compost, and miscellaneous gaseous byproducts of decomposition. Organic matter is lost as carbon (C) compounds decompose, creating CO_2 and volatile compounds such as ammonia (NH_3) that evaporate into the environment. Many biological transformations and products occur in the composting process, mediated by a variety of microorganisms, inhabiting diverse microenvironments (Epstein, 1997; Poincelot, 1975). Although organisms decompose some organic materials, they continue to create new organic compounds from the products of decomposition. Elements such as nitrogen (N) and sulfur (S) combine with other elements, changing repeatedly between soluble and insoluble forms (Miller, 1993). Soluble elemental forms are subject to microbial use or possible leaching. Other physical and chemical processes are also at work, affecting the porosity, water- and nutrient-holding capacity, conductivity, pH, and other properties that may influence either the composting process or potential uses of the product. See the chapter by Day and Shaw in this book for more details about the biological, chemical, and physical processes involved in composting.

Oxygen is provided to the composting materials via aeration (see following section). Aeration mechanisms can be very effective, but they are not perfect (Epstein, 1997). In reality, part of the decomposition also occurs anaerobically (i.e., without O_2). Anaerobic processes contribute to the overall decomposition of the composting materials. However, excessive anaerobic decomposition is undesirable during composting because it results in incomplete degradation and odors (Miller, 1993). Providing well-aerated conditions minimizes the odors associated with anaerobic processes and completes the decomposition of partially degraded anaerobic byproducts such as organic acids, which can contribute to phytotoxicity when the compost is ultimately used (Epstein, 1997).

In addition to O_2, the organisms need moisture, a balance of nutrients, and favorable temperatures and pH (Table 3.1). Abundant moisture is required for the metabolic activities of the organisms. Water provides a medium for biochemical reactions and transport of nutrients and organisms (Miller, 1991). However, excessive water interferes with aeration by filling pore spaces and by increasing the weight of the composting materials, causing them to compact and lose pore space (Das and Keener, 1996). The ideal balance of moisture generally falls in the range of 50 to 60% (wet basis). Depending of the materials and composting method, moisture contents of 40 to 70% are tolerable. Below 40%, the composting rate slows due to lack of moisture. Above an upper limit of 60 to 70%, aeration is extremely difficult (Haug, 1993). Proper moisture content is established initially by adding water to dry feedstocks or by mixing dry feedstocks and wet feedstocks together. Mixing diverse materials is common because it also improves the physical characteristics of the composting substrate and its nutrient balance. Once the composting process starts, high temperatures and air movement drive evaporation, which generally reduces the compost moisture content over time. In outdoor systems, rainfall can reduce or even reverse this drying process, as can metabolic water produced as a byproduct of decomposition. During the composting process, moisture levels can be maintained by adding water to materials that have dried excessively. If materials become too wet from precipitation and metabolic water, they can be aerated or agitated to promote drying.

Table 3.1 Preferred Conditions for Composting

Condition	Reasonable Range[z]	Preferred Range
Carbon to nitrogen (C:N) ratio	20:1–40:1	25:1–30:1
Moisture content	40–65%[y]	50–60%
O_2 concentrations	Greater than 5%	Much greater than 5%
Particle size (diameter in mm)	3–13	Varies[y]
pH	5.5–9.0	6.5–8.0
Temperature (°C)	43–66	54–60

[z] Recommendations for rapid composting. Conditions outside these ranges can also yield successful results.

[y] Depends on the specific materials, pile size, and weather conditions.

From Rynk et al., 1992. *On-Farm Composting Handbook.* NRAES, Ithaca, New York. With permission.

Although most organic materials provide all of the nutrients needed by the microorganisms, the nutrient content is often not balanced for optimal composting. Ordinarily, nutrients are managed by providing balanced proportions of two primary nutrients, C and N. In some cases other nutrients, such as phosphorus (P), have been found to be limiting (Brown et al., 1998). Nevertheless, it is commonly assumed that with a reasonable carbon to nitrogen (C:N) ratio, other required nutrients are available in sufficient quantities (Rynk et al., 1992). An ideal C:N ratio is considered to be in range of 25:1 to 30:1 (Epstein, 1997). With ratios above 30:1, composting may be limited by lack of N. With lower C:N ratios, excess N is converted to NH_3, which is subject to loss via volatilization and leaching. In practice, the C:N ratio of the feedstocks is not as crucial as this implies. Composting takes place effectively at C:N ratios from 20:1 to 50:1 and even higher (Horwath et al., 1995). Also, the biochemical processes depend not only on the concentrations of nutrients available but also on their biological availability, a factor that is more difficult to account for. This is particularly true for C, which is often confined in biologically resistant organic compounds (Epstein, 1997; Lynch, 1993).

Microorganisms must be able to access the nutrients in organic substrates for composting to proceed. Composting organisms colonize the surface of solid particles where water and O_2 are present; here they can hydrolyze organic compounds into more degradable soluble forms. With more substrate surface area, more organic material is accessible to the organisms. Because smaller particles provide greater surface area, decomposition increases with smaller particle size (Golueke, 1972). However, very small particles reduce the size of the pores in the composting materials, which hinders aeration. Therefore, particle size is another factor to balance during composting. An ideal particle size is impractical to specify because it depends on the feedstocks, stage of the process, aeration system, and many other dynamic factors. Generally, a mix of coarse and fine particles in the range of 3 to 50 mm works well (Gray and Biddlestone, 1974; Rynk et al., 1992).

The heat produced during composting is a direct result of the biological activity (Miller, 1991). Therefore, the composting temperature indicates the performance and stage of the composting process (Golueke, 1972). Microbial activity early in the process quickly raises the temperature to thermophilic levels, typically reaching 40 to 60°C. As the easily degradable organic materials are consumed, gradually the activity slows and the temperature drops back to mesophilic levels. Near the end of the process, only the more resistant organic materials remain. Microbial activity slows to a constant level, with temperatures remaining near ambient. Thermophilic temperatures are not necessary to composting. The process takes place readily at lower mesophilic temperatures (10 to 40°C). In fact, some transformations are thought to be carried out solely by mesophilic organisms (Rynk et al., 1992). However, thermophilic temperatures provide the advantages of faster biochemical reaction rates and more effective destruction of weed seeds and pathogens. Extremely high temperatures, above 60°C, are detrimental to the composting organisms, retard the process, and may impair the compost quality (Finstein and Hogan, 1993; Hoitink et al., 1993).

Environmental and health regulations require certain feedstocks, especially biosolids, to be composted at thermophilic temperatures for a prescribed amount of time (unless pathogen destruction is established by an alternative means). For example, the U.S. Environmental Protection Agency (U.S. EPA) regulations regard composting as a "Process to Further Reduce Pathogens" (PFRP) if temperatures are maintained above 55°C for 3 consecutive days in a static pile or in-vessel composting system (U.S. EPA, 1985). In a turned windrow system, 55°C must be maintained for 15 days with a minimum of five turns in that time. A wider range of uses are allowed for biosolids that meet PFRP conditions. In some jurisdictions, other feedstocks (e.g., yard trimmings) are required to be composted at PFRP conditions if the compost is offered for sale or public use (CIWMB, 2000).

III. FUNCTIONS AND MECHANISMS OF AERATION

The composting method determines how aeration occurs. Aeration is a crucial and inherent component of composting. It provides the O_2 needed for aerobic biochemical processes, and also removes heat, moisture, CO_2, and other products of decomposition. In fact, during most of the composting period, the amount of aeration required for cooling greatly exceeds the amount required for removing moisture or supplying O_2 (Finstein et al., 1986; Haug, 1993; Kuter, 1995). Thus, the need for aeration is more often determined by temperature than by O_2 concentration.

Although there are many variations, aeration generally takes place either passively or by forced air movement. Passive aeration, often called natural aeration, takes place by diffusion and natural air movement. Forced aeration relies on fans to move air through the mass of composting materials. A possible third mode of aeration being developed is a system that injects nearly pure O_2 into a closed composting reactor (Rynk, 2000c, 2000d).

A. Passive Aeration

Passive or natural aeration is driven by at least three mechanisms: molecular diffusion, wind, and thermal convection. Oxygen diffuses into material because there is more O_2 outside than within. Similarly, CO_2 diffuses outward. Although molecular diffusion is constantly at work to correct concentration imbalances, the process is slow and probably does not have a major effect on aerating piles (Haug, 1993; Miller, 1991). In exposed outdoor locations, wind can be a significant factor in O_2 transfer. Many composting site operators have observed gusts of wind causing puffs of steam to spout from piles. The influence of wind on aeration of open piles has not been widely documented.

Thermal convection is probably the greatest driving force for passive aeration in most composting systems (Haug, 1993; Lynch and Cherry, 1996a; Randle and Flegg, 1978). The heat generated during composting increases the temperature of gases with the materials, which in turn decreases their density. The warm gases rise out of the composting mass, create a vacuum, and cool fresh air enters. The aeration rate is determined by the temperature difference between the interior gases and the ambient air plus the air flow resistance of the composting media. Therefore, the keys to obtaining reliable passive air movement are generating heat to drive thermal convection and establishing a porous physical structure within the composting materials (Rynk et al., 1992). Porosity is particularly important. Dense, wet mixtures such as are common in sludge composting can reduce or eliminate the potential for convective O_2 transfer (Finstein et al., 1980).

Most composting systems that rely on passive aeration commonly include periodic agitation or "turning" of the materials. Although turning charges the materials with fresh air, the air introduced by turning is quickly consumed by the composting process (Epstein, 1997; Haug, 1993). The longer lasting effect of turning on aeration may be to rebuild the pore spaces in the material, which are crucial to diffusion and convection. However, there is evidence that this effect can be shorter lived as well (McCartney and Chen, 1999; Michel et al., 1996). The functions and effects of turning are discussed in more detail in the subsection covering the turned windrow method of composting.

B. Forced Aeration

With forced aeration, air is supplied mechanically, via fans and associated ducts, plenums and control devices — the aeration delivery and control system. There are innumerable possible combinations of aeration and control strategies and equipment configurations. Materials can be aerated by individual fans each turning on and off independently, or by a group of fans feeding a common manifold with valves controlling air flow to individual piles, pile sections, bins, or vessels. Air can be provided by positive pressure, forcing air into the distribution network and up through the materials (Stentiford, 1996), or by negative pressure, sucking air through the materials from the exterior and into the distribution network. Air movement is generally more efficient with positive pressure because pressure loss is favorable (Finstein and Hogan, 1993; Kuter, 1995). However, negative pressure offers the advantage of allowing the exhaust gases to be easily collected for odor treatment. It is also possible to use both strategies, reversing the direction of air flow at different stages of the composting process (Stentiford and Pereira Neto, 1985). Negative pressure might be primarily used early in the process when it is more important to capture gases for odor treatment. Periodically reversing the direction of air flow also helps to correct moisture and temperature gradients that occur in static composting systems (Stentiford and Pereira Neto, 1985). Some contained composting systems recycle a portion of the exhaust air to retain moisture and heat within the composting environment (Panter et al., 1996).

Depending upon the composting systems, forced aeration can be continuous and then increased as needed, or intermittently turned on and off as needed. Continuous aeration can reduce the required air flow rate. It also reduces the fluctuation of temperature and O_2 levels that occur over time. However, continuous aeration can cause gradients within the composting environment, leading to excessive drying and a permanently cool zone in the area where air enters (Citterio et al., 1987). This may be a concern if PFRP is required. Forced aeration is typically controlled based on the temperature within the composting materials. With many composting systems, temperature activates the aeration devices directly, via a temperature feedback control system of sensors, electronic components, and computer programs. Aeration is activated or increased when the process temperature surpasses a temperature set point. In other systems, aeration is determined by a time cycle that is adjusted either manually or automatically according to process temperature. Even with a direct temperature feedback control system, a timer is often required to activate aeration at regular intervals to maintain aerobic conditions when temperatures remain below the set point, especially during the initial and final stages of composting (Finstein et al., 1983). Aeration rates and intervals normally vary with the stage of composting

(Lenton and Stentiford, 1990). Composting systems may include several temperature zones, each requiring slightly different air flow rates and temperature set points.

IV. COMPOSTING AS A SYSTEM OF OPERATIONS

A composting system is a set of distinct operations that together produce compost products from raw feedstocks. A system also includes the associated infrastructure such as buildings, equipment, and utilities. What is often called the composting method is only part of the system, but it is the principal part. The biological transformations that take place occur primarily via the composting method. Some systems may employ two methods of composting to accommodate different stages of composting.

Supporting operations in the composting system *physically* alter the materials and materials handling operations move the materials through the system. Common supporting operations include: feedstock receiving and inspection, sorting and separation, feedstock storage, grinding or size reduction, mixing, curing, product storage, product blending, bagging and shipping (Figure 3.1). Not all of the operations listed above are necessary in all systems. Whether a given operation is used or not depends on the feedstocks, the composting method employed, and the intended use and markets for the compost. Grinding, mixing, screening, odor treatment, and curing are the most common supporting operations.

Grinding speeds the composting process by reducing particle size. It also improves materials handling and can facilitate other operations like mixing, prescreening, and sorting. Grinding is a necessary operation for large woody feedstocks like brush, tree branches, and construction and demolition wastes. It is frequently used for other materials as well including straw, mixed solid wastes, leaves, vegetation, and food. Hammermills, shear shredders, and rotating drums can all be effective in reducing the size of feedstocks prior to composting (Diaz et al., 1982; Richard, 1992).

When two or more separate materials are composted together, *mixing* of feedstocks prior to composting usually takes place as a separate operation. However, for methods that include regular agitation such as windrows and agitated beds, it may be sufficient to load feedstocks together in rough proportions and rely on the agitation during composting to create a homogeneous blend. When mixing is performed as a separate operation it is accomplished with bucket loaders, batch mixing equipment using augers or paddles, continuous pug mills, or windrow turners (Higgins et al., 1981; Rynk, 1994). Many batch mixing machines are truck mounted and equipped with conveyors that can construct windrows and piles as the mixed feedstocks are unloaded.

Screening is most often used to remove large particles (e.g., sticks, rocks, wood chips) from finished compost to improve the appearance and performance of the compost. Screening may not be necessary if the feedstocks do not contain large particles or if large particles are tolerable in the compost. Large particles are generally acceptable for compost that is used for mulch, erosion control, and many soil amendment applications. Screening is also used to recover large particles of bulking agent for reuse in the composting process and for separating particular materials

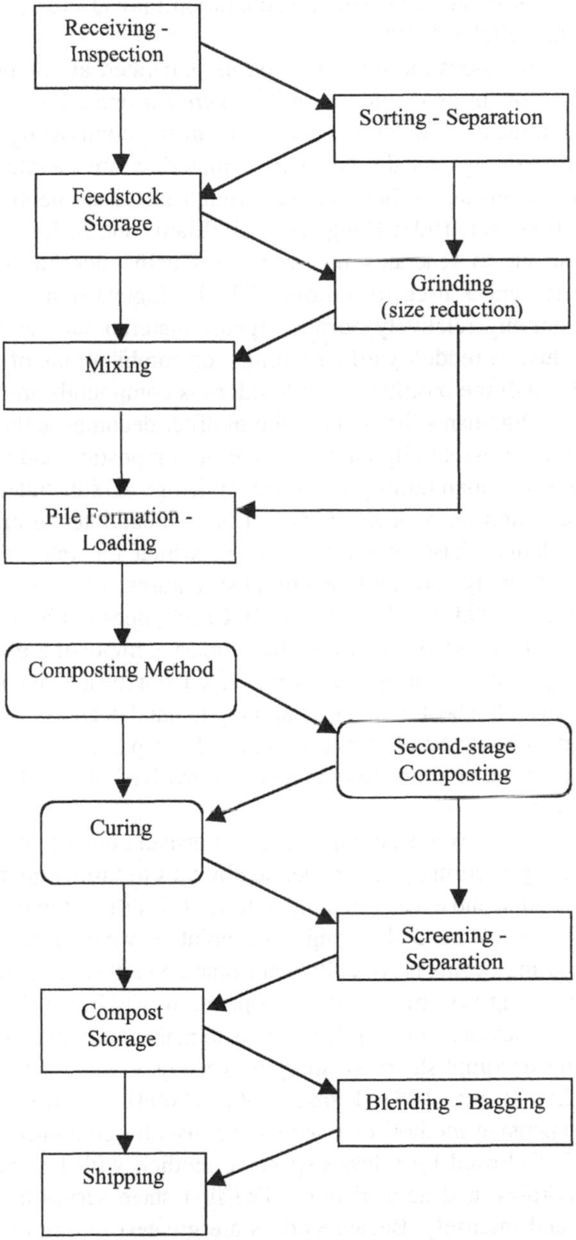

Figure 3.1 Flow chart of common operations in a typical composting system.

from raw feedstocks (e.g., removing soil from yard trimmings prior to grinding). The most common types of screens used at composting facilities are trommel screens (Rynk, 1994). Shaking, vibrating, and rotating disc screens are also used for size classification. If plastic or similar contaminants need to be separated from the compost, an air classification or ballistic separation may also be used as a compost refining technique (Richard, 1992).

Due to concerns about the impact of odors generated at composting facilities, many composting facilities include an *odor treatment* operation in the composting system. Odor treatment captures exhaust air from the composting aeration system and/or air from buildings on the site where materials are handled, processed, or stored. The odor-laden air is then passed through physical-chemical or biological treatment units (Dunson, 1993; Haug, 1993; Williams and Miller, 1993). Chemical units expose the air to reactants that remove specific odorous compounds. For example, sulfuric acid is used to remove NH_3. Biological units are large beds, or "biofilters," containing relatively porous organic material such as wood, compost, peat moss, sawdust, shredded yard trimmings, or combinations of these materials. As air passes through the biofilter, volatile odorous compounds are adsorbed on the organic particles. Organisms that inhabit the biofilter decompose the adsorbed compounds. A biofilter is essentially another stage of composting and therefore it must be managed as such, maintaining adequate moisture, C substrate, and O_2 levels. Additional information on odor management is included later in this chapter.

Curing is a latter phase of composting in which the rate of decomposition declines to a slow steady pace and the compost matures at low mesophilic (<40°C) temperatures (Haug, 1993; Rynk et al., 1992). Curing does not have to be a separate operation, but can be an integral part of the primary composting method. However, there are advantages in separating the curing stage from the active composting stage. As compost matures, the heat generation and O_2 demand decrease substantially. This generally permits a lower level of management. By separating the curing step, it is possible to reduce time or simplify management involved in monitoring and manipulating the process.

Curing typically involves stacking partially finished compost in piles that aerate passively. Increasingly, curing piles are aerated by fans to further guard against anaerobic conditions, odors, and excessive temperatures. It is difficult to draw the boundary between active composting and curing. The point at which active composting is terminated and curing begins is typically determined by either temperature or time. In continuous composting systems, the curing operation may begin simply because the compost must be removed from the pile or vessel to make room for incoming materials.

Some systems accomplish composting in two stages, with each stage employing a different method (Haug, 1993; Rynk, 2000c, 2000d). The first stage is a more intensive and expensive method of composting, usually contained within a vessel or building. It is followed by a less expensive method with less containment like static piles, windrows, and aerated piles. The first stage serves to get the process started quickly and intensely. Because odors are greatest during the early phase of composting, the first stage isolates the materials from the surroundings and captures odors. When the partially composted materials are delivered to the next stage, the potential for odor and the attraction to pests has been greatly reduced.

For more information about composting systems and operations that support compost production, refer to Dougherty, 1999; Haug, 1993; Kuter, 1995; Richard, 1992; Rynk, 1994; Rynk et al., 1992; The Compost Council, 1994; and Willson et al., 1979.

V. CHARACTERIZATION OF COMPOSTING METHODS

The composting method provides the conditions for microorganisms to convert raw feedstocks into compost. The method dictates how the composting materials are aerated, contained, and moved through the system. There are numerous ways to characterize composting methods including the degree of containment (open vs. in-vessel), mode of aeration (passive vs. forced), use of agitation (static vs. turned), and the physical progression of materials through the composting process (batch vs. continuous). Table 3.2 defines these and other terms that are used to classify and identify composting methods.

Table 3.2 Terms Used to Classify Composting Methods and Systems

Open: Materials are composted in freestanding piles or windrows (i.e., long narrow piles). Materials may be stacked in simple bins that are not fully enclosed. Systems may be enclosed within a building but the composting environment is otherwise not controlled.

In-vessel or contained: Materials are composted within reactors or vessels. Most methods employ forced aeration and some means of agitation. Environment surrounding the composting materials is controlled. Examples of reactors include aerated steel containers, large polyethylene tubes, vertically oriented rectangular and cylindrical reactors, and various enclosed bin configurations. Horizontal agitated bins are usually considered within this category.

Passive or natural aeration: Relies only on natural air movement as means of aeration. Driving mechanisms include diffusion, wind, and thermal convection.

Forced aeration: Employs fans or blowers and an air distribution network of vents or pipes to deliver air to composting materials.

Static: Materials are composted without regular agitation or turning. Some infrequent and irregular turning when piles are moved or combined.

Agitated or turned: Materials are agitated or turned regularly at intervals ranging from every day to every 2 months. A variety of agitation mechanisms may be used.

Batch: Feedstocks are composted in identifiable batches, usually with little or no change in physical location through the process. After a batch is formed and starts to compost, no new material is added to the composting unit.

Continuous: Feedstocks physically move through the compost system in a nearly continuous fashion. Movement corresponds with progressive stages of decomposition. Compost is removed and new feedstock is added regularly and frequently.

Modular: Materials are composted in multiple, often relatively small units or modules. Each module may represent an individual batch of material. Modules can be freestanding piles or windrows, bins, or enclosed reactors. As the number of modules increases, the system approaches continuous operation.

Due to differences in cost and management, composting methods are broadly categorized by whether or not the composting materials are physically contained. Two categories are typically used: (1) "open" methods that provide little or no containment and (2) "in-vessel" methods that contain composting materials in a reactor or vessel. The distinction between open and in-vessel composting is not sharp. Several methods can be considered in either category. Nevertheless, these categories provide a reasonable framework for describing the various commercial composting methods that are used. Therefore, individual composting methods are discussed within these broad categories. Table 3.3 summarizes the methods covered in this chapter.

Table 3.3 Summary of Selected Commercial-Scale Composting Methods

Method and Description
Open Methods
Turned windrows: Long narrow piles that are regularly turned and aerated passively.
Passively aerated static piles: Freestanding piles that are turned infrequently or not at all and aerate passively without aeration aids.
Static piles and windrows with assisted passive aeration (e.g., Passively Aerated Windrow System — PAWS; and Naturally Aerated Static Piles — NASP): Static windrows and piles with passive aeration aids such as perforated pipe and aeration plenums.
Aerated static piles and bins: Freestanding piles or simple bins with forced aeration and no turning.
Aerated and turned piles, windrows, and bins: Freestanding piles or windrows, or simple bins with forced aeration system. Materials are turned regularly or occasionally.
In-Vessel or Contained Methods
Horizontal agitated beds: Materials are composted in long narrow beds with regular turning, usually forced aeration, and continuous movement.
Aerated containers: Materials are contained in variety of containers with forced aeration.
Aerated-agitated containers: Commercial containers that provide forced aeration, agitation, and continuous movement of materials.
Silo or tower reactors: Vertically oriented forced aerated systems with top to bottom continuous movement of materials.
Rotating drums: Slowly rotating horizontal drums that constantly or intermittently tumble materials and move them through the system.

A. Open Composting Methods

With methods that would be considered open, the materials are composted in freestanding piles or windrows. In some cases, materials are placed in simple two- or three-sided bins. Composting may take place outdoors or under the cover of a building. The defining feature of open composting methods is that they do not control the environment surrounding the composting materials. Examples of open composting methods include turned windrows, passively aerated static piles, and forced aerated static piles and bins.

1. Turned Windrow Method

Turned windrow composting may be the most common method practiced (Haug, 1993; Rynk et al., 1992). Since this method typifies large-scale composting, it is frequently the standard by which other methods are compared (Dougherty, 1999; Golueke, 1972; Hay and Kuchenrither, 1990; Kuter, 1995; Rynk et al., 1992).

Figure 3.2 Turned windrow composting facility.

Turned windrow composting forms materials into long narrow piles or windrows (Figure 3.2). Windrows essentially aerate by passive means — diffusion, wind, and convection. To supplement passive aeration, windrows are turned on a regular basis. Turning is simply a thorough agitation of the materials. It is accomplished with bucket loaders or special windrow-turning machines. In practice, the number of turnings and time between turnings varies greatly, ranging from 3 or 4 turnings over 6 to 12 months to 40 turnings in a 2-month period.

Turning mixes and blends feedstocks; homogenizes materials in the windrow; releases trapped gases and heat; distributes water, nutrients, and microorganisms throughout the windrow; and exchanges material from the cool oxygenated environment at the surface of the windrow with material from the warmer O_2-poor areas near the core. Depending on the feedstocks and aggressiveness of turning equipment, turning also reduces particle size.

It is often said that turning aerates the windrow. This is true, but only to a limited extent. Although turning does add fresh air and O_2, microorganisms consume the O_2 within hours (Epstein, 1997). Between turnings, windrows must aerate passively to remain aerobic. Another belief is that turning fluffs materials in the windrow, increasing porosity, reducing density, and making passive aeration more effective. Michel et al. (1996) showed that this is not necessarily the case. In experiments with windrow turning of yard trimmings, the bulk density of the window materials increased after turning — the windrow became more compacted after turning. Most likely the effect of turning on bulk density depends on the feedstocks and stage of composting. With loose brittle material like leaves, turning decreases particle size and increases bulk density. With material that is already dense, like manure or nearly finished compost, turning does reduce the bulk density. However, the effect may last for only several days as the turned material again settles and becomes compacted

(McCartney and Chen, 1999). Therefore, without very frequent turning, the effect of turning on aeration may be small. The turning equipment is also a factor. Some compost turners aggressively shred materials in the windrow while other turners have more of a tossing effect. Turning with bucket loaders is also more likely to decrease bulk density. Although its effect on aeration may be limited, turning still advances the composting process. It charges the windrow with fresh air and generally invigorates the process (Michel et al., 1996; Rynk et al., 1992).

Windrows are managed according to the goals and preferences of facility operators and managers (Golueke, 1972; Rynk et al., 1992). Thus, in some systems windrows are turned infrequently when operators have time and the weather is good. At the other extreme, turnings take place almost daily based on measurement of CO_2 concentrations and temperature. A common strategy is to turn windrows based on temperature patterns. For example, a windrow may be turned when temperature falls or rises to prescribed levels or if temperatures consistently decrease over a certain number of consecutive days (Figure 3.3). This management strategy typically results in turnings occurring daily to every 2 days during the first 2 or 3 weeks of composting, followed by weekly turnings for another 6 to 8 weeks. Again, windrow management varies greatly among facilities because goals, resources, feedstocks, and management philosophies also vary greatly among facilities.

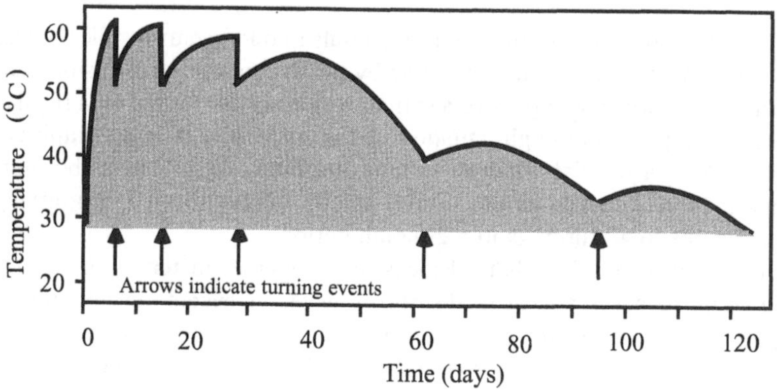

Figure 3.3 Typical relationship of windrow temperatures with turning. Turning events are indicated by vertical arrows.

Many facilities turn windrows with a bucket loader because loaders are already available. Bucket loaders turn by repeatedly lifting and dropping materials while rebuilding the windrow. Turning with a loader can be time consuming. Normally, slightly less than 1 min is required to turn one bucket-load of material (Rynk, 1994). Thus a loader with a bucket that holds 1 m^3 can process roughly 100 m^3 per hour. In contrast, special windrow turning machines handle several thousand m^3 per hour. Windrow turners use a variety of mechanical devices to agitate and rebuild windrows, including flails or paddles mounted on a rotating shaft, augers, or inclined conveyors. Nearly all turners have a housing that shapes the windrow. Some turners are self-propelled, while others require a tractor for power and/or travel. A few include nozzles and a water tank so water can be added during turning.

Windrow dimensions vary with the materials being composted, the turning equipment, and the characteristics of the site. Generally, windrows range from 1 to 5 m high and 4 to 7 m wide (Rynk et al., 1992). Windrows higher than 5 m are discouraged because they are difficult to aerate and also increase the risk of fire from spontaneous combustion. Windrows can be of any length allowed by the site layout and movement of materials. Usually, turning equipment dictates the windrow cross-sectional dimensions. When a bucket loader is used for turning, the reach of the loader determines the maximum windrow height and width. Most bucket loaders can build windrows 3 to 4 m high. Windrow turning machines create windrows that are much smaller. The largest windrow turners can handle windrows up to 3 m high and 7 m wide. However, most turners form windrows that range from 1 to 2 m high and 4 to 5 m wide.

The composting feedstocks also affect windrow size. Large windrows are difficult to aerate, so materials that are dense and resist air penetration should be composted in smaller windrows, less than 2 m high. Bulky materials, like deciduous leaves or bark, may be composted in larger windrows, up to 4 m high. Larger piles are sometimes used, but increase the risk of odor as well as spontaneous combustion from excessively high temperatures (Rynk, 2000a, 2000b). When windrows are turned with a loader, the size can be adjusted to suit composting conditions. For example, windrow size may be increased during the winter to retain more heat.

The turned windrow method of composting is a simple and flexible approach to composting. It easily accommodates a wide range of feedstocks, scales of operation, financial resources, equipment, and management strategies. It is a proven, successful method of composting. One of the disadvantages is that windrows and space for turning activities occupy a large area and, therefore, are expensive to enclose within a building. A second disadvantage is that aerobic conditions are not always maintained within windrows. Occasionally the windrows are anaerobic and odorous in the interior. Because odors are released during turning, windrows may not be acceptable for some locations.

2. Passively Aerated Static Piles

The passively aerated static pile method of composting is a low management approach used for slowly decomposing feedstocks like deciduous leaves, brush, bark, wood chips, and some farm residues. It is described by a number of guidelines and manuals concerning the recycling of these particular feedstocks (BioCycle, 1989; Massachusetts Department of Environmental Protection, 1988; Richman and Rynk, 1994; Rynk et al., 1992; Strom and Finstein, 1985).

The passively aerated static pile method takes a patient approach to composting, relying on little more than passive aeration, natural decomposition, and time to produce compost piles. Little manipulation of the process takes place. Several feedstocks may be combined and mixed to adjust moisture, porosity, bulk density, and/or C:N ratio, but once a pile is formed it is left undisturbed for months. Typically piles are combined after they shrink and then they are turned occasionally with a bucket loader (e.g., every 1 to 3 months). Often, turning is performed only when piles are moved within the site. Each turning reduces the time required to produce compost. Piles need to be turned at least once or twice during the composting cycle. Otherwise, the composted material is inconsistent and contains visible remains of the feedstocks. If necessary, the consistency can be further improved by shredding or screening the compost.

Most static piles are simply freestanding piles. They tend to be relatively large, usually ranging from 2 to 5 m high. The equipment forming the piles determines the maximum height. Normally the width of a freestanding pile is slightly less than twice its height. Pile length does not affect the process. It is determined by materials handling preferences and the constraints of the site. The large pile size conserves heat and moisture, allowing decomposition to continue steadily but slowly. Piles may be covered with a layer of finished compost to filter odors or a tarp to exclude precipitation. A few operations build extremely large extended piles, exceeding 10 m in height and 30 m in width. Piles of this size are the exception. The extreme size not only greatly lengthens the composting period but also risks the development of fires from spontaneous combustion (Rynk, 2000a, 2000b). In smaller volume applications, piles are sometimes enclosed in bins to more neatly contain the materials, segregate batches, or allow materials to be stacked higher in a narrow space. A common configuration is a series of three-sided bins housed within an open-sided building (Figure 3.4).

Figure 3.4 Passively aerated material contained in covered composting bins.

Because static piles depend on passive air movement only, maintaining aerobic conditions is a challenge. Oxygen diffusion and air movement are restricted by the large mass of compacted materials. Therefore, O_2 concentrations within the piles tend to be very low (Lopez-Real and Baptisa, 1996). Normally, aerobic conditions exist only within 1 to 2 m from the pile surface and perhaps along air channels that form within the pile. As the porosity of composting feedstocks increases, aerobic conditions penetrate deeper. Also, O_2 distribution improves if the feedstocks are uniform (e.g., well-mixed). Nevertheless, with most large static piles, much of the decomposition occurs at low O_2 concentrations. This lengthens the time required for composting and increases the chance of odors.

In large piles, passive aeration is practical only with porous feedstocks that decompose at a slow to moderate rate. Therefore, static piles are most often used for coarse feedstocks like dry leaves, shredded brush, bark, and wood chips. Intense odors are not normally a problem with these materials because of their slow rate of decomposition. However, odors can be strong when piles are disturbed.

Unless the composting site is isolated from neighbors or the piles can be kept small and porous enough to maintain aerobic conditions, static piles are not appropriate for materials which tend to decompose quickly, like grass clippings, biosolids, food residues, and manure. Under anaerobic conditions, these materials produce foul odors that can travel considerable distances, especially if the pile is disturbed. Manure is sometimes allowed to compost anaerobically in passive piles because odors are acceptable in some agricultural settings. However, this situation is changing. For many farms, manure cannot be composted in static piles without odor problems (see odor management section of this chapter).

The key factor in composting by the passively aerated static pile method is time. Composting occurs slowly in static piles because conditions are largely anaerobic and because the materials rarely receive the benefits of turning. Large passive piles turned several times require a year to produce mature compost. With extremely large piles and no turning, 3 years or more may be needed. Because of the long composting period, a second key factor is space. The site must have a large enough area to hold 1 to 3 years of partially composted feedstock, although the area requirement is reduced somewhat by the large pile size.

In general, the static pile method is a potentially economical and successful approach to composting if time and space are available, and if odors are not critical. A 1-year composting cycle is practical for many feedstocks, especially those that are generated almost entirely at a particular time of year, such as deciduous tree leaves. In this case, the material can remain composting on the site for a year. Then, after a year, the compost can be moved off-site to make room for the next lot of incoming feedstocks.

3. Methods Using Assisted Passive Aeration of Static Piles

To overcome the limitations of passive aeration, techniques have been developed that assist air movement through static windrows and piles. The most effective way to promote passive aeration is to create a composting medium that is porous. However, a porous medium is not always possible to achieve. Also good porosity, by itself, may not be enough, especially with highly degradable feedstocks like grass clippings, food, or manure. Therefore, to increase air flow and improve O_2 distribution, some composting methods employ aeration aids such as pipes and air plenums.

Prominent within this category is a method known as the Passively Aerated Windrow System (PAWS). Researchers with Agriculture Canada developed the

PAWS method in the late 1980s for composting fish processing residuals and farm manures. Several modifications and variations of the original PAWS technique have since evolved, but the basic concept remains the same (Lynch and Cherry, 1996a, 1996b; Mathur, 1995, 1997; Mathur and Richards, 1999; Mathur et al., 1988, 1990).

The PAWS method (and its variations) involves relatively short (1 to 3 m high) windrows that are not turned but have some deliberate means of delivering air without using fans. There are four basic features: (1) a homogenous and relatively porous mixture of composting feedstocks; (2) a delivery system for passive air flow; (3) a base layer of stable absorbent material like straw or compost; (4) and an exterior layer (approximately 14 cm thick) of stable coarse material that retains heat, moisture, odors and NH_3. These features are common to most PAWS variations, although the base and cover layers may not be used in all cases. Other modifications include box-like containers, variations in the air delivery system (discussed later), fabric covers, and varying windrow dimensions.

The original PAWS method delivers air to the windrows through a series of perforated PVC pipes running parallel to one another across the width of the windrow (Figure 3.5). The pipes are the same as those used for septic systems. Each pipe is 10 cm in diameter and, along its length, it contains two rows of holes, 1.25 cm in diameter spaced 7.6 cm apart. One row of holes is situated at the 10 o'clock position, the other at 2 o'clock.

Figure 3.5 Passively aerated windrow system (PAWS) with perforated PVC pipes (courtesy of Sukhu Mathur).

A variation of the original method replaces the pipes with a concrete platform which has a hollow core. The core serves as an air channel or plenum. Holes or slits in the top of the concrete platform direct air from the plenum into the windrow above. The permanent concrete platform eliminates the inconvenience of handling pipes when constructing and breaking down windrows. It is strong enough to support materials handling equipment such as a small bucket loader (Patni, 2000).

The most recent modification of the PAWS approach is called the Natural Aeration Static Pile (NASP) method (Mathur and Richards, 1999). The NASP method eliminates the pipes and concrete platform and relies solely on a layer of porous base material as an air plenum. The thickness of the base layer is approximately 45 cm. The base can be any coarse material which facilitates air movement such as wood chips, bark, or straw. Windrows used in the NASP variation tend to be higher, 2 to 3 m high, compared to the 1 to 2 m windrows typical of the PAWS method.

The key to all variations of the PAWS method is generating high temperatures to drive thermal convection. This method depends on the heat of composting to cause warm gases to rise out of the top of the windrow and cool O_2-rich air to replace it at the base. The purpose of the pipes, concrete platform, and porous plenum is to direct and distribute the incoming air. Research examining the PAWS method, and more recently NASP, demonstrates that these passive aeration techniques are sufficient to produce and maintain thermophilic temperatures with a wide variety of feedstocks including fish, several types of livestock manure, sawmill and paper pulp, food residues, and yard trimmings (Mathur, 1997). The required composting time varies with the feedstock but generally ranges from 3 to 6 months (Mathur and Richards, 1999; Rynk et al., 1992).

4. Aerated Static Piles

The aerated static pile method of composting was developed in the 1970s by researchers with the U.S. Department of Agriculture (USDA) for composting biosolids from wastewater treatment facilities (Willson, 1980). Forced aeration was employed to improve aeration, reduce the processing time, and reduce the odors associated with composting of biosolids. It has become the archetypal method for composting biosolids. It is used with biosolids more than any other type of feedstock (Goldstein and Gray, 1999). The method has generated numerous variations, particularly in terms of process control, but the basic approach remains the same (Finstein et al., 1985, 1987a, 1987b, 1987c; Haug, 1993; Kuter, 1995; Rynk et al., 1992; Singley et al., 1982).

The aerated static pile method relies on fans to aerate and ventilate the composting materials. No turning or agitation takes place except for what occurs incidentally when materials are moved. Piles are constructed on top of a system of aeration vents that supplies and distributes air through the composting materials. Forced aeration provides O_2, cools the pile; and removes water vapor, CO_2, and other products of decomposition.

The components of an aerated static pile include the air distribution network, a coarse porous base layer, the composting materials, an outer layer of stable material, and the air delivery and control system. The air distribution network underlies the pile. Several air distribution techniques have been used. The most common is perforated PVC or polyethylene pipe that rests on the composting pad. The pipe has holes along its length for delivering air to, or collecting air from, the pile above. Because the pipe interferes with materials handling and is often damaged when the pile is taken apart, many facilities have constructed durable air vents embedded in the composting pad. One approach is to cut air channels in the concrete pad. The channels can contain the aeration pipe or serve as an air plenum with slotted or spaced covers to distribute the air to the material above. Some facilities use air ducts buried beneath the pad. Spaced along the duct, vertical risers or orifices extend up to the pad surface. Aeration pipe and embedded channels are normally covered with a mound of wood chips or some other coarse material. The porous material helps prevent the air holes from clogging and further spreads the air exiting or entering the pipe or channel. Facilities that use vertical orifices typically do not use the base covering because high pressure at the orifices keeps the holes clear and distributes the air.

The mixture of composting feedstock is stacked over the air distribution and porous base. The feedstock mixture is capped with a layer (18 to 30 cm thick) of stable material, usually compost. This exterior layer insulates the pile; separates flies and other pests from feedstocks; and helps retain odors, NH_3, and water.

Air is supplied to the aerated static pile via fans and associated piping and control devices — the aeration delivery and control system. There are innumerable possible combinations of aeration and control strategies and equipment configurations. Piles can be aerated by individual fans each turning on and off independently, or by a group of fans feeding a common manifold with valves controlling air flow to individual piles or pile sections. Air can be provided by positive or negative pressure, continuously or intermittently. The process variables that control the amount of airflow can also vary widely. The original USDA system was based on O_2 levels, but today aerated static pile aeration systems are commonly controlled by temperature or by a timer that is adjusted according to temperature levels.

Aerated static piles range in height from 2 to 4 m. The height is limited by equipment forming the piles and by the weight of the materials compacting the lower portion. Pile lengths commonly range from 40 to 80 m. Length is limited by the variation in air distribution along the aeration pipe and ducts. Otherwise, aerated static piles vary in size from small individual piles 3 to 6 or 10 m wide to very large extended piles that may exceed 50 m in width (Figure 3.6). Single piles have a typical triangular or parabolic shaped cross-section. As new feedstock is generated, it is added to the end. One or two rows of aeration pipe run the length of the pile.

Extended aerated stated piles are large and roughly rectangular in shape. New feedstock is added to the side in batches or cells. New cells are constructed by stacking feedstocks against the side of the previous cell. With extended piles, a row of aeration pipe is placed at the base of each cell, or ducts are spaced at regular intervals in the pad. Air valves are opened as the rows are covered with new feedstock. Extended piles are commonly used because they use space efficiently. Frequently they are enclosed within a building. The aerated static pile method of composting can also be accomplished in bins rather than freestanding piles. It is common for materials to be stacked on aeration pipes between partitions in a building.

Because the materials in aerated static piles are not turned, feedstocks must be well mixed before being placed in the pile. The mix also must be relatively porous and have good structure to resist compaction and settling. For composting biosolids,

Figure 3.6 Extended aerated static pile composting method.

good structure is usually achieved by adding wood chips as an amendment or bulking agent. Other bulking agents used include deciduous leaves, mixed yard trimmings (ground), sawdust, and finished compost (Naylor, 1996). A rigid bulking agent like wood chips not only supports the pile physically, but also creates channels for air to penetrate the materials as they compost. However, the formation of air channels can also be a drawback of this method. Excessive channeling leads to short circuits in the air flow. This leaves some sections of the pile without sufficient O_2 and air movement, while others sections become dried out. As a result, materials within the pile compost inconsistently. A good combination of feedstocks, thorough blending of feedstocks, proper design of the aeration system, and occasional agitation minimize channeling.

With a mixture of biosolids and wood chips, composting by the aerated static pile method typically takes 3 to 6 weeks, followed by a curing period of 1 to 2 months (Kuter, 1995; Rynk et al., 1992). The composting time should be similar for manure and grass clippings. Less degradable feedstocks like mixed yard trimmings require a longer period. Without the physical agitation of turning, wood chips and other large woody bulking agents remain almost intact in the finished compost. In most cases, the compost is screened to improve its consistency and to recover wood chips for reuse as bulking material.

The aerated static pile is a well practiced and proven approach to composting. Like the windrow method, it is technically simple, but it requires less area and accomplishes aeration with more certainty than passively aerated methods, including turned windrows. Because it is more space-efficient, extended aerated piles can be enclosed within a building. The primary shortcoming is the fact that it is a static system. Thus, it requires special attention to the type and amount of amendments used, and it can suffer from compaction, short circuiting of air, and inconsistent decomposition within a batch of compost.

5. *Methods that Combine Turning and Forced Aeration of Open Piles and Windrows*

Several composting facilities employ methods that combine attributes of turned windrows and aerated static piles. The intent is to gain the advantages of forced aeration while avoiding the disadvantages of a static system. Three methods in particular exemplify this strategy. Although they lack generally recognized names, these methods can be referred to as forced-aerated windrows, turned extended aerated piles, and turned aerated bins.

Forced-aerated windrows are built like typical windrows but over an aeration channel or plenum recessed in the composting pad (Hay and Kuchenrither, 1990). As with the standard windrow method, the windrows are turned periodically with a conventional windrow turner. Between turnings, fans push or pull air through the windrow via the plenum. A windrow must be kept centered over the plenum to maintain even air flow on both sides of the windrow. Because the aeration system provides O_2 and temperature control, turning is less frequent than required for standard windrows.

The turned extended aerated pile technique resembles the previously described extended aerated static pile in all respects except that the pile is regularly turned (Schoenecker and McConnell, 1993). With each turning, the pile is shifted to one side in successive sections by a turner with a side-discharging conveyor. Compost is always removed from the pile on the same side and fresh feedstocks are always added to the opposite side. Starting at the side containing the oldest material, or the compost, the turner moves along the edge of the pile and picks up the compost in a slice equivalent to the width of the turner. The side-discharge conveyor places the compost directly into a truck bed. Next, the turner moves to the adjacent, newly exposed edge of the pile, picks up a slice, and deposits it over the strip of floor space previously occupied by the harvested compost. The turner progresses through the remaining slices, shifting each one over toward the harvested side of the pile. Each slice roughly covers the floor area vacated by the previous slice. Eventually, the entire pile is moved over by a distance equivalent to the width of the harvested compost. Finally, new feedstock is placed along the side of the pile, in the space left vacant when the last slice shifted over. The air distribution system is recessed in the floor beneath the pile. It operates in the same manner as aeration for an extended aerated static pile. Facilities that employ this technique have used specially designed turners and conventional elevating face turners fitted with side-discharging conveyors.

The turned aerated bin approach is a minor variation of the aerated static pile method. A series of three-sided bins are used for composting, each with aeration plenums recessed in the floor (Alix, 1998). A batch of mixed feedstocks is placed in the first bin. It is aerated and managed in the same manner as a conventional aerated static pile. After 1 to 2 weeks, a bucket loader or conveyor moves the batch from the first bin into the adjacent bin. The batch aerates in the second bin for several days to several weeks until it is shifted again to the next bin in the series. This is repeated two to five times until the material is harvested as compost from the last bin. After the compost is harvested, in succession, bins are emptied and refilled with a batch from the adjacent bin. When a batch is shifted between bins it can be mixed and adjusted for moisture if necessary. Two or more batches can be combined to compensate for shrinkage in volume. This technique agitates and mixes the composting materials, breaks up air channels and clumps of poorly composted feedstocks, and generally allows for improved physical manipulation of the materials. The series of bins also allows for finer design and control of the aeration system. Each bin represents a different aeration zone that can be sized and managed based on stage of decomposition taking place in the bin. Composting also can be accomplished in nearly the same manner by unloading materials from a bin, turning, and then reloading them into the same bin (Croteau et al., 1996).

Combining forced aeration and turning improves conditions and generally speeds the composting process, compared to systems that are only turned or only aerated with fans. However, it also requires both the task of turning and the requirements of an aeration system. The methods just discussed relate to systems that, practically speaking, involve open piles and windrows. Agitated beds and several other in-vessel composting systems provide both agitation and forced aeration as an inherent part of their design.

B. In-Vessel or Contained Systems

In-vessel composting systems encompass a diverse group of methods that contain the materials and attempt to control the composting environment within a bin, reactor, or vessel. Perhaps they are better described as "contained" composting methods, but the term "in-vessel" has taken hold within the industry. Many in-vessel systems are commercially developed, with features that are unique to the commercial design. Generic in-vessel methods include agitated beds, modular aerated containers and tunnels, aerated-agitated containers, silo or tower reactors, and rotating drums. Descriptions of in-vessel technologies are provided by numerous references (Haug, 1993; JG Press, 1986; Kuter, 1995; Rynk, 2000c, 2000d; Rynk et al., 1992; U.S. EPA, 1994).

1. Horizontal Agitated Beds

The horizontal agitated bed method of composting is prevalent because it provides the advantages of forced aeration (usually), frequent turning, plus automated turning and materials handling. Because the composting materials are partly contained, the agitated bed method is normally considered within the category of in-vessel composting. However, it can reasonably be considered as a category in itself or even grouped with open systems like windrows and aerated static piles. Agitated bed systems are available from numerous commercial companies, each employing their own particular system design features and turning equipment (Haug, 1993; Rynk, 2000c, 2000d).

Agitated bed composting systems contain the composting materials in a long narrow horizontal bed formed within a channel created by concrete walls (Figure 3.7). The top of the bed is open, so the system is usually housed in a building. A turning machine agitates and moves the materials in the bed. The turner moves down the bed automatically (i.e., without an operator), usually guided by rails on the walls of the channel. Agitated bed systems operate in continuous, plug-flow mode. Feedstocks are loaded in the front end of the channel. Compost is removed from the opposite end. Starting at the compost discharge end, the turner moves toward the front or loading end. With each pass, the turner displaces material a set distance toward the back of the channel where the materials are discharged as compost. Depending on the turner, material is shifted 2 to 4 m with each turning. This displacement distance plus the length of the bed and the turning frequency determine composting period in the bed (generally 10 to 28 days).

Figure 3.7 Horizontal agitated bed composting system with a single bed (courtesy of Karin Grobe).

Turners function similarly in all of the commercial systems, although the turner design differs among the systems. Most turners agitate with flails or paddles on a rotating shaft and use a conveyor to displace the composting material backwards. Another type of turner is a rotating cylinder that carries material backward as it rotates down the bed. Turner dimensions match the size of the beds. Normally, the depth of the material in the bed decreases gradually from the loading end to the discharge end because the materials shrink in volume as they compost. With some turners, the displacement can be adjusted to build up the height of the bed toward the back end of the channel.

The dimensions of individual beds vary among the commercial systems, with depths ranging from 1 to 3 m and widths of 2 to 4 m. Bed lengths typically range from roughly 50 to 100 m. Most applications use multiple channels and one or two turning machines that are transported among channels. Single-channel applications are occasionally used for smaller scale applications, usually for composting manure on farms. Rather than separate multiple channels, some commercial systems use a single bed, 7 to 13 m wide. An overhead crane supports and moves the turner through the bed in strips. The turner otherwise works in the same manner as other agitated bed turners.

In most applications, forced aeration is provided through the floor of the channel. Air flow is controlled by temperature or a time sequence adjusted according to temperature levels. Usually the channels are divided into three to seven aeration zones along their length. Zones nearest the loading end require a higher rate of aeration. Most aerated system use positive pressure, forcing air up through the bed. If odor treatment is needed, air within the building is captured by an exhaust fan and directed to a biofilter. There is some interest in switching to negative pressure aeration for improved odor control.

2. *Aerated Containers*

Aerated containers are fully enclosed and covered containers of various materials and dimensions (Figure 3.8). The types of containers used vary from solid-waste roll-off boxes to flexible polyethylene bags (Rynk, 2000c, 2000d). Aerated containers typically compost feedstocks in batches. Therefore, they are modular. Additional containers are added to accommodate successive batches. Applications employ from 2 to over 40 containers depending on the scale of operation and the container system.

Except for a few small passively aerated containers that are used for very low-volume applications, aerated containers rely on fans to supply O_2 and remove heat, moisture, CO_2, and other process gases. Several containers can be aerated from a single fan system by connecting individual containers to an air distribution header. In most cases, air is introduced at the base of the material from a pipe or plenum in the floor and flows up through the composting mass into a headspace at the top of the container. In other cases, air flows in the opposite direction, from the headspace to the plenum. Some systems periodically reverse the airflow to correct temperature and moisture gradients. Exhaust is usually collected and passed through a biofilter for odor treatment. Aeration may be controlled by time or temperature, depending on the system. Some systems employ sophisticated aeration controls that include temperature feedback control, moisture and O_2 monitoring, and air recirculation.

Aerated containers are essentially static systems. No agitation or turning takes place within the container. Therefore, materials must be well mixed prior to loading the containers. Most systems allow for the containers to be emptied partway through the process. After emptying, the materials are mixed and adjusted externally, and then reloaded into the container for continued composting (or delivered to a second composting system like windrows). Emptying the container provides an opportunity to add more feedstock or combine materials from several containers to compensate for the shrinkage. Because the exercise of emptying and reloading containers requires labor and expense, it is not practiced in many cases.

Figure 3.8 Interior of an aerated-container composting reactor (courtesy of Jim McNelly).

Common examples of commercial-scale aerated containers include modified steel solid waste roll-off boxes, long polyethylene bags, and specially constructed tunnels. Several commercially available systems use corrosion-resistant roll-off containers modified with a tight fitting lid, an air distribution plenum in the floor, and air vents in the headspace. Individual units range in capacity from 15 to over 50 m^3. They are typically 2 to 3 m high and wide and from 3 to 15 m long. These containers are moveable and modular. They are loaded with conveyors or bucket loaders through the lid and emptied from the end by tipping with a roll-off truck or tipping equipment. The tipping system allows materials to be emptied and remixed externally as described previously. As containers are filled, they are connected to the air delivery manifold and materials compost as a batch. Valves govern the airflow to individual containers. Aeration and process control strategies for these units can be highly technical, involving computer control, monitoring of several process parameters, and variation in airflow rate and direction.

Tunnel composting systems, commonly used by the mushroom industry, are finding use for composting other feedstocks as well (Panter et al., 1996). Tunnels are similar in scale and operation to the roll-off container systems, except they usually are not intended to be moved. They may be constructed on site with concrete or prefabricated with corrosion-resistant steel. They provide controlled forced aeration through a floor plenum. One characteristic of tunnel systems is their ability to monitor and adjust internal air conditions by recirculating internal air with some outside air blended in. Tunnels are loaded from one end and operate in batch mode after the tunnel is fully loaded. Multiple tunnels are used to attain a nearly continuous operation. Unlike the top loaded roll-off containers, tunnels cannot be aerated until they are fully loaded.

The polyethylene bag composting system is a different example of an aerated container. It is essentially an aerated static pile in a bag. The bags used are similar to those used on farms for making silage but, for composting, aeration pipe is inserted to create an aerobic environment inside the bag. Bags range in size from 1.5 to 3 m in diameter and up to 65 m in length. Special equipment loads the bags with premixed feedstock to a prescribed density. Aeration pipe is rolled out beneath the feedstock mix as it is pushed into the bag. After the bag is loaded, the pipe is connected to a positive-pressure blower which pushes air through the composting pile. Ports in the side of the bag serve as exhaust vents and allow access for temperature probes. A bag can be temporarily sealed and aerated before it is completed full. This system does not accommodate agitation or transport of an intact container. When the compost is finished, or ready for the next stage of processing, the bag is sliced open and the contents are removed with a bucket loader.

3. Aerated-Agitated Containers

Several commercially available composting systems combine forced aeration and internal agitation. However, at present only one is applicable to the production of compost on a large scale (Rynk, 2000c, 2000d).

In this system, the composting materials are supported on perforated stainless steel trays that advance in sequence through an enclosed aeration chamber or tunnel. An external hydraulic ram pushes an empty tray into the tunnel to be loaded with the feedstock mixture from an overhead hopper. As the new tray enters the tunnel, it nudges the preceding trays along and the last tray is discharged. At the discharge point, augers unload compost from the exiting tray. Within the tunnel, air is forced through the trays from a plenum below. Air is recirculated and eventually exhausted to a biofilter, which is an integral part of the unit. Two aeration or temperature zones exist. Higher temperatures are maintained in the first zone for pathogen destruction. Inside the tunnel, as a tray moves from the first zone to the next, the compost is agitated and, if necessary, water can be added during agitation. One self-contained unit incorporates the entire system — hopper, tunnel, aeration, agitation, augurs, and biofilter. Generally, system capacity is determined by the size and number of the units used. Individual units range in throughput capacity from less than 100 kg to over 30 Mg per day.

4. Silo or Tower Reactors

Silo type composting systems use one or more vertically oriented vessels, or silos, in which composting materials move through the silo from top to bottom (U.S. EPA, 1994). The movement starts when an auger or other unloading mechanism removes a section of compost from the bottom of the silo. The materials above shift downward and then a fresh mixture of feedstocks is added to the space created at the top of the silo. Although the materials within the silo move, they are not well agitated. Therefore feedstocks must be well mixed before loading. Material typically remains in the silo from 10 to 30 days, depending on the application. Often, material removed from the silo moves to a second silo or a secondary composting system.

Nearly all silo systems are aerated with fans. Normally air is introduced at the bottom and moved up through the materials in the silo. The air is collected at the top and exhausted, usually to a biofilter. As the air moves up through the deep bed of material it gathers heat, moisture and CO_2. Therefore, the air loses its effectiveness as a means of providing O_2 and cooling, so it is difficult to maintain uniform composting conditions across the entire depth. Attempting to overcome this deficiency, some systems use a set of narrow aeration ducts inserted down into the silo to introduce fresh air at different depths within the silo.

One small commercial silo system relies on passive aeration (Rynk, 2000c, 2000d). This system contains the composting materials in silos formed by tall narrow wire-mesh cages. The cages are arranged in series like slices in a loaf of bread. A 10-cm air space separates adjacent cages and provides a channel for passive air flow and O_2 diffusion. Cages are approximately 3 m high, 7 m long, and 1.2 m wide. Therefore, the core of the composting mass is only about 0.5 m from the air space that surrounds the cage. Materials move vertically through the system in a silo fashion with feedstock loaded at the top of the cages and compost removed from the bottom. Because the cages are made of wire mesh, the system is not totally enclosed.

The main advantage of silo systems is the small surface area, or footprint, afforded by the vertical orientation. However, vertical stacking of materials also creates aeration challenges due to the greater depth of material in the silo and the compaction that occurs when materials are stacked high.

5. Rotating Drum Reactors

Rotating drum composting digesters have been used for many years, primarily for large-scale municipal solid waste (MSW) facilities (Richard, 1992) and, at a much smaller scale, for backyard composting (Dickson et al., 1991). Recently, several versions of commercial-scale rotating drums have emerged that are appropriate for applications between these extremes (Rynk, 2000c, 2000d). Although the commercial drum systems differ in size, design details and process management, they share the basic technique of promoting decomposition by tumbling material inside an enclosed reactor.

Drums are mounted horizontally, usually at a slight incline (Figure 3.9). They slowly rotate either continuously or intermittently, tumbling the material inside. The tumbling action mixes, agitates, and generally moves material through the drum. Feedstocks are loaded at one end of the drum and compost is removed at the opposite end. Various loading and unloading devices are used, depending on the specific system. Loading with conveyors and unloading by gravity are common. Some of the large drums have internal partitions that separate the drum into compartments and define distinct batches of materials. Doors in the partitions allow material to be transferred from one compartment to the next.

Figure 3.9 Small-scale rotating drum composting reactor.

In regard to the composting process, the key function of the rotation is to expose the material to fresh air, to add O_2, and to release heat and gaseous products of decomposition. To deliver the fresh air and remove the gaseous products, forced aeration is usually but not always provided. In some cases, the short drums can obtain sufficient aeration by passive air exchange through the openings in the ends. When forced aeration is used, air is directed from the compost or discharge end of the drum to the opposite end, where the feedstocks are loaded.

The largest drum systems are used at facilities composting diverse feedstocks like MSW. These drums are 3 m or more in diameter and over 50 m long. Several drums may be used in parallel. Smaller scale drum systems are targeted for source-separate feedstocks like livestock manure, animal mortalities, yard trimmings, and food residuals. These units range from 1.5 to 3 m in diameter and 3 to 15 m in length.

Rotating drum composting reactors have always been associated with very short retention times, typically 3 to 5 days. In practice, drums have served primarily as a first stage of composting. Material removed from the drums is usually finished in windrows, aerated piles, or another secondary composting system. The drums start the feedstocks composting quickly and evenly in a controlled high-temperature environment. Drums are particularly effective at homogenizing heterogeneous mixtures like MSW. However, depending on the mixture and drum design, some systems have experienced insufficient aeration leading to organic acid formation and a severe drop in pH. This can be problematic, since in several applications, drums are the only stage of active composting. In some cases the composting time is extended to several weeks, which should allow aerobic degradation of any acids as aeration demand drops over time. In other cases, the compost is cured after only 3 or 5 days of composting in the drum. Such an abbreviated composting period deserves caution.

The material discharged may be appropriate for some uses, such as direct land application in the winter or autumn. However, analysis of compost maturity parameters suggests that a period of several weeks is necessary to achieve a compost product that is mature enough for general horticultural use (Epstein, 1997; Golueke, 1972).

VI. ODOR MANAGEMENT

Odor is among the greatest threats to successful operation of composting facilities, and is a major contributor to many facility closures (Miller, 1993). Odor problems are often misdiagnosed, and some purported solutions can actually make the problem worse. A good understanding of the nature and types of odors associated with composting is important to their effective control.

One of the difficulties with odor is that it is a subjective and quite variable trait. Although significant advances in odor quantification continue to occur, there will always be wide variation between two people's perception of the same odor phenomenon. Not only are there differences in sensitivity among individuals, the same odor that smells bad to one person may be tolerable or even pleasurable to another. In addition to significant psychological effects (Schiffman et al., 1995), odors can affect both cognitive performance and physiological response (Lorig, 1992; Ludvigson and Rottman, 1989). Accepting and addressing the differences in perception and health impacts of odors is a critical part of responding to a community's odor concerns.

Anaerobic decomposition is the most common source of malodorous compounds at composting sites. Despite all attempts to maintain aerobic conditions, during the active phase of composting there are likely to be pockets of anaerobic activity within large particles and in clumps where compaction or high moisture creates resistance to airflow. Anaerobic odors include a wide range of volatile organic acids, N-containing compounds including NH_3 and amines, ketones, phenols, terpines, alcohols, and S compounds (Eitzer, 1995; Epstein, 1997; Miller, 1993). Most of these compounds are intermediate products of decomposition, and require an aerobic metabolic pathway to complete their degradation.

Not all malodors are produced anaerobically. Ammonia is a common odor that can be produced either aerobically or anaerobically. Protein degradation in a low C:N mixture generates NH_3 in excess of microbial growth requirements. Under low pH conditions this NH_3 remains in solution as the aqueous ammonium ion (NH_4^+), but under high pH (>8.5) it is in the form of more volatile NH_3 gas (Figure 3.10). The partitioning into NH_3 is also encouraged by high temperatures, and compounded by high airflow rates that insure high concentration gradients at the gas/solute interface. Ammonia gas is lighter than air and not as persistent as some of the strictly anaerobic odors such as sulfides and organic acids, so although NH_3 may be the predominant odor in the immediate vicinity of low C:N composting, it rarely is a significant problem off site.

Odors can be generated in almost any component of a composting system. In facilities accepting readily degradable materials like food scraps and grass, the odors

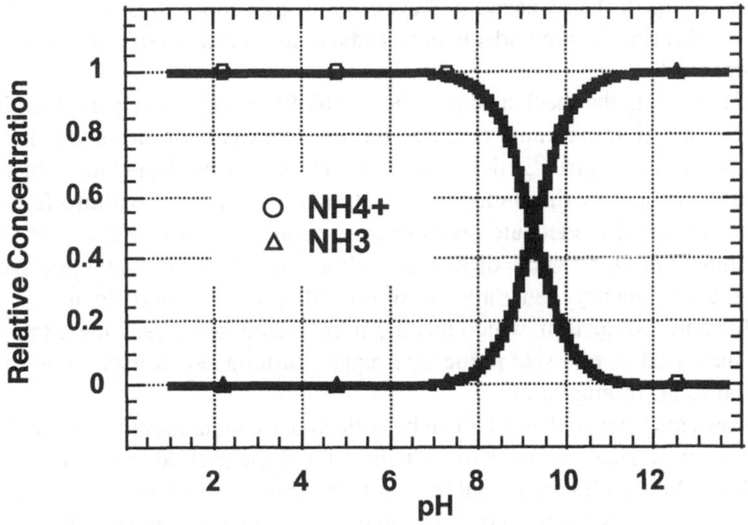

Figure 3.10 Effect of pH on equilibrium between gaseous ammonia (NH_3) and ammonium ion (NH_4^+).

may be arriving with the incoming materials. An enclosed tipping area with negative air pressure is one approach to collecting these emissions. The initial stages of composting are also likely to be a major odor source, especially when composting highly degradable materials, a low C:N ratio mixture (< 25:1), or a mixture that contains more than about 60% moisture. Quantifying these risk factors is imprecise because of the wide range of composting feedstocks and mixture characteristics, but if the O_2 supply is inadequate to keep up with demand — or unable to thoroughly penetrate the composting matrix — anaerobic odors are likely to result.

Odors can also be generated during the later stages of composting. Anaerobic conditions may occur when material is moved from active composting to a curing area, either because forced aeration is no longer being used, or because the pile size has been increased beyond the ability of diffusion or convection to supply adequate O_2. Heavy precipitation in an outdoor curing pile can saturate the pores and reduce air movement, also resulting in anaerobic conditions. Odor in a curing pile is particularly problematic as it can affect both the site neighbors and potential customers. Aside from the obvious impact on the aesthetics of the final product, many of the odorous organic acids can also be phytotoxic to plants.

Although it is important to minimize anaerobic conditions by maintaining good process control, complete odor prevention is not always successful at even well-managed composting facilities. If odors do occur, there are a number of interventions possible before those odors impact a sensitive neighbor. Many anaerobically produced odorous compounds still contain considerable amounts of energy, and if they pass through an aerobic zone they can be further decomposed. Adsorption and microbial degradation of odors is actively encouraged in biofilters (described later) but also occurs *in-situ* near the surface of composting piles. *In-situ* biofiltration can be enhanced by covering freshly decomposing materials with a layer of older, stabilized compost. Unmanaged and even unrecognized by most facility operators, *in-situ* biofiltration is often adequate to reduce odors to tolerable levels, especially in unturned piles.

Understanding the mechanisms of *in-situ* biofiltration can help explain the common misconception that more frequent turning reduces odor emissions. If one calculates the stoichiometric O_2 demand at typical rates of biodegradation, many compost mixtures consume the amount of O_2 introduced by turning within a few minutes (Richard, 1997). If inadequate porosity or excessive pile size are constraining the replenishment of O_2 by wind, diffusion, and/or convection, turning a pile even daily (the highest frequency generally possible) still provides more than 23 hours of anaerobic odor production, which turning immediately releases. Unless the porosity can be increased or pile size reduced, frequent turning can worsen odor emissions compared to no turning at all.

For systems where exhaust air can be collected, treatment is possible by chemical, thermal, or biological mechanisms. Chemical and thermal odor treatment methods are well established for industrial applications, but to date their implementation at composting facilities has been rare. For classical chemical treatment, the complexity of composting odors would require multistage scrubbers with acid and alkali recovery systems (Dunson, 1993). Thermal treatment requires large energy inputs when odor concentrations are relatively low, making it expensive to treat the large volumes of mildly odorous air found at many composting facilities (Dunson, 1993).

Biological odor treatment via biofilters has been the most common approach taken at composting facilities. A biofilter uses moist organic materials to adsorb and then biologically degrade odorous compounds. Cooled and humidified compost process air is typically injected through a grid of perforated pipes into a bed of filtration media. The materials that have been used for biofilter construction include compost, soil, peat, and chipped brush and bark, sometimes blended with a biologically inert material such as gravel to maintain adequate porosity. Biofilter bed depths typically range from 1 to 1.5 m deep, with shallower beds subject to short circuiting of gas flow and deeper beds more difficult to keep uniformly moist. Biofilters have been shown to be effective at treating essentially all of the odors associated with composting, including NH_3 and a wide range of volatile organic compounds (including S compounds and amines).

The principal design criterion for biofilters is the airflow rate per unit surface area of the biofilter. Literature values for biofilter airflows range from 0.005 to 0.0030 $m \cdot s^{-1}$ (1 to 6 cfm per ft^2) (Lesson and Winer, 1991; Naylor, 1996) and are typically 0.015 to 0.02 $m \cdot s^{-1}$ (3 to 4 cfm per ft^2). Over a 1 to 1.5 m path length, a typical residence time for the gas is 45 to 60 s (Naylor, 1996). For the purpose of selecting the biofilter blower, the backpressure expected across the biofilter at this airflow rate is usually in the range of 20 to 120 mm H_2O per m depth (0.22 to 0.9 inches H_2O per ft), although the pressure drop can be considerably higher through dense composts and soil (Shoda, 1991).

Moisture and porosity are also essential to maintaining an effective biofilter. Moisture content should be between 40 and 60%, while air-filled porosity should be in the range of 32 to 50% (Naylor, 1996). Unsaturated air coming out of the compost building dries the biofilter, and rewetting from the surface is generally not

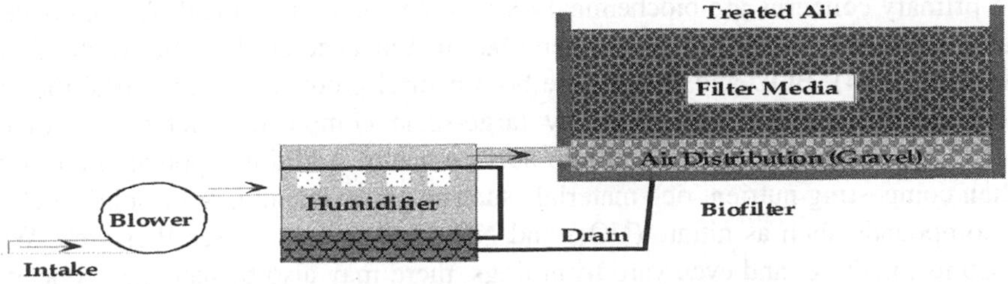

Figure 3.11 Diagram of biofilter components with humidification.

uniform. A simple humidification scheme can help to maintain the correct moisture and porosity as illustrated in Figure 3.11.

Several other approaches to odor mitigation and treatment can be used at composting sites. A number of facilities use odor masking agents, which attempt to cover up malodors with a more desirable odor. Unfortunately, masking agents are rarely as persistent as the original malodorous compounds, and thus may prove ineffective downwind of the site. Chemical, bacterial, and enzymatic deodorants have a mixed track record (Miner, 1995; Ritter, 1981), although some have recently been demonstrated as effective in reducing manure odor emissions (Bundy and Hoff, 1998).

Vegetative buffers can provide both a pleasing visual screen and a multitude of surfaces for particulate odor filtration, gaseous odor adsorption, and then microbial odor degradation (Miner, 1995). Trees also affect wind velocity and direction in their immediate vicinity, which may be used to advantage if odor problems are associated with a particular wind pattern.

If odors ultimately escape the composting facility they can travel significant distances downwind. Many anaerobic odors are relatively persistent. Sulfur compounds eventually oxidize, while precipitation is the likely fate of highly soluble compounds like NH_3 and organic acids (Miller, 1993). The fate and transport of odors is complex and subject to considerable local variation. When siting and designing large facilities, or if a significant odor problem has already developed, site-specific odor dispersion models can be used to evaluate alternative sites and treatment options (Epstein, 1997; Haug, 1993; Miner, 1995).

VII. SITING AND ENVIRONMENTAL PROTECTION

Composting has long been viewed as an environmentally beneficial activity. To maintain that positive reputation it is essential that compost facilities consider and mitigate any adverse environmental impacts. Water quality protection can be accomplished at most composting facilities by proper attention to siting, ingredient mixtures, and compost pile management.

The results of several water quality monitoring studies indicate that outdoor windrow composting can be practiced in an environmentally sound manner (Cole, 1994; Richard and Chadsey, 1994; Rymshaw et al., 1992). However, there are a few aspects of this process that can potentially create problems. For leaf composting, the primary concerns are biochemical oxygen demand (BOD) and phenol concentrations found in water runoff and percolation (Lafrance et al., 1996; Richard and Chadsey, 1994). BOD and phenols are both natural products of decomposition, but the concentrated levels generated by large-scale composting dictate that runoff should not be directly discharged into surface water. Additional potential concerns when composting nutrient rich materials such as grass, manure, or biosolids include N compounds such as nitrate (NO_3) and NH_3, and in some cases P as well. With biosolids, manure, and even yard trimmings, there may also be pathogen concerns. Although important, these concerns are readily managed, and can be mitigated through careful facility design and operation.

A. Facility Design

Selecting the right site is critical to many aspects of a composting operation, from materials transport and road access to neighborhood relations. From an environmental management perspective, the critical issues are soil type, slope, and the nature of the buffer between the site and surface- or groundwater resources. Soils can impact site design in a variety of ways. If the soils are impermeable, groundwater is protected from NO_3 pollution, but runoff is maximized, which must be managed to prevent BOD, P, and pathogens from entering surface waters. On the other hand, highly permeable soils reduce the runoff potential but may allow excessive NO_3 infiltration to groundwater. Intermediate soil types may be best for sites that are operated on the native soil. For some large facilities, or those handling challenging feedstocks, a working surface of gravel, compacted sand, oiled stone, or even asphalt or concrete may be appropriate. Such surfaces can improve trafficability during wet seasons, but the surface or groundwater quality issues remain.

The buffer between the site and surface- or groundwater resources is the first line of defense against water pollution. Deep soils, well above the seasonally high water table, can filter solid particles and minimize NO_3 migration. Horizontal buffers can filter and absorb surface runoff, and can be enhanced by specially designed grass filter strips.

Site design issues that may impact water quality include the selection of a working surface (native soil or an improved surface), exclusion of run-on to the site by surface diversions, possible drainage of wet sites, and the possible provision of roofs over some or all of the composting area to divert precipitation and to keep compost or waste materials dry. In all but fully roofed sites, surface runoff may need to be managed as described later. Slope of the site and surface drainage to either divert uphill water away from the site or collect site runoff for management should be considered in the design process.

A number of factors combine to determine the quality of water running off compost sites. One obvious factor which is often overlooked is the excess water running onto the site from upslope. Diversion ditches and berms that divert water around the site minimize the runoff that needs to be managed. Siting the facility on a soil with moderate to high permeability also significantly reduces the runoff generated on the site. For the runoff which remains, alternatives to surface discharge include such simple technologies as soil treatment, filter strips, or recirculation, so that sophisticated collection and treatment systems should not be needed.

These simple, low-cost treatment strategies have proven effective for a variety of wastewaters and organic wastes (Loehr et al., 1979). Soil treatment forces the percolation of water through the soil profile, where these organic compounds can be adsorbed and degraded. Vegetative filter strips slow the motion of runoff water so that many particles can settle out of the water, while others are physically filtered and adsorbed onto plants. Recirculation involves pumping the runoff water back into the compost windrows, where the organic compounds could further degrade and the water would be evaporated through the composting process. This last option should work very well during dry summer or early fall weather, when water often needs to be added, but would not be appropriate if the moisture content of the compost was already high.

B. Operations

The day to day operation of the composting site offers considerable opportunities to minimize water quality impacts. The proper selection, mixing, and management of materials can help control overall runoff, BOD, and pathogen and nutrient movement. Assuring appropriate moisture and C:N ratios throughout the composting process can be very effective at limiting these pollutants.

NO_3 is most easily controlled by maintaining an appropriate C:N ratio in the composting mixture. Raw materials should normally be blended to approximately 30:1 or greater C:N ratio by weight. The ratio between these key elements is based on microbial biomass and energy requirements. Inadequate N (a high C:N ratio) results in limited microbial biomass and slow decomposition, while excess N (a low C:N ratio) is likely to leave the composting system as either NH_3 (odors) or NO_3 (water pollution). In a N-limited system, microorganisms efficiently assimilate NO_3, NH_3 and other N compounds from the aqueous phase of the compost, thus limiting the pollution threat.

The ideal C:N ratio depends on the availability of these elements to microbial decomposition. C availability is particularly variable, depending on the surface area of particles and the extent of lignification of the material. Composting occurs in aqueous films on the surfaces of particles, so greater surface area increases the availability of C compounds. Lignin, because of its complex structure and variety of chemical bonds, is resistant to decay. For both of these reasons the C in large wood chips is less available than that in straw or paper, so greater quantities of wood chips would be required to balance a high N source like manure.

The data from experimental studies indicate low C:N ratio mixtures can generate NO_3 levels above the groundwater standard (Cole, 1994; Rymshaw et al., 1992). Much of this NO_3 in runoff and leachate infiltrates into the ground. Although microbial assimilation and denitrification may somewhat reduce these levels as water passes through the soil, these processes have a limited effect and are difficult to control. Proper management of the C:N ratio is perhaps the only practical way to limit NO_3 contamination from the site short of installing an impermeable pad and water treatment system.

The other important factor to consider when creating a composting mixture is water content. Excess water, in addition to increasing the odor potential via anaerobic decomposition, increases the runoff and leachate potential of a composting pile during rainfall events.

With both C:N ratios and moisture content, the optimum water and N levels for rapid composting may create a greater than necessary water pollution threat. Increasing the C:N ratio from 30:1 to 40:1 and decreasing the water content from 60 to 50% may slow down decomposition somewhat, but can provide an extra margin of safety in protecting water quality.

Once the materials are mixed and formed into a compost pile, windrow management becomes an important factor. Windrows should be oriented parallel to the slope, so that precipitation landing between the windrows can move freely off the composting area. This will minimize windrow saturation. Pile shape can have a considerable influence on the amount of precipitation retained in a pile, with a flat or concave top retaining water and a convex or peaked shape shedding water, particularly in periods of heavy rain. These effects are most pronounced when the composting process is just starting or after a period of dry weather. In the early phases of composting, a peaked windrow shape can act like a thatched roof or haystack, effectively shedding water. Part of this effect is due to the large initial particle size, and part is due to waxes and oils on the surfaces of particles. Both of these initial effects diminish over time as the material decomposes. During dry weather the outer surface of even stabilized organic material can become somewhat hydrophobic, limiting absorption and encouraging runoff.

If a pile does get too moist, the only practical way to dry it is to increase the turning frequency. The clouds of moisture evident during turning release significant amounts of water, and the increased porosity that often results from turning increases diffusion and convective losses of moisture between turnings. This approach can be helpful during mild or warm weather, but caution must be exercised in winter when excessive turning can cool the pile.

C. Runoff Management

Implementation of the preventative measures described previously can considerably reduce the water pollution threat. However, some facilities may require additional management of runoff from the site. The runoff pollutants of primary concern are BOD and P, largely associated with suspended solids particles. Pathogenic cysts may either be absorbed on particles or be free in solution, and the relative significance is not adequately researched. Four readily available strategies exist to help control these pollutants: (1) vegetative filter strips, (2) sediment traps or basins, (3) treatment ponds, and (4) recirculation systems.

The simplest runoff management strategy is the installation of a vegetative filter strip. Vegetative filter strips trap particles in dense surface vegetation. Grasses are commonly used and must be planted in a carefully graded surface over which runoff can be directed in a thin, even layer. Suspended particles flowing slowly through the grass attach to plants and settle to the soil surface, leading to a significant reduction in nutrients, sediment, and BOD levels (Dillaha et al., 1989; Magette et al., 1989).

Sediment traps operate by settling dense particles out of the runoff. Particles settle by gravity during passage through a basin of slowly moving water. This approach can be particularly effective for removing P associated with sediment (NRCS, 1992). Because much of the BOD and N in compost site runoff is associated with light organic particles, the effectiveness of this approach may be somewhat limited. However, it helps limit sediment movement off the site, and can be a useful adjunct to either a vegetative filter strip or a treatment pond, enhancing the effectiveness of each.

During dry periods of the year, compost runoff can be recirculated to the compost piles themselves, or alternately used to irrigate cropland or pasture. The nutrients as well as moisture can thus serve a useful purpose, either by supplying needed moisture to the compost windrows or by providing nutrients and water to crops. However, a recirculation system requires both a pumping and distribution system and adequate storage capacity for prolonged wet periods. Although this approach offers a closed system, which appears ideal for pathogen control, care may need to be taken to separate runoff from fresh feedstocks (especially manure) to avoid contaminating finished compost or crops.

Storage requires the construction of a pond, which can also be used to treat the waste (Figure 3.12). Ponds can be designed for aerobic or facultative treatment of runoff water. In either case microorganisms continue the decomposition process started in the compost pile, but in an aqueous system. As the organic material stabilizes, the BOD levels drop. Pathogen levels are also expected to drop, although the rate depends on seasonal temperature variations and slows during winter in unfrozen portions of a pond. To be effective, ponds must be designed to contain the runoff from major storm events, with an adequate residence time for microbial stabilization. Details of pond design vary with climate, runoff characteristics, and pond effluent requirements. The Natural Resources Conservation Service (NRCS, 1992) has considerable expertise in adapting treatment systems to the local situation.

All these treatment options help remove N and P as well as BOD and pathogens. Sediment basins and ponds settle out particulate matter, which includes bound nutrients such as P. However, these sedimentation mechanisms do not remove nutrients or BOD as well as soil adsorption and crop uptake in a land treatment system. For N removal, vegetative filter strips and irrigation systems can both be effective, and either is enhanced by alternating flow pulses with rest periods. Phosphorus removal is most efficient under aerobic conditions, and irrigation systems generally show higher removal rates than vegetative filter strips, although either can be effective. Although little is currently known about the effectiveness of these approaches in destroying the pathogens of concern, increased opportunities for adsorption, desiccation, and other forms of environmental and microbiological stress are integral to the physical and biological treatment processes described. An appropriate combination of these removal mechanisms can be designed to address the pollution parameters of local concern.

Water quality protection at a composting site can be accomplished through proper site design, operations, and runoff management. Composting facilities vary widely in size, materials processed, and site characteristics, and all these factors will affect the design of appropriate preventative measures. Although the available evidence is

Figure 3.12 Water retention pond at a composting site. Hay bales filter out sediment in runoff.

limited, current indications are that runoff from composting windrows has BOD and nutrient levels comparable to low-strength municipal wastewaters. Land treatment systems that have proven effective for these other wastewaters are likely to be effective for windrow composting facilities as well.

REFERENCES

Alix, C.M. 1998. Retrofits curb biosolids composting odors. *BioCycle* 39(6):37–39.

BioCycle. 1989. *The BioCycle Guide to Yard Waste Composting*. JG Press, Emmaus, Pennsylvania.

Brown K.H., J.C. Bouwkamp, and F.R. Gouin. 1998. The influence of C:P ratio on the biological degradation of municipal solid waste. *Compost Science and Utilization* 6(1):53–58.

Bundy, D.S. and S.J. Hoff. 1998. The testing procedures and results of pit additives tested at Iowa State University, p. 270–282. In: C. Scanes and R. Kanwar (eds.). *Animal Production Systems and the Environment: An International Conference on Odor, Water Quality, Nutrient Management and Socioeconomic Issues. Proceedings Volume 1: Oral Presentations*.

CIWMB (California Integrated Waste Management Board). 2000. *California Code of Regulations*, Title 14, Division, 7, Chapter 3. Website: http://www.ciwmb.ca.gov/Regulations/Title 14/default.htm. Verified 24 June 2000.

Citterio, B., M. Civilini, A. Rutili, A. Pera, and M. de Bertoldi. 1987. Control of a composting process in bioreactor by monitoring chemical and microbial parameters, p. 633–642. In: M. de Bertoldi, M.P. Ferranti, P. L'Hermite, and F. Zucconi (eds.). *Compost: Production, Quality and Use*. Elsevier Applied Science, London, United Kingdom.

Cole, M.A. 1994. Assessing the impact of composting yard trimmings. *BioCycle* 35(4):92–96.

Croteau, G., K. May, and M. Schaan. 1996. Costs and benefits of on-site organics composting. *BioCycle* 37(5):65–68.

Das, K. and H.M. Keener. 1996. Process control based on dynamic properties in composting: moisture and compaction considerations, p.116–125. In: M. de Bertodli, P. Sequi, B. Lemmes, and T. Papi (eds.). *The Science of Composting*. Blackie Academic & Professional, London, United Kingdom.

Diaz, L.F., G.M. Savage, and C.G. Golueke. 1982. *Resource Recovery from Municipal Solid Wastes — Volume 1, Primary Processing*. CRC Press, Boca Raton, Florida.

Dickson, N., T.L. Richard, and R.E. Kozlowski. 1991. *Composting to Reduce the Waste Stream: A Guide to Small Scale Food and Yard Waste Composting*. Natural Resource, Agriculture, and Engineering Service (NRAES), Ithaca, New York.

Dillaha, T.A., R.B. Reneau, S. Mostaghimi, and D. Lee. 1989. Vegetative filter strips for agricultural nonpoint source pollution control. *Transactions of the American Society of Agricultural Engineers* 32(2):513–519.

Dougherty, M. (ed.). 1999. *Field Guide to On-Farm Composting*. Natural Resource, Agriculture, and Engineering Service (NRAES), Ithaca, New York.

Dunson, J.B., Jr. 1993. Control of odors by physical-chemical approaches, p. 242–261. In: H.A.J. Hoitink and H. Keener (eds.). *Science and Engineering of Composting: Design, Environmental, Microbiological, and Utilization Aspects*. Renaissance Publications, Worthington, Ohio.

Eitzer, B.D. 1995. Emissions of volatile organic chemicals from municipal solid waste composting facilities. *Environmental Science and Technology* 29(4):896–902.

Epstein, E. 1997. *The Science of Composting*. Technomic Publishing, Basel, Switzerland.

Finstein, M.S. and J.A. Hogan. 1993. Integration of composting process microbiology, facility structure and decision making, p. 1–23. In: H.A.J. Hoitink and H. Keener (eds.). *Science and Engineering of Composting: Design, Environmental, Microbiological, and Utilization Aspects*. Renaissance Publications, Worthington, Ohio.

Finstein, M.S., J. Cirello, S.T. MacGregor, F.C. Miller, and K.M. Psarianos. 1980. *Sludge Composting and Utilization: Rational Approach to Process Control*. US-EPA Project C-340-678-01-1. Accession No. PB82-13623, National Technical Information Service, Springfield, Virginia.

Finstein, M.S., F.C. Miller, P.F. Strom, S.T. MacGregor, and K.M. Psarianos. 1983. Composting ecosystem management for waste treatment. *Bio/Technology* 1:347–353.

Finstein, M.S., F.C. Miller, F.C. MacGregor, and K.M. Psarianos. 1985. *The Rutgers Strategy for Composting: Process Design and Control*. Rutgers University, New Brunswick, New Jersey.

Finstein, M.S., F.C. Miller, and P.F. Strom. 1986. Waste treatment composting as a controlled system, p. 362–398. In: H.J. Rehm and G. Reed (eds.). *Biotechnology: A Comprehensive Treatise in 8 Volumes – Volume 8: Microbial Degradations*. VCH Verlagsgesellschaft, Weinheim, Federal Republic of Germany.

Finstein, M.S., F.C. Miller, J.A. Hogan, and P.F. Strom. 1987a. Analysis of EPA guidance on composting sludge. Part I: biological heat generation and temperature. *BioCycle* 28(1):20–26.

Finstein, M.S., F.C. Miller, J.A. Hogan, and P.F. Strom. 1987b. Analysis of EPA guidance on composting sludge. Part II: biological process control. *BioCycle* 28(2):42–47.

Finstein, M.S., F.C. Miller, J.A. Hogan, and P.F. Strom. 1987c. Analysis of EPA guidance on composting sludge. Part III: oxygen, moisture, odor, pathogens. *BioCycle* 28(3):38–44.

Goldstein, N. and K. Gray. 1999. Biosolids composting in the United States. *BioCycle* 40(1):63–75.

Golueke, C.G. 1972. *Composting: A Study of the Process and Its Principles*. Rodale Press, Emmaus, Pennsylvania.

Gray, K.R. and A.J. Biddlestone. 1974. Decomposition of urban waste, p. 743–775. In: C.H. Dickinson and G.J.F. Pugh (eds.). *Biology of Plant Litter Decomposition, Volume 2*. Academic Press, London, United Kingdom.

Haug, R.T. 1993. *The Practical Handbook of Compost Engineering*. CRC Press, Boca Raton, Florida.

Hay, J.C. and R.D. Kuchenrither. 1990. Fundamentals and applications of windrow composting. *Journal of Environmental Engineering* 116:746–763.

Higgins, A.J., V. Kasper Jr., D.A. Derr, M.E. Singley, and A. Singh. 1981. Mixing systems for sludge composting. *BioCycle* 22(5):18–22.

Hoitink, H.A.J., M.J. Boehm, and Y. Hadar. 1993. Mechanisms of suppression of soilborne plant pathogens in compost, p. 601–621. In: H.A.J. Hoitink and H. Keener (eds.). *Science and Engineering of Composting: Design, Environmental, Microbiological, and Utilization Aspects*. Renaissance Publications, Worthington, Ohio.

Horwath, W.R., L.F. Elliot, and D.B. Churchill. 1995. Mechanisms regulating composting of high carbon to nitrogen ratio grass straw. *Compost Science and Utilization* 3(3):22–30.

JG Press, Inc. 1986. *The BioCycle Guide to In-Vessel Composting*. Emmaus, Pennsylvania.

Kuter, G.A. (ed.). 1995. *Biosolids Composting*. Water Environment Federation, Alexandria, Virginia.

Lafrance, C., P. Lessard, and G. Buelna. 1996. Évaluation de la filtration sur tourbe et compost pour le traitmement de l'effluent d'une usine de compostage de résidus verts. *Canadian Journal of Civil Engineering* 23:1041–1050.

Lenton, T.G. and E.I. Stentiford. 1990. Control of aeration in static pile composting. *Waste Management Research* 8:299–306.

Lesson, G. and K.A.M. Winer. 1991. Biofiltration: An innovative air pollution control technology for VOC emissions. *Journal of the Air and Waste Management Association* 41(8):1045–1054.

Loehr, R.C., W.J. Jewell, J.D. Novak, W.W. Clarkson, and G.S. Friedman. 1979. *Land Application of Wastes, Volume II*. Van Nostrand Reinhold, New York.

Lopez-Real, J. and M. Baptisa. 1996. A preliminary study of three manure composting systems and their influence on process parameters and methane emissions. *Compost Science and Utilization* 4(3):71–82.

Lorig, T.S. 1992. Cognitive and noncognitive effects of odour exposure; electrophysiological and behavioural evidence, p. 161–173. In: S. Van Toller and G.H. Dodd (eds.). *The Psychology and Biology of Perfume*. Elsevier Applied Science, London, United Kingdom.

Ludvigson, H.W. and T.R. Rottman. 1989. Effects of ambient odors of lavender and cloves on cognition, memory, affect and mood. *Chemical Senses* 14:525–536.

Lynch, J.M. 1993. Substrate availability in the production of composts, p. 24–35. In: H.A.J. Hoitink and H. Keener (eds.). *Science and Engineering of Composting: Design, Environmental, Microbiological, and Utilization Aspects*. Renaissance Publications, Worthington, Ohio.

Lynch, N. J. and R.S. Cherry. 1996a. Design of passively aerated compost piles: vertical air velocities between pipes, p. 973–982. In: M. de Bertodli, P. Sequi, B. Lemmes, and T. Papi (eds.). *The Science of Composting*. Blackie Academic & Professional, London, United Kingdom.

Lynch, N. J. and R.S. Cherry. 1996b. Winter composting using the passively aerated windrow system. *Compost Science and Utilization* 4(3):44–52.

Magette, W.L., R.B. Brinsfield, R.E. Palmer, and J.D. Wood. 1989. Nutrient and sediment removal by vegetated filter strips. *Transactions of the American Society of Agricultural Engineers* 32(2):663–667.

Massachusetts Department of Environmental Protection. 1988. *Leaf Composting Guide*. Boston, Massachusetts.

Mathur, S.P. 1995. Natural aeration static pile (based on Agriculture Canada's PAWS technology). In: *Composting Technologies and Practices – A Guide for Decision Makers*. The Composting Council of Canada, Ottawa, Ontario, Canada.

Mathur, S.P. 1997. Composting processes, p. 154–196. In: A.M. Martin (ed.). *Bioconversion of Waste Materials to Industrial Products*. Blackie Academic and Professional/Chapman & Hall, London, United Kingdom.

Mathur, S.P. and R.W. Richards. 1999. Large-scale generation of hygenic, high-quality compost from beef feedlot manure and sawmill residues in natural aeration static piles (NASP) without pipes, plenums or envelope, p. 36. In: P. Warman and T. R. Munro-Warman (eds.). *Abstracts of the International Composting Symposium, September, 1999*. Coastal BioAgresearch, Ltd., Truro, Nova Scotia, Canada.

Mathur, S.P., N.K. Patni, and M.P. Levesque. 1988. *Composting of Manure Slurries with Peat Without Mechanical Aeration*. Canadian Society of Agricultural Engineers, Annual Conference Paper No. 88–123. 13 pages.

Mathur S.P., N.K. Patni, and M.P. Levesque. 1990. Static pile, passive aeration composting of manure slurries using peat as a bulking agent. *Biological Wastes* 34:323–333.

McCartney, D.M. and H. Chen. 1999. Physical modeling of composting using biological load cells: effect of compressive settlement on free air space and microbial activity, p. 37. In: P. Warman and T. R. Munro-Warman (eds.). *Abstracts of the International Composting Symposium, September, 1999*. Coastal BioAgresearch, Ltd., Truro, Nova Scotia, Canada.

Michel, F.C., Jr., L.J. Forney, A. J.-F. Huang, S. Drew, M. Czuprenski, J.D. Lindeberg, and C.A. Reddy. 1996. Effects of turning frequency, leaves to grass ratio, and windrow vs. pile configuration on the composting of yard trimmings. *Compost Science and Utilization* 4(1):26–43.

Miller, F.C. 1991. Biodegradation of solid wastes by composting, p. 1–31. In: A.M. Martin (ed.). *Biological Degradation of Wastes*. Elsevier Applied Science, London, United Kingdom.

Miller, F.C. 1993. Minimizing odor generation, p. 219–241. In: H.A.J. Hoitink and H. Keener (eds.). *Science and Engineering of Composting: Design, Environmental, Microbiological, and Utilization Aspects*. Renaissance Publications, Worthington, Ohio.

Miner, R.J. 1995. *A Review of the Literature on the Nature and Control of Odors from Pork Production Facilities*. National Pork Producers Council, Des Moines, Iowa.

Natural Resources Conservation Service (NRCS). 1992. *National Engineering Handbook Part 651: Agricultural Waste Management Field Handbook*. U.S. Department of Agriculture (USDA), Washington, DC. http://www.ftw.nrcs.usda.gov/awmfh.html. Verified 24 June 2000.

Naylor, L.M. 1996. Composting, p. 193–269. In: M.J. Girovich (ed.). *Biosolids Treatment and Management: Processes for Beneficial Use*. Marcel Dekker, New York.

Panter, K., R. DeGarmo, and D. Border. 1996. A review of features, benefits and costs of tunnel composting systems in Europe and in the USA, p. 983–986. In: M. de Bertodli, P. Sequi, B. Lemmes, and T. Papi (eds.). *The Science of Composting*. Blackie Academic & Professional, London, United Kingdom.

Patni, N. 2000. Videotape interview. In: R. Rynk (producer). *The Future of Agricultural Composting, A Video Workshop*. Compost Education and Resources for Western Agriculture, University of Idaho, Moscow, Idaho.

Poincelot, R.P. 1975. *The Biochemistry of Composting*. The Connecticut Agricultural Experiment Station, New Haven, Connecticut.

Randle, P. and P.B. Flegg. 1978. Oxygen measurements in a mushroom compost stack. *Scientia Horticulturae* 8:315–323.

Richard, T.L. 1992. Municipal solid waste composting: physical and biological processing. *Biomass and Bioenergy* 3(3-4):163–180.

Richard, T.L. 1997. Course notes on composting aeration design. http://www.ae.iastate.edu/Ae573_ast475/composting-aeration_design_not.htm

Richard, T.L. and M. Chadsey. 1994. Environmental impact assessment, p. 232–237. In: *BioCycle* staff (eds.). *Composting Source Separated Organics*. J.G. Press, Emmaus, Pennsylvania. Also published in 1990 as: Environmental monitoring at a yard waste composting facility. *BioCycle* 31(4):42–46.

Richman, B. and R. Rynk. 1994. *Recycling Yard Debris*. Panhandle Health District, Couer d'Alene, Idaho.

Ritter, W.F. 1981. Chemical and biochemical odor control of livestock wastes: A review. *Canadian Agricultural Engineering* 23:1–4.

Rymshaw, E., M.F. Walter, and T.L. Richard. 1992. *Agricultural Composting: Environmental Monitoring and Management Practices*. Department of Agricultural and Biological Engineering, Cornell University, Ithaca, New York.

Rynk, R. 1994. Composting equipment, p. 147–179. In: C.W. Heuser and P.E. Heuser (eds.). *Recycling and Resource Conservation, A Reference Guide for Nursery and Landscape Industries*. Pennsylvania Nurserymen's Association, Harrisburg, Pennsylvania.

Rynk, R. 2000a. Fires at composting facilities: Part I. *BioCycle* 41(1):54–58.

Rynk, R. 2000b. Fires at composting facilities: Part II. *BioCycle* 41(2):58–62.

Rynk, R. 2000c. Review of contained composting systems: Part I. *BioCycle* 41(3):30–36.

Rynk, R. 2000d. Review of contained composting systems: Part II. *BioCycle* 41(4):67–72.

Rynk, R.F., M. van de Kamp, G.B Willson, M.E. Singley, T.L. Richard, J.J. Kolega, F.R. Gouin, L.L. Laliberty, D. Kay, D.W. Murphy, H.A.J. Hoitink, and W.F. Brinton. 1992. *On-Farm Composting Handbook*. Natural Resource, Agriculture, and Engineering Service (NRAES), Ithaca, New York.

Schiffman, S.S., E.A. Sattely Miller, M.S. Suggs, and B.G. Graham. 1995. The effect of environmental odors emanating from commercial swine operations on the mood of nearby residents. *Brain Research Bulletin* 37(4):369–375.

Schoenecker, B. and A. McConnell. 1993. Composting yard waste under cover and over air. *BioCycle* 34(1):44–45.

Shoda, M. 1991. Methods for the biological treatment of exhaust gases, p. 1–30. In: A.M. Martin (ed.). *Biological Degradation of Wastes*. Elsevier Applied Science, London, United Kingdom.

Singley, M., A. J. Higgins, and M. Franklin-Rosengaus. 1982. *Sludge Composting and Utilization — A Design and Operating Manual*. Rutgers University, New Brunswick, New Jersey.

Stentiford, E.I. 1996. Composting control: principles and practice, p. 49–59. In: M. de Bertodli, P. Sequi, B. Lemmes, and T. Papi (eds.). *The Science of Composting*. Blackie Academic & Professional, London, United Kingdom.

Stentiford, E.I. and T.J. Pereira Neto. 1985. Simplified systems for refuse/sludge compost. *BioCycle* 26(5):46–49.

Strom, P.F. and M.S. Finstein. 1985. *Leaf Composting Manual for New Jersey Municipalities*. Rutgers University, New Brunswick, New Jersey.

The Compost Council. 1994. *Compost Facility Operating Guide*. Alexandria, Virginia.

United States Environmental Protection Agency (U.S. EPA). 1985. *Composting of Municipal Wastewater Sludges*. EPA/625/4-85-016. Office of Wastewater Management, Cincinnati, Ohio.

United States Environmental Protection Agency (U.S. EPA). 1994. *In-Vessel Composting of Municipal Wastewater Sludge*. EPA/625/8-89/016. Office of Wastewater Management, Cincinnati, Ohio.

Williams, T.O. and F.C. Miller. 1993. Composting facility odor control using biofilters, p. 262–281. In: H.A.J. Hoitink and H. Keener (eds.). *Science and Engineering of Composting: Design, Environmental, Microbiological, and Utilization Aspects*. Renaissance Publications, Worthington, Ohio.

Willson, G.B. 1980. *Manual for Composting Sewage Sludge by the Aerated Static Pile Method*. U.S. Environmental Protection Agency (U.S. EPA), Cincinnati, Ohio.

Willson, G.B., J.F. Parr, and J.L. Thompson. 1979. Evaluation of mixers for blending sewage sludge with wood chips, p. 48–54. In: *Municipal and Industrial Sludge Composting*. Information Transfer Inc., Rockville, Maryland.

CHAPTER 4

Compost Quality Attributes, Measurements, and Variability

Dan M. Sullivan and Robert O. Miller

CONTENTS

1-56670-460-X/01/$0.00+$.50

I. WHAT IS COMPOST QUALITY?

Compost testing is used by both compost producers and users. This chapter is designed to assist compost producers and users in collecting representative compost samples, requesting compost analyses, and in understanding and interpreting compost analytical data.

The most critical compost quality factors depend on the planned compost end use (Table 4.1). For most applications, plant growth response is the ultimate indicator of compost quality. Compost nutrient content, especially plant-available nitrogen (N), is most important for field crops where compost is applied as a supplement or replacement for other nutrient sources. The pH and soluble salt content of compost is a key characteristic where compost is used at high rates, such as in potting media or compost sold to the general public. Successful marketing to the general public requires a mature, well-decomposed, dark brown to black compost with an earthy odor. Compost maturity and biological stability are most important for compost use in potting media, for bagged products, and for compost-mediated disease suppression. Consistent particle size is needed for soilless potting media and other high-value applications such as amendment of athletic turf or golf greens.

Table 4.1 Relative Importance of Compost Quality Measurements for Horticultural Applications

Kind of Quality Measurement	Target Compost Use			
	Greenhouse or Nursery Crops	Sales to General Public; Bulk or Bagged	Soil Amendment for Vegetable and Fruit Crops	Mulch
Plant growth response	++	++	++	–
Nutrient content	–	+	+	–
pH and soluble salts	++	++	+	–
Man-made inerts	++	++	+	+
Sensory: color and odor	+	++	–	+
Maturity and biological stability	++	++	+	–
Particle size	++	+	+	+

Note: –, +, ++ indicates low, medium, and high importance for specified compost use.

II. COMPOST QUALITY SPECIFICATIONS OR GUIDELINES

Compost industry organizations have recently adopted suggested compost specifications for a variety of uses. These specifications are usually based on consensus among experts. They are often not specific enough for a given end use in a particular location. The guidelines developed by the U.S. Composting Council (Table 4.2) were evaluated by several groups of professional end-users in the Seattle, WA area (E & A Environmental Consultants and Stenn, 1996). Horticultural professionals considered the guidelines too general to apply to their specific situations. The study concluded that the U.S. Composting Council guidelines are best used by compost producers as a minimum quality standard.

Table 4.2 Suggested Compost Quality Guidelines for Horticultural Applications[z]

Quality Parameter	Soil Amendment for Turf, Vegetable Crops, or Planting Beds	Potting Media	Landscape Mulch
Particle size	Passes 25 mm screen	Passes 13 mm screen	Passes 10 mm screen
Soluble salts	Maximum in soil blend of 2.5 to 6 dS m^{-1} depending on crop	Maximum in mixed media of 3 dS m^{-1}	Must report
Stability	Stable to highly stable	Highly stable	Moderately to highly stable

[z] Adapted from U.S. Composting Council, 1996. Other quality parameters suggested by the Council are the same across horticultural compost use categories: Nutrient content, water-holding capacity, bulk density, and organic matter content must be reported; must pass germination and growth screening; and must not exceed Part 503 limits for trace element concentrations (U.S. EPA, 1993). Moisture content ("as-is" basis) should be 35–55%, and pH from 5.5 to 8.0.

Voluntary or regulation-driven compost quality assurance programs have recently emerged in North America (California Compost Quality Council, 1999; Compost Council of Canada, 1999) and in Europe (Bildingmaier, 1993; Verdonck, 1998). Generally, there are rigid standards for essential quality parameters such as minimum organic matter content, maximum levels of trace elements, maximum levels of man-made inerts, and freedom from human pathogens. Beyond these minimal standards, most compost quality assurance programs have a list of other parameters that must be reported for the product.

A few quality assurance programs include periodic verification of product quality by a third-party laboratory or an oversight agency. In voluntary programs, the compost producer obtains the right to advertise with an organizational "seal of approval" (California Compost Quality Council, 1999; Woods End Research Laboratory, 1999a). However, only a few of the parameters that may be important to high-value horticultural use (Inbar et al., 1993) are evaluated in most current quality assurance programs.

A major problem in compost guideline development or the development of quality assurance standards for compost is the difference in perspective between researchers, compost producers, and compost users (E & A Environmental Consultants and Stenn, 1996). Research studies typically focus on how the use of a specific compost product affects the growth of specific plant species in a particular application. Compost users and producers have much broader information needs. Typically, they are interested in efficient methods for compost handling, how compost can be used on a variety of soils and plant species, and how compost use affects other crop maintenance activities (e.g., fertilization, disease and weed control). Guidelines and quality assurance standards will continue to improve as more experience is gained on compost use in different environments.

One of the first steps towards standardization of compost quality is the standardization of laboratory analysis procedures. The U.S. Composting Council has developed a comprehensive publication describing procedures for compost sampling and testing, *Test Methods for the Examination of Composting and Compost* (TMECC; Leege and Thompson, 1997). The format for TMECC is designed primarily for laboratory use. Quick tests for approximation of compost product quality are also included. Detailed instructions are given for carrying out each test, using a format similar to that used by the American Society for Testing and Materials.

Most of the chemical and physical test methods listed in TMECC were adapted from existing standard methods for soil and plant material analysis and are unlikely to change significantly with time. Many of the biological methods for assessing compost stability and maturity were recently developed by researchers, and are likely to be refined as they are adopted by the compost industry. The current version of TMECC (Leege and Thompson, 1997) is undergoing extensive peer review by laboratory personnel, compost users, scientists, and regulatory officials. Future editions of TMECC will reflect the collective expertise of the peer review group. In this chapter, we will frequently reference TMECC methods from the 1997 edition.

III. COMPOST SAMPLING

Compost sampling is perhaps the most critical phase of compost analysis. A compost sample that accurately represents the compost product is essential. Best results from compost testing come from carefully planned sampling.

Deciding what tests are needed and what laboratories will do the analysis is the first step in designing a sampling plan. For evaluation of horticultural use potential, compost tests can be performed by a laboratory that routinely does analyses for other organic growing media. Other tests, such as those required by regulation (e.g., human pathogens or trace elements), should be performed by a laboratory that specializes in such testing. Some agricultural soil and plant tissue testing laboratories can perform many of the horticultural and environmental tests.

We suggest working backwards from the interpretation of test results to determine when and how to sample. If compost is purchased, tell the supplier what components of compost quality are essential for the intended use. Discuss how and when the compost is sampled, to make sure the analysis reflects "as delivered" quality. If one is producing compost, compost test results can be used to adjust the composting process to meet one's specific needs. To assist in producing quality compost, a producer may want to sample compost feedstocks and actively composting piles, as well as the finished compost.

The generalized sampling protocol described in Table 4.3 is applicable to samples collected for all analyses except for microbiological analyses. A sterile sample collection and preservation technique is needed for microbiological testing (U.S. EPA, 1992). Composite sampling, where individual samples are combined into one sample submitted to a laboratory, is the recommended protocol for representing average compost quality. When information is needed on the variability of compost analyses within a pile, a variety of other sampling techniques can be used (Leege and Thompson, 1997).

Table 4.3 Generalized Protocol for Sampling Compost from Windrows

Sample size: A 12 L compost sample is usually needed for a complete chemical, physical, and biological analysis. Check with your laboratory for optimal sample size for the requested analyses.
Number of sampling locations: Randomly select six locations along the length of the windrow.
Subsample collection: At each location along the windrow, collect three subsamples of equal volume to represent a cross section of the compost pile. Expose the center of the large piles using a front end loader or other equipment. Collect at least a total of 18 subsamples (6 locations ×3 subsamples per location) to represent a windrow. Mix the three subsamples from each sampling location in a 15 L plastic bucket.
Sample mixing and volume reduction: Empty the six composite "location samples" on a large plastic tarp. Mix all samples together on the tarp. Reduce sample size by repeated mixing, quartering and subsampling. Final sample volume to submit to the laboratory = 12 L.
Sample containers and preservation: Transfer a 12 L blended compost sample to three 4 L zippered plastic freezer bags. Cool sample to 4°C with ice or refrigeration. Ship in a plastic pail with blue ice packs. The sample should arrive at the laboratory within 24 to 48 hours.

Adapted from Leege and Thompson, 1997.

The best time to collect a composite sample is immediately after a pile has been thoroughly turned or mixed. Within days or hours after turning, a pile develops gradients in moisture, aeration, biological stability, and bacterial populations. Even after turning, piles may not be thoroughly mixed, so many small samples from different locations in the pile must be combined to reflect average compost quality.

The most common sampling situations are sampling from windrows or sampling from curing piles. For windrow sampling, it is important to take samples from random locations representing the entire length of the windrow. This is especially important when windrows are built gradually from end to end, and may have substantial variation in compost feedstocks and processing time. Curing piles are often extremely variable in moisture, maturity, and bulk density. Frequently, curing piles are very large and contain material from several active composting piles, and are not turned or mixed. In sampling large static windrows and curing piles, it is essential to break into the center of the pile with a front-end loader or other equipment to obtain a representative sample.

IV. PHYSICAL PROPERTIES OF COMPOSTS

A. Moisture Content

Compost moisture content is easily determined, but may fluctuate widely due to differences in feedstocks, processing, and storage conditions. Moisture content can be expressed on a weight or volume basis. Moisture is most often expressed as a fraction of total compost weight (Table 4.4). As moisture content increases, dry matter per unit weight decreases. Moisture content may also provide some understanding of processing or storage conditions. Composts with moisture contents of less than 35% may not have been fully stabilized due to low moisture, or may have been stored for excessively long periods leading to moisture loss. Composts with less than 35% moisture are often dusty and unpleasant to handle.

B. Bulk Density

Bulk density, the weight per unit volume of compost, is affected by moisture content, inorganic (ash) content, particle size distribution, and the degree of decomposition. Bulk density is used to convert nutrient analyses from dry weight to an "as-is" basis.

Bulk density on an as-is basis (Table 4.4) mainly indicates water content. Most composts with an as-is moisture content of 35 to 55% will have a bulk density of 500 to 700 kg m^{-3} (about 900 to 1200 lb per yd^3).

Bulk density on a dry weight basis is an indicator of particle size and ash content. Dry bulk density usually increases with composting time as ash content increases and as particle size is reduced by decomposition, turning, and screening (Raviv et al., 1987). The dry bulk density of compost is most important when compost comprises a large proportion of the growing media (e.g., potting media). As bulk density increases, drainage and air-filled porosity of growing media are reduced, and water-holding capacity is increased.

Compost users use bulk density and moisture analyses to calculate volume-based application rates (e.g., m^3 compost per 100 m^2) that are approximately equal to a given compost dry weight per unit area (e.g., kg dry matter per m^3). The measurement of as-is bulk density in the laboratory (Table 4.4) simulates a small pile of compost. Compost in big piles, or packed into a truck, may have higher bulk density values.

C. Water-Holding Capacity

Water-holding capacity is the amount of water held in pores after gravitational loss for a specified time. This test is used to assess the utilization of compost for potting media. Water-holding capacity (Table 4.4) is a measure of the water retained by a compost sample after free drainage for 4 h. This procedure is container specific. Water retention after free drainage is strongly affected by the height of the measurement vessel (Inbar et al., 1993).

Table 4.4 Common Analyses for Compost Physical Properties

Analysis	TMECC[z] Method Number	Metric Units	Common Field Units	Laboratory Procedure Comments
Gravimetric moisture content	7.01A&B	g water per g of "as-is" compost	% w/w "as-is"	Dry weight of compost sample measured at 70°C. Moisture content can be calculated from total solids content: Moisture content (%) = 100 – total solids (%). Moisture contents for soils are usually expressed in different units (g water per g of dry soil).
Bulk density	7.01A&B	g compost per cm³ of "as-is" compost	lb per cubic yard	A reproducible method for packing compost in the measurement vessel (2000 cm³ beaker) is essential for consistent results. This measurement is used to calculate other physical properties on a volume basis.
Gravimetric water-holding capacity	7.01A&B	g water per g of "saturated and drained" compost	% w/w "as-is"	Water held after free drainage for 4 h in a 2000 cm³ beaker with perforated bottom. This procedure overestimates water-holding capacity of compost in the field because some saturated compost will occur at the bottom of the beaker. Data from this procedure can be used to calculate total porosity and air-filled porosity.
Particle size	5.01-B	% passing sieve (dw[y])	% passing sieve (dw)	Percentage (by dry weight) which passes a given sieve mesh opening (e.g., less than 12 mm). Nested sieving yields particle size distribution.
Man-made inerts	5.01-B	g inerts per g compost (dw)	% (dw)	Visual sorting process. Sample size small because the procedure is time consuming. Includes glass, plastic, rubber, and metal. Usually does not include rocks. Plastics may be a small amount by weight but be a visual concern.

[z] TMECC: *Test Methods for the Examination of Composting and Compost* (Leege and Thompson, 1997).
[y] Dry weight basis.

Water-holding capacity measurements are of limited importance for field compost use. Composts applied to soil, even at high rates, may not increase the net amount of water that is readily available to plants between soil matric potentials of –0.2 and –0.8 bars (Chang et al., 1983). Compost addition to soil increases net water availability at matric potentials near saturation (0 to –0.2 bars; Chang et al., 1983, McCoy, 1992), but this water drains away rapidly in a field soil.

D. Particle Size and Man-Made Inerts

Particle size provides a number of critical indicators for the potential user. Large particles (e.g., those retained by 12 mm screen) prevent efficient spreading for some field applications. Screening can remove larger compost particles, but it is difficult to remove small particles. Small particle size may also limit use for applications such as potting mixes or golf greens, where rapid drainage is important. Too many fine compost particles are undesirable in a mulch product, because they can retain enough water to promote weed seed germination.

Man-made inerts, such as glass or plastics, are seldom a problem except for composts derived from municipal solid waste (MSW). Plastics can be a problem with urban yard debris composts, especially if grass is collected in plastic bags.

V. CHEMICAL PROPERTIES OF COMPOSTS

A. Total Organic Carbon (C)

The total organic C concentration of a compost is an indicator of its organic matter concentration. Total organic C is generally measured by two laboratory methods: (1) combustion (Method 9.08-B in Table 4.5) and (2) Walkley–Black (Schulte, 1988). The combustion method relies on high-temperature furnace oxidation and subsequent direct measurement of C by an infrared detector. Combustion is the preferred procedure because it is more accurate and precise than the Walkley–Black determination. The Walkley–Black method provides an estimate of organic C, based on partial chemical oxidation of total organic C. The Walkley–Black test is calibrated for soil organic matter, which is not completely similar to compost organic matter. Another disadvantage of the Walkley–Black method is its use of dichromate, a chemical classified as a hazardous waste.

Both the combustion and Walkley–Black methods do not discriminate between organic and inorganic C (e.g., carbonates). Testing for inorganic C is recommended for composts that have a saturated paste pH above 7.3, or composts that have been amended with alkaline materials such as lime.

B. Volatile Solids (Volatile Organics)

The volatile solids (volatile organics) method estimates organic and ash concentrations. The portion of the sample lost in high-temperature combustion (550°C) estimates organic matter; the portion remaining after combustion is ash. Because organic matter is not determined directly, the volatile solids content of a sample is only approximately equal to its organic matter content. The volatile solids estimate includes nonorganic matter sources of weight loss, including rubber, plastic, and "bound" water. This method is also referred to as loss-on-ignition (LOI).

Table 4.5 Common Analyses for Compost Chemical Properties[z,y]

Analysis	TMECC[z] Method Number or Other Reference	Metric Reporting Units[x]	Common Units[x]	Laboratory Procedure Comments
Total organic carbon (TOC)	9.08-B	mg kg^{-1}	%	Measures TOC utilizing a combustion furnace and infared detector. Organic C via combustion can be inaccurate for high pH composts that contain a lot of inorganic C as carbonate. The sample size used by different commercial combustion analyzers varies from 0.1 to 2.0 g. A larger sample size usually increases analytical precision and accuracy. This measurement for organic C is preferred for estimating C for C:N ratio. Total C and N analysis can be done simultaneously with some instruments.
Volatile solids[w]	9.08-A	mg kg^{-1}	%	Sample is preheated to remove moisture, weighed, placed in 550°C furnace and then reweighed. Weight loss is "volatile solids" or "volatile organics". Material remaining after ignition is ash. Compost C is approximately 50% of volatile solids content (rough estimate).
Cation exchange capacity (CEC)	8.03-B	cmol (+) per kg	meq 100 g^{-1}	Sample is saturated with a cation such as NH_4^+, Na^+ or Ba^{2+}. CEC is measured by the replacement technique. Compost CEC varies with pH. CEC determined at pH 7 is adequate for most composts.
Total nitrogen	8.09-A 8.09-D	mg kg^{-1}	ppm or %	Measures sum of inorganic plus organic N forms. Two acceptable methods: Total Kjeldahl (TKN) or combustion with infared detector. Some Kjeldahl methods do not include measurement of nitrate-N.
Inorganic nitrogen	8.09-B 8.09-C	mg kg^{-1}	ppm	Inorganic N includes ammonium N (NH_4-N), ammonia N (NH_3-N), and nitrate N (NO_3-N). A number of colorimetric methods are suitable. Nitrate is most accurately determined by the cadmium reduction method. NH_3-N can be determined by the ion electrode method. Sample inorganic N concentrations can change rapidly with sample drying or unrefrigerated storage.
pH	8.07-A 8.07-B	—	—	Saturated paste or volume addition methods. Saturated paste extract useful for other tests (see below). Adding large volumes of water changes pH. Usually, pH by volume addition is 0.1 to 0.3 units higher than saturated paste pH.
Electrical conductivity (EC)	Gavlak et al., 1994[y]	dS m^{-1}	mmhos cm^{-1}	EC estimates soluble salt concentrations. EC determined on saturated paste extract. Sample is saturated with water, vacuum filtered, and EC of extract is measured (usually with a conductivity probe). Extract also used for determination of some elements like Cl and B.

[z] TMECC: *Test Methods for the Examination of Composting and Compost* (Leege and Thompson, 1997).
[y] *Plant, Soil and Water Reference Methods for the Western Region* (Gavlak et al., 1994).
[x] Dry weight basis.
[w] Volatile solids test is sometimes called "loss-on-ignition" (LOI) or "volatile organics." Volatile solids are equal to "biodegradable volatile solids" when sample does not have significant quantities of plastics and rubber.

C. Cation Exchange Capacity

Cation exchange capacity (CEC) is a measure of the capacity of compost to hold exchangeable cations such as potassium (K), calcium (Ca), magnesium (Mg), and sodium (Na), to negatively charged surfaces. Sources of negative charges in compost include dissociation of acidic functional groups found in organic matter (e.g., OH, COOH). As pH increases, the CEC of organic matter increases. Most composts have a pH of 6 to 8, which is similar to that used in most CEC test methods (pH 7).

CEC test methods recommended for compost (Table 4.5) saturate the compost with a single cation such as NH_4^+, Na^+ or barium (Ba^{2+}), then subsequently displace and determine the saturating cation.

Compost CEC measurements are used in formulating potting media for container plants, and as an indicator of compost maturity (Table 4.6). A potting medium with a relatively high CEC provides more buffering capacity against changes in pH and is more easily managed in container plant production. Compost CEC increases with composting time, as the compost organic matter becomes more humified.

D. Total Nitrogen (N)

The total N content of composts can vary substantially based on feedstocks, processing conditions, curing, and storage (see "Chemical Indicators of Maturity" in this chapter and chapter 14 in this book for more interpretive information).

Total N is the sum of inorganic + organic N forms in compost. Total N is measured by two laboratory methods, total Kjeldahl and combustion.

For the Kjeldahl method, strong acid is added to digest the sample, and ammonium-N (NH_4-N) in the digested sample is subsequently determined via colorimetric analysis. Some Kjeldahl procedures do not include nitrate-N (NO_3-N) in the total N determination. For most composts, omitting NO_3 from total N analyses is insignificant, since composts usually contain less than 0.2% NO_3-N.

The combustion method is a direct measurement of total N. The sample is oxidized in a high-temperature furnace, and N is determined by an infared detector. The combustion method is generally more accurate and precise than the Kjeldahl method. Samples containing large quantities of lignin will have lower quantities of N by the Kjeldahl method versus the combustion method. The Kjeldahl method does not digest N present in heterocyclic ring compounds like those found in lignin; the combustion method detects all N forms.

E. Inorganic Nitrogen (N)

Inorganic N includes NH_4-N, ammonia-N (NH_3-N), and NO_3-N. A number of extractants and colorimetric determination methods are acceptable for NH_4 and NO_3 analyses (Gavlak et al., 1994).

Laboratory procedures for NH_4-N, and NH_3-N are identical. Laboratories differ in how they report test results. Usually, soil testing labs report NH_4-N, while environmental laboratories report NH_3-N. From a chemical perspective, NH_4-N is the

Table 4.6 Sensory and Chemical Indicators of Compost Maturity

Method	TMECC[z] Method Number	Trend During Composting	Suggested Value for Mature Compost	Comments
		Sensory Indicators		
Color	9.03A	Darkens	Black to very dark brown	Subjective. Feedstock dependent
Odor	9.03A	Foul anaerobic odor to earthy odor	Earthy, soil-like, no odor	Subjective. Not very sensitive for composts during curing stage
		Chemical Indicators		
Volatile solids reduction	9.10	Decrease	45 to 60+%	Feedstock dependent. Only measurable by compostercompost producer. Calculation is based on the initial ash content of the feedstock mixture (TMECC 9.10-A; Stentiford et al.and Pereira-Neto,1985)
Cation exchange capacity (CEC)	8.03	Increase	> 60 meq.100 g^{-1} volatile solids for MSW composts (Harada et al., 1981)	Maximum CEC in mature compost depends on the feedstocks.
C:N ratio	9.02A	Decrease or increase depending on C:N of feedstocks	Mature compost: 15 to 20:1. Composts with C:N ratios above 25 to 30:1 usually immobilize inorganic N.	Ratio is meaningful for assessing maturity for composts derived from high C:N mixtures (initial C:N ratio > 25:1).
Inorganic N	9.02C	NH_4 decrease; NO_3 increase	Mature composts contain more NO_3-N than NH_4-N.	Dry, unstable compost piles can give high nitrate NO_3 values. Rewetting of dry, immature compost can result in rapid loss of NO_3 via denitrification.

[z] TMECC: *Test Methods for the Examination of Composting and Compost* (Leege and Thompson, 1997).

more accurate representation. NH_3 is usually a very small proportion of compost NH_4-N + NH_3-N.

Sample preservation techniques and holding time can affect inorganic N test results. NH_4 and NO_3 can change rapidly due to drying and unrefrigerated storage. Poorly stabilized or immature composts will often contain significant quantities of NH_4-N that can be rapidly lost to the air during handling and storage. It is best to rapidly freeze samples that will be submitted for NH_4 or NO_3 analysis.

The form and amount of N present in inorganic forms can be a useful indicator of compost maturity (see "Chemical indicators of compost maturity" in Table 4.6). Compost inorganic N is also important as an estimate of plant-available N supplied with the compost.

F. Acidity/Alkalinity (pH)

The pH range for most finished composts is from 6.0 to 8.0. The final pH of the compost is highly dependent on the feedstock, the compost process, and the addition of any amendments. Excessive acidity or excessive alkalinity can injure plant roots, inhibiting plant growth and development. Compost feedstocks such as wood may be quite acidic, while others (e.g., lime-treated biosolids) may be a significant source of alkalinity.

Where compost accounts for sizable portions of a potting medium mix, attention must be paid to matching the final pH of the potting medium to plant requirements. In potting media, compost pH can be increased by lime addition, and reduced by elemental sulfur (S) addition. Some composts with high pH may be unsuitable for acid-loving plants because of the difficulty in lowering compost pH with elemental S. To be rapidly effective in reducing pH, elemental S must be of very fine particle size (Marfa et al., 1998). As compost CEC increases, the amount of lime or elemental S needed to change the pH also increases.

Compost pH is measured by two methods in the laboratory, saturated paste and volume addition. For the paste method, water is added to the sample until its moisture content just exceeds water-holding capacity. Then, pH is measured by immersing an electrode into the paste. The volume method involves mixing a specified volume of compost with a specified volume of water (e.g., 1:1 or 1:2 compost to water). Then, pH is measured by immersing the pH electrode into the slurry mixture. Compost pH determined by the volume method usually results in a value 0.1 to 0.3 pH units higher than that determined by the saturated paste method. Traditionally, the saturated paste method has been used to assess compost for landscape applications, and the volume addition method has been used for potting media assessment.

G. Electrical Conductivity (Soluble Salts)

Salinity is estimated from measurement of electrical conductivity (EC) (Table 4.5). Like pH measurement, soluble salts can be measured via saturated paste or volume addition methods.

Electrical conductivity does not provide information on the type of salts present. Some cations or anions are nutrients such as Ca, Mg, sulfate-S (SO_4-S), or NO_3-N. Salts containing Na, chloride (Cl) or boron (B) can be toxic to plants at elevated concentrations. These elements are usually determined in a saturated paste extract (Table 4.5) or volume addition extract.

High salt contents in compost affect seed germination and root health. Crops differ widely in salt tolerance (California Fertilizer Association, 1990). Some vegetable crops, such as beans (*Phaseolus vulgaris* L.) and onions (*Allium cepa* L.) are highly sensitive to salts. Repeated application of high-salt composts can lead to soil salinity build-up in field soils in arid climates. Composts containing Cl at over 10 meq L^{-1} of a saturated paste extract may limit the growth of grapes (*Vitis* spp.), and B contents in excess of 1.0 mg L^{-1} of a saturated paste extract may affect sensitive crops such as beans.

H. Phosphorus (P), Potassium (K), Calcium (Ca), Magnesium (Mg), and Micronutrients

Total P, K, Ca, and Mg are determined by total digestion of the compost in strong acid, with subsequent analysis by atomic absorption spectrometry or inductively coupled plasma spectrometry.

Only a portion of the total P, Ca, and Mg in a compost sample will be plant available. Essentially all of total compost K is plant-available. The exchangeable (plant available) fraction of total K, Ca, and Mg can be determined via a soil test procedure called "exchangeable bases." Determination of exchangeable bases, including Na, is recommended for some composts (e.g., compost derived from beef feedlot manure). High quantities of exchangeable Na may indicate water infiltration problems. In these instances, analysis of exchangeable Ca and Mg concentrations will determine if there is a need to amend the compost with gypsum.

Soil test methods for extractable P, such as the Bray (dilute acid-flouride), Olsen (bicarbonate), and Mehlich 3 (ammonium nitrate, ammonium flouride, EDTA, and HNO_3) methods are sometimes performed by laboratories on compost samples. Interpretation of these soil test methods for compost samples is difficult, because the tests were primarily designed for predicting plant growth responses on mineral soils.

Micronutrient analyses (i.e., zinc [Zn], manganese [Mn], iron [Fe], and copper [Cu]) are sometimes of value when composts are used in potting media. The usual test method involves saturation of the compost with an 0.005 M DTPA (diethylene triamine pentaacetic acid) extraction solution, filtration of the extract, and subsequent analysis for the metals of interest (Whitney, 1998). Composts containing more than 25 mg kg^{-1} of Zn and 2.5 mg kg^{-1} of B via DTPA extraction may have a detrimental impact on plant growth.

VI. EVALUATING COMPOST MATURITY AND STABILITY

Compost maturity and stability are critical for compost use in potting media, for bagged products, and for compost-mediated disease suppression. Maturity is a general term describing fitness of the compost for a particular end use, and stability refers to the resistance of compost organic matter to degradation. Mature composts are ready to use; they contain negligible or acceptable concentrations of phytotoxic compounds like NH_3 or short-chain organic acids. The more stable the compost, the less shrinkage occurs during container plant production. Stable composts remain cool when bagged. Different degrees of compost stability are needed for control of specific plant diseases (Hoitink et al., 1997; Hoitink chapter in this book).

The development of a "mature compost" is a continuous process. The first phase, rapid composting, is characterized by high temperatures (55 to 75°C), a supply of readily decomposable organic matter, and rapid rates of organic matter decomposition by thermophilic bacteria. Weed seeds and most fungi and bacteria are killed during rapid composting. The second phase, curing, begins when the supply of readily decomposed organic matter becomes limiting. During curing, pile temperatures are lower (< 40°C) and the compost is recolonized by mesophilic bacteria and fungi. The third phase, maturity, is the most subjective. By our definition, a compost is considered mature when it has cured long enough for a particular end use.

Maturity measurements have a number of purposes. First, indicators of maturity are used by compost producers to evaluate the success of the composting process. From a processor standpoint, processing compost for the minimum time necessary decreases cost and increases product volume. Second, maturity indicators are sometimes incorporated into minimum product standards by government agencies or compost industry organizations. From a regulatory standpoint, a single measurement that is rapid, reproducible, and accurately reflects product quality is desirable. Unfortunately, several tests are often needed to characterize maturity. Often, the most reliable tests are those that are the slowest, most expensive, or least available. Third, maturity measurements are sometimes used by compost users as a check on compost quality for their particular application. Our discussion here focuses on the horticultural compost user, apart from regulatory considerations.

Compost maturity can be evaluated by sensory, chemical, stability, or phytotoxicity methods (Tables 4.6, 4.7, and 4.8). Sensory and chemical methods are the simplest and most readily available. They evaluate maturity indirectly, and are all somewhat feedstock dependent. They rely on correlations between measured parameters and compost respiration rate or plant growth response. Compost stability, as measured by respirometry or self-heating, describes the relative stability of organic C compounds present in the compost. Standards for compost stability are applicable across a wide range of compost feedstocks. Phytotoxicity tests are often the most difficult tests to standardize and interpret, because of the many variables involved in plant response to compost.

A. Sensory Indicators of Maturity

Evaluation of compost color and odor are reasonable screening methods for rejecting composts that have obvious problems. A compost with a foul anaerobic odor is unlikely to be rated as mature by any other test. A standardized matrix for color and odor evaluation is available (Leege and Thompson, 1997; Method 9.03A in Table 4.6). Compost color darkens during composting, and is strongly affected by feedstocks. Mature yard trimmings composts are usually dark black in color, while manure composts usually attain a more brownish color when mature.

B. Chemical Indicators of Maturity

A wide variety of chemical indicators of compost maturity have been proposed (Chen and Inbar, 1993; Henry and Harrison, 1996; Jimenez and Garcia, 1989). We describe the most widely used chemical indicators here and in Table 4.6.

Table 4.7 Laboratory and Field Methods for Assessing Compost Stability

Method	TMECC[z] Method Number or Other Reference	Trend During Composting	Suggested Value for Mature Compost	Comments
Specific oxygen uptake rate (SOUR); moist compost	9.09B	Decrease	Very stable < 0.5, stable 0.5–1.5, mod. unstable 1.5–3.5, unstable 3.5–6.0 mg O_2 g VS^{-1} h^{-1}	Requires specialized apparatus. Not widely available at commercial laboratories. Affected by compost moisture and sample preconditioning. Short duration test (60 to 90 min). Requires volatile solids (VS) determination.
Specific oxygen uptake rate (SOUR); compost slurry	Lasaridi and Stentiford, 1998a, 1998b	Decrease	Very stable < 0.5, stable 0.5–1.5, mod. unstable 1.5–3.5, unstable 3.5–6.0 mg O_2 g VS^{-1} h^{-1}	Only respiration measurement not affected by compost moisture content. Reported to give similar data to TMECC 9.09B with greater precision. Method is widely available, since it is adapted from a wastewater procedure for biological oxygen demand (BOD). Requires computer-assisted control of O_2 inputs and measurements of dissolved O_2. Test duration 20 h.
CO_2 evolution (trapped in KOH or NaOH)	9.09C	Decrease	Very stable < 2, stable 2–8, mod. unstable 8–15, unstable 15–40 mg CO_2-C g VS^{-1} d^{-1}	Standard vessel size is 4L with air renewal every 24 h, temperature 35°C. Sample preconditioned for 72 h. Requires volatile solids (VS) determination.
CO_2 evolution (colorimetric gel – Solvita™)	Woods End Reseach Laboratory, 1999b	Decrease	Semiquantitative with eight colorimetric categories corresponding to raw, active, and finished compost. Color categories cover the range from 2 to 30 mg CO_2-C per g compost-C per day	For on-site testing. Test provides a semiquantitative assessment of CO_2 evolution rate. Uses a closed vessel (125 mL) for a fixed time period (4 h) with a specified volume of compost. Test done at ambient temperature with no sample preconditioning. Calibrated by manufacturer with relative scale. Colorimetric gel has limited shelf life. The 1999 version of the Solvita™ kit also includes a colorimetric test for ammonia (NH_3).
Dewar self-heating	9.11	Decrease	Maximum self-heating in 2 to 9 day test: 0–20°C finished; 20–40°C active; 40°C fresh compost (Brinton et al., 1995)	Simple, apparatus and interpretation. Simulates natural heating process in a compost pile. Measurements in "field units": heat output per unit volume. Compost moisture affects test result. Self-heating data roughly correlated to O_2 uptake and CO_2 evolution data for some composts.
Pile reheating	State of Florida regulations	Decrease	Mature compost will not reheat more than 20°C above ambient temperature upon standing (Ozores-Hampton et al., 1998)	Affected by pile size, porosity, and moisture content.

[z] TMECC: *Test Methods for the Examination of Composting and Compost* (Leege and Thompson, 1997).

Table 4.8 Methods for Assessing Phytotoxic Substances in Compost

Method	TMECC[z] Method Number	Trend During Composting	Suggested Value for Mature Compost	Comments
Seed germination and root elongation	9.05	Increase	Germination index (Zucconi et al., 1985) using garden cress[y] > 60%. Other procedures: germination index similar to that of a mature compost produced with similar feedstocks.	Plant species vary in sensitivity to compost extracts. Garden cress test too sensitive for many compost end uses. Composts with high salt concentrations inhibit germination of some seeds at all stages of curing.
Short-chain organic acids (volatile fatty acids)	9.12	Decrease	Acetic acid conc. > 300 mg kg^{-1} inhibited garden cress seed germination (DeVleeschauwer et al., 1981).	Unstable compost contains short chain C organic acids such as acetic, butryric, and propionic acids that are phytotoxic. Direct determination of short-chain organic acids is expensive, requiring gas or ion chromatography. Generally not a sensitive test during curing.

[z] TMECC: *Test Methods for the Examination of Composting and Compost* (Leege and Thompson, 1997).
[y] Garden cress = *Lepidium sativum* L.

1. Organic Matter

Volatile solids, an estimate of compost organic matter, decrease during composting. Typically, about half of the initial organic matter is lost during composting. CEC generally increases as the compost matures (Chen and Inbar, 1993). This measurement is most meaningful for comparisons within a particular class of feedstocks (e.g., cattle manure composts). Some organic materials have a relatively high CEC prior to composting (Casale et al., 1995). A minimum CEC of 60 meq·100 g^{-1} of compost volatile solids (ash-free basis) has been proposed as a target for mature MSW composts (Harada et al., 1981).

2. Carbon and Nitrogen

Compost total N, carbon to nitrogen (C:N) ratio, and inorganic N concentrations are often more related to feedstocks than to maturity. For this discussion, maturity with respect to N cycling occurs when the compost can be incorporated into growth media without causing excessive immobilization of N or NH_3 toxicity. A variety of maturity indicators can be derived from measurements of compost C and N (Table 4.6).

Potential problems with N are associated with particular feedstocks (chapter 14 in this book). Nitrogen immobilization is a major problem for immature composts derived from low N content feedstocks such as MSW (Jimenez and Garcia, 1989; Ozores-Hampton et al., 1998). Plants grown in composts that immobilize N are often yellow and stunted because of N deficiency. For high N feedstocks such as manures or biosolids, N availability is highest in immature compost. As composting proceeds, inorganic N and readily mineralizable N is lost as NH_3, or incorporated into complex organic forms (Pare et al., 1998). Immature manure or biosolids composts with NH_4-N concentrations above 1000 mg kg^{-1} can produce enough water-soluble NH_3 to be toxic to plant roots (Barker, 1997). The potential for NH_3 toxicity is primarily a concern for composts or compost-amended media that have a pH greater than 7.5 to 8.0.

Ideal compost feedstock mixtures have an initial C:N ratio of about 30:1, decreasing to less than 20:1 as the composting process proceeds. The use of C:N ratio is based on the C:N ratio of stable soil organic matter, which usually ranges from 10 to 15:1. If cured for an extended period, compost C:N will approach that of soil organic matter. For many composting systems, the C:N ratio is not a sensitive indicator of maturity (Forster et al., 1993; Lasaridi and Stentiford, 1998b). For example, in compost production systems with pH > 7.5, the C:N ratio may change very little during composting, since C loss as CO_2 and N loss as NH_3 occur simultaneously.

The amount or ratio of NH_4-N and NO_3-N is another simple chemical indicator of maturity. NH_4-N is often highest in the early stages of composting, declining as compost stability increases. The lower respiration rates that occur in mature compost are more favorable for NO_3 production via nitrification and less favorable for NO_3 loss via denitrification. Also, nitrification is strongly inhibited at temperatures above 40°C. NH_4 and NO_3 concentrations are strongly affected by drying and re-wetting in immature composts (Grebus et al., 1994).

C. Compost Stability as a Maturity Indicator

Compost stability is one aspect of compost maturity. Stability, as measured by respirometry or self-heating, describes the relative stability of organic C compounds present in the compost. Standards for compost stability are applicable across a wide range of compost feedstocks (Frost et al., 1992; Haug, 1993).

1. Respirometry

Respirometry is the measurement of O_2 consumed or CO_2 released by a sample. It is used to estimate biological activity in a sample. The measured respiration rate can be used to estimate the rate of compost weight loss over time, and to estimate compost maturity.

Measurement of O_2 and CO_2 from air samples taken directly from an actively composting pile can provide data to guide pile aeration requirements (Haug, 1993). However, such measurements cannot be considered maturity measurements because the time of air contact with the compost is unknown.

It is important to understand what units the laboratory uses to report the compost respiration rate. The most commonly accepted units (Table 4.7) base the respiration rate on the amount of volatile solids or the amount of organic C present in the sample. Such units allow comparison per unit of organic matter or C. Compost respiration rates and organic matter contents can be used to estimate "shrinkage" of a compost via organic matter decomposition. For example, for a compost with 50% organic matter (25% C) and a respiration rate of 2 mg CO_2-C per g compost C per day, the rate of product loss via decomposition is approximately 0.1% per day.

There is great variation in the technology used to measure compost respiration rates. Test procedures range from quantitative to qualitative. Most respiratory procedures include a 2 to 3 day sample preconditioning step to achieve uniform moisture (about 50% total solids) and a compost microbial population dominated by mesophilic microorganisms. A recently proposed adaptation of the specific oxygen uptake rate (SOUR) test used in wastewater analysis (Lasaridi and Stentiford, 1998a, 1998b) does not require sample preconditioning or moisture adjustment.

Most respirometric procedures require a standardized temperature (25 to 35°C) and repeated measurements over time to determine respiration rate (Table 4.7). Since the compost sample produces heat, a water bath is often required to hold temperature constant. The simplest of the quantitative respiration measurements is CO_2 evolution rate measured by alkaline trapping. Carbon dioxide trapped in KOH is determined via titration (Method 9.09C in Table 4.7). Measurements of O_2 consumption using Clark-type polarographic electrodes require repeated measurements every 10 min for at least 90 min (Frost et al., 1992). Therefore, O_2 uptake measurements are usually coupled with a datalogger or a computer (Iannotti et al., 1994), or reported as a unitless O_2 uptake index (Grebus et al., 1994). Neither CO_2 evolution nor O_2 consumption measurements of compost respiration rate are currently widely available at commercial laboratories.

A rapid semiquantitative procedure, the Solvita™* test, uses a colorimetric gel determination of CO_2 evolution (Woods End Research Laboratory, 1999b). The Solvita procedure does not rigidly control compost temperature and moisture. The sample is not "preconditioned" prior to testing. The measured respiration rate is estimated per unit volume of as-is compost at ambient temperature. The interpretive scale provided has eight categories ranging from "raw" compost (categories 1 to 2), "active" compost (categories 3 to 6), and "finished" compost (categories 7 to 8). Raw compost is poorly decomposed and probably phytotoxic, and finished compost is ready for most uses. The Solvita test is being used in connection with agency compost specifications for maturity in Washington State, Texas, California, Minnesota, Maine, and Illinois in the U.S., and in Germany and Denmark (Woods End Research Laboratory, 1999a). Eighteen states in the U.S. are currently reviewing the Solvita procedure for inclusion in compost testing protocols.

2. Dewar Self-Heating Test

This test is a standardized procedure for measurement of compost heat production (Brinton et al., 1995; Method 9.11 in Table 4.7). It is an indirect measurement of respiration rate. Moist compost is placed in an insulated vacuum bottle, and the rise in temperature is recorded over a 2 to 9 day period. The maximum temperature increase over ambient is used for interpretive purposes. The test is simple to perform,

* Registered Trademark of Woods End Research Laboratory, Inc., Mt. Vernon, Maine.

but time consuming. Unlike short-term O_2 or CO_2 respirometry, the Dewar test allows development of a natural succession of compost microflora similar to that which occurs in a compost pile. Therefore, sample preconditioning is not as critical for this test. Also, compost samples often reach a self-limiting temperature in the Dewar procedure, which also simulates the natural behavior of compost piles.

There is debate about the proper level of compost moisture for the Dewar test (Brinton et al., 1995). Earlier guidance was to dry compost to 30% moisture, which is below the optimum for microbial activity. Current guidance is to moisten compost to the optimum range for microbial activity, usually above 50% moisture. However, at higher moisture levels, more heat is needed for a given rise in temperature; water addition increases the heat capacity of the compost sample.

Dewar self-heating test values (Method 9.11 in Table 4.7) are correlated with quantitative measurements of respiration (Woods End Research Laboratory, 1999b). Raw compost via the Dewar test corresponds with a respiration rate of greater than 20 mg CO_2-C per g compost-C per day. Finished compost via the Dewar test has an approximate respiration rate of less than 4 mg CO_2-C per gram compost-C. Active compost via the Dewar test has an approximate respiration rate of 8 to 20 mg CO_2-C per g compost-C per day.

D. Phytotoxicity as a Maturity Indicator

Composts can contain a variety of phytotoxic substances that inhibit or prevent plant growth. Phytotoxicity tests are most interpretable when the test duplicates or represents a specific compost end use.

Standardized germination and growth tests evaluate a combination of phytotoxic factors in compost including NH_3, soluble salts, short-chain organic acids, and pH (Leege and Thompson, 1997; Method 9.05 in Table 4.8). Growth of most plant species and cultivars is inhibited with highly unstable composts (Garcia et al., 1992; Keeling et al., 1994; Zucconi et al., 1981a, 1981b). As compost becomes more stable, variation in plant species susceptibility to phytotoxic factors becomes more important.

Germination and growth tests directly estimate the plant growth inhibition by compost under specified environmental conditions. Most tests are semiquantitative, with test scores grouped into two to four inhibition categories, such as none, mild, strong, and severe inhibition of germination and growth. Tests require 1 to 14 days depending on the method. Tests using compost extracts are usually more rapid and reproducible than direct seeding tests, but require additional time for extract preparation. Compost extracts must be prepared aseptically via millipore filtering to remove bacteria and to prevent rapid degradation of short-chain organic acids.

The choice of plant species can have a large effect on germination and growth test results when the compost is high in soluble salts. Very stable composts with high salt concentrations may inhibit germination of some plant species (Iannotti et al., 1994). We recommend using seeds with higher salt tolerance (California Fertilizer Association, 1990) when evaluating composts with elevated soluble salts.

Short-chain organic acids resulting from decomposition of organic matter can inhibit or reduce seed germination and root growth. Organic acids responsible for growth inhibition include acetic, butyric, propionic, and valeric acids (Brinton, 1998;

Liao et al., 1994). These acids also produce the foul odor associated with compost that has been decomposing anaerobically. They are produced as a natural byproduct of the early stages of organic matter decomposition. As compost matures, the short-chain organic acids are lost via decomposition. These compounds can be determined quantitatively with sophisticated laboratory gas or ion chromatography procedures (Brinton, 1998; Liao et al., 1994). Brinton (1998) reported mean short-chain organic acid concentrations of 4385 mg kg^{-1} and a range of 75 to 51,474 mg kg^{-1} for 626 compost samples from across the U.S. Phytotoxic concentrations of acetic acid can be as low as 300 mg kg^{-1} (DeVleeschauwer et al., 1981).

Composts may have one or more quality problems that impose limitations on their use (Table 4.9). Most quality problems can be traced to either the compost feedstocks or the composting process. Reducing compost application rates or allowing additional time for compost stabilization can minimize most of the common quality problems.

Table 4.9 Diagnosis and Management of Potential Plant Production Problems in Compost-Amended Media

Problem	Impact of Composting Feedstocks and Process	Compost Analytical Characteristics	Compost Use Suggestions
Nitrogen deficiency	Reported problems for composted MSW and woody debris, and some yard trimmings composts. Higher compost stability or higher N feedstocks needed to overcome problem.	Compost C:N ratio greater than 25-30:1 NO_3-N <100 ppm (mg kg^{-1}) High respiration rate[z].	Allow additional time for compost stabilization. Apply additional N fertilizer with compost.
Ammonia toxicity	Unstable composts especially those with pH > 8.	NH_4-N >1000 ppm (mg kg^{-1}) and C:N < 20:1. High respiration rate	Allow additional time for compost stabilization. Reduce pH to 7. Provide aeration to enhance conversion to nitrate.
Short chain organic acids	Unstable composts. Reported for many feedstocks.	Compost phytotoxic in germination test. High respiration rate	Allow additional time for compost stabilization. Aerate compost to speed decomposition of short-chain organic acids.
Soluble salts	Feedstocks are the source of salts. Elevated salts often associated with composted manure and grass clippings. Composted paper or cardboard can elevate boron concentrations.	E.C. >3 dS m^{-1} in growing media. Compost phytotoxic in germination test Above 10 meq Cl L^{-1} of saturated paste extract Above 1 mg B L^{-1} of saturated paste extract	Leach compost with water before seeding or planting. Avoid use on sensitive crops.

[z] High respiration rate using a stability assessment procedure for CO_2 evolution, O_2 uptake, or self-heating. See Table 4.7 for stability assessment options.

VII. VARIABILITY IN COMPOST ANALYTICAL DATA

The compost testing methods outlined in this chapter are valuable tools for product quality assessment. Laboratory data are most valuable when one is familiar with the accuracy and precision of the data (how closely it reflects reality). This section describes how to choose a laboratory to perform analyses, and what variability is commonly observed in chemical laboratory analysis procedures. There are very limited published data on the variability of compost physical and biological tests; such tests likely have variability considerably greater than listed here for the chemical tests (Tables 4.10 and 4.11).

Table 4.10 Analytical Variability for a Chicken Manure Compost Sample Analyzed by 42 Commercial Laboratories

		Mean	Relative Standard Deviation (%)[y]	
Analysis	Units[z]	All Laboratories	Intralaboratory[x]	Interlaboratory
pH (saturated paste)	none	7.8	1	3
pH (1:2 v/v)		8.0	1	2
Conductivity	dS m^{-1}	7.9	11	22
Total N (combustion)	%	1.1	5	6
Total N (Kjeldahl)	%	1.1	5	5
Total organic C (TOC)	%	19.6	6	9
Volatile solids (LOI)	%	46.0	10	12
Total P	%	1.0	9	17
Total K	%	1.0	10	15
Total Ca	%	4.4	7	17
Total Mg	%	0.4	7	15
Total S	%	0.3	11	21
Total Zn	mg kg^{-1}	221.0	9	11
Total B	mg kg^{-1}	30.1	13	30
Total Cu	mg kg^{-1}	103.0	10	19
Total As	mg kg^{-1}	14.9	19	35
Total Cd	mg kg^{-1}	1.0	23	149
Total Pb	mg kg^{-1}	9.7	12	60
Total Se	mg kg^{-1}	0.4	32	86

[z] Dry matter basis.
[y] Relative standard deviation = standard deviation/mean × 100.
[x] Intralaboratory precision for three analyses of the same sample.

From personal communication, R.O. Miller, Soil and Crop Sciences Dept., Colorado State University, Fort Collins, CO. Data from Western States Proficiency Testing program, 3rd Quarterly Report, Sept. 1997. Laboratories participating in the proficiency testing program received a subsample of a large bulk sample. With permission.

We recommend selecting a laboratory that has compost testing experience and performs the test methods routinely. Generally, any laboratory that performs compost tests several times each month is sufficient. Preference should be given to testing laboratories that participate in a compost analysis proficiency testing program or a sample exchange program. One example is the Compost Analysis Proficiency (CAP)

Table 4.11 Analytical Variability for Two Compost Samples Analyzed by Six Commercial Laboratories[z]

		Chicken Manure Compost		Yard Trimmings Compost	
Compost Analysis	**Units**	**Mean**	**RSD[y] (%)**	**Mean**	**RSD[y] (%)**
pH		6.6	10	6.9	5
Conductivity	dS m^{-1}	25	34	7	36
Total N	%	3.55	12	1.18	16
Total P	%	2.2	16	0.2	15
Total K	%	2.8	9	0.6	37
Volatile solids	%	70	16	37	8

[z] Adapted from Granatstein, 1997. Laboratories received a subsample of a large bulk sample. Laboratories were not told what method to use, or informed that they were part of a "study."
[y] Relative standard deviation (interlaboratory) = standard deviation/mean × 100.

program coordinated by the Utah State University Analytical Laboratory (Logan, UT, USA). Proficiency testing programs provide a check on laboratory data quality on a regular basis (usually every 3 months). Ask the laboratory to provide their results from the proficiency testing program. Compare their analytical values to the mean or median value for all laboratories participating in the proficiency program.

The quality of laboratory data for a specific test has two components, accuracy or bias, and precision. Bias is the deviation of a laboratory analysis from its true value, and precision describes the reproducibility of a test value. Bias is assessed using a standard reference sample with known analytical values. Precision can be assessed via repeated analysis of a single well-blended sample.

Tables 4.10 and 4.11 illustrate intralaboratory and interlaboratory precision for well-blended compost samples. Precision between multiple laboratories (interlaboratory) is generally higher than that within a single laboratory (Table 4.10). Sampling error, the failure to collect a truly representative sample, is not included in the compost analytical data presented in Tables 4.10 and 4.11.

The precision of laboratory data is method dependent (Table 4.10). For example, the pH saturated paste test method may have an intralaboratory precision of 1.3%, while that of total N is 4.5% and that of total arsenic (As) is 18.5%.

VIII. COMPOST QUALITY IN THE FUTURE

This chapter reflects the growing state of compost quality evaluation. Compost quality testing is becoming a more predictable and routine process as compost use expands, and as analytical methods tailored specifically to compost are developed. The development of guidelines, regulations, and quality assurance programs for compost quality is also spurring improvements in compost analysis. However, the quantity of compost analyses performed by commercial laboratories is still very small compared to the quantity of analyses performed for soil or plant tissue. The recent initiation of a cooperative compost-testing program, the Compost Analysis

Proficiency (CAP) program coordinated by the Utah State University Analytical Laboratory, reflects increasing interest in compost analyses.

The greatest current research activity is in the area of rapid determination of compost stability and maturity parameters. Regulations and user demand for mature or stable compost are pushing the standardization of these tests forward.

The development of interpretive statements based on compost test data is still an art. The interpretation of test data must consider the needs of the compost user and must integrate chemical, physical, and biological properties of the compost. Even with reliable compost analytical data, expert opinions can differ substantially. Recommendations for compost application rates, adjustments in cultural practices (e.g., irrigation, fertilization, pest control), and determination of "acceptable" quality are based on understanding of interactions. Different interactions may occur with each crop, soil, or growing medium, and with other components of the horticultural production or marketing system. Refining recommendations for compost quality for specific applications will continue to provide a challenge for the future.

REFERENCES

Barker, A.V. 1997. Composition and uses of compost, p. 140–162. In: J.E. Rechcigl and H.E. MacKinnnon (eds.). *Agricultural Uses of By-products and Wastes.* American Chemical Society, Washington, D.C. ACS Symposium Series 668.

Bildingmaier, I.W. 1993. The history of the development of compost standards in Germany, p. 536–550. In: H.A.J. Hoitink and H.M. Keener (eds.). *Science and Engineering of Composting: Design, Microbiological and Utilization Aspects.* Renaissance Publications, Worthington, Ohio.

Brinton, W.F. 1998. Volatile organic acids in compost: production and odorant aspects. *Compost Science and Utilization* 6(1):75–82.

Brinton, W.F., E. Evans, M.L. Droffner, and R.B. Brinton. 1995. Standardized test for compost self-heating. *BioCycle* 36(11):64–69.

California Compost Quality Council (CCQC). 1999. *CCQC Registered Compost.* [Online]. Available at http://www.crra.com/ccqc/. Verified July 15, 1999. California Compost Quality Council, San Francisco, California.

California Fertilizer Association. 1990. *Western Fertilizer Handbook, Horticulture Edition.* Interstate Publishers, Danville, Illinois.

Casale, W.L., V. Minassian, J.A. Menge, C.J. Lovatt, E. Pond, E. Johnson, and F. Guillemet. 1995. Urban and agricultural wastes for use as mulches on avocado and citrus and for delivery of microbial biocontrol agents. *Journal of Horticultural Science* 70:315–332.

Chang, A.C., A.L. Page, and J.E. Warneke. 1983. Soil conditioning effects of municipal sludge compost. *Journal of Environmental Engineering* 109:574–583.

Chen, Y. and Y. Inbar. 1993. Chemical and spectroscopical analyses of organic matter transformations during composting in relation to compost maturity, p. 551–600. In: H.A.J. Hoitink and H.M. Keener (eds.). *Science and Engineering of Composting: Design, Microbiological and Utilization Aspects.* Renaissance Publications, Worthington, Ohio.

Compost Council of Canada. 1999. *Setting the Standard. A Summary of Compost Standards in Canada.* [Online]. Available at http://www.compost.org/standard.html. Verified June 30, 1999. Compost Council of Canada, Toronto, Ontario, Canada.

DeVleeschauwer, D., O. Verdonck, and P. Van Assche. 1981. Phytotoxicity of refuse compost. *BioCycle* 22(1):44–46.

E & A Environmental Consultants and H. Stenn. 1996. *Compost End-Use Guidelines Development Project.* Report CM-96-1. Clean Washington Center, Department of Trade and Economic Development, Seattle, Washington.

Forster, J.C., W. Zech, and E. Wurdinger. 1993. Comparison of chemical and microbiological methods for the characterization of the maturity of composts from contrasting sources. *Biology and Fertility of Soils* 16:93–99.

Frost, D.I., B.L. Toth, and H.A.J. Hoitink. 1992. Compost stability. *BioCycle* 33(11):62–66.

Garcia, C., T. Hernandez, F. Costa, and J.A. Pascual. 1992. Phytotoxicity due to the agricultural use of urban wastes. Germination experiments. *Journal of the Science of Food and Agriculture* 59:313–319.

Gavlak, R.G., D.A. Horneck, and R.O. Miller. 1994. *Plant, Soil and Water Reference Methods for the Western Region.* University of Alaska-Fairbanks. Western Regional Extension Publication 125.

Granatstein, D. 1997. Lab comparison study completed. In: D. Granatstein (ed.) *Compost Connection for Northwest Agriculture* 4:1–4 (May 1997). [Online]. Available at http://csanr.wsu.edu. Verified July 15, 1999. Washington State University Cooperative Extension, Center for Sustaining Agriculture and Natural Resources, Pullman, Washington.

Grebus, M.E., M.E. Watson, and H.A.J. Hoitink. 1994. Biological, chemical and physical properties of composted yard trimmings as indicators of maturity and plant disease suppression. *Compost Science and Utilization* 2(1):57–71.

Harada, Y., A. Inoko, M. Tadaki, and T. Izawa. 1981. Maturing process of city refuse compost during piling: application of composts to agricultural land. *Soil Science and Plant Nutrition* 27:357–364.

Haug, R.T. 1993. *The Practical Handbook of Compost Engineering.* Lewis Publishers, Boca Raton, Florida, pp. 307–384.

Henry, C.L., and R.B. Harrison. 1996. Compost fractions in compost and compost maturity tests, p. 51–67. In: *Soil Organic Matter: Analysis and Interpretation.* Soil Science Society of America, Madison, Wisconsin. Special Publication 46.

Hoitink, H.A.J., A.G. Stone, and D.Y. Han. 1997. Suppression of plant diseases by compost. *HortScience* 32:184–187.

Iannotti, D.A., M.E. Grebus, B.L. Toth, L.V. Madden, and H.A.J. Hoitink. 1994. Oxygen respirometry to assess stability and maturity of composted municipal solid waste. *Journal of Environmental Quality* 23:1177–1183.

Inbar, Y., Y. Chen, and H.A.J. Hoitink. 1993. Properties for establishing standards for utilization of composts in container media, p. 668–694. In: H.A.J. Hoitink and H.M. Keener (eds.). *Science and Engineering of Composting: Design, Microbiological and Utilization Aspects.* Renaissance Publications, Worthington, Ohio.

Jimenez, E.I. and V.P. Garcia. 1989. Evaluation of city refuse compost maturity: a review. *Biological Wastes* 27:115–142.

Keeling, A.A., I.K. Paton, and J.A. Mullett. 1994. Germination and growth of plants in media containing unstable refuse-derived compost. *Soil Biology & Biochemistry* 26:767–772.

Lasaridi, K.E. and E.I. Stentiford. 1998a. A simple respirometric technique for assessing compost stability. *Water Research* 32:3717–3723.

Lasaridi, K.E. and E.I. Stentiford. 1998b. Biological parameters for compost stability assessment and process evaluation. *Acta Horticulturae* 469:119–128.

Leege, P.B. and W.H. Thompson (eds.). 1997. *Test Methods for the Examination of Composting and Composts.* [Online]. Available at http://www.edaphos.com. Verified July 15, 1999. U.S. Composting Council, Amherst, Ohio.

Liao, P.H., A. Chen, A.T. Vizcarra, and K.V. Lo. 1994. Evaluation of the maturity of compost made from salmon farm mortalities. *Journal of Agricultural Engineering Research* 58:217–222.

Marfa, O., J.M. Tort, C. Olivella, and R. Caceres. 1998. Cattle manure compost as substrate II- Conditioning and formulation of growing media for cut flower cultures. *Acta Horticulturae* 469:305–312.

McCoy, E.L. 1992. Quantitative physical assessment of organic materials used in sports turf rootzone mixes. *Agronomy Journal* 84:375–381.

Ozores-Hampton, M., T.A. Obreza, and G. Hochmuth. 1998. Using composted wastes on Florida vegetable crops. *HortTechnology* 8:130–137.

Pare, T., H. Dinel, H. Schnitzer, and S. Dumontet. 1998. Transformation of carbon and nitrogen during composting of animal manure and shredded paper. *Biology and Fertility of Soils* 26(3):173–178.

Raviv, M., S. Tarre, Z. Geler, and G. Shelef. 1987. Changes in some physical and chemical properties of fibrous solids from cow manure and digested cow manure during composting. *Biological Wastes* 19:309–318.

Schulte, E.E. 1988. Recommended soil organic matter tests, p. 29–31. In: W.C. Dahnke (ed.). *Recommended Chemical Soil Test Procedures for the North Central Region.* North Dakota State University, Fargo. North Dakota Agricultural Experiment Station Bulletin 499 (revised).

Stentiford, E.I. and J.T. Pereira-Neto. 1985. Simplified systems for refuse/sludge composts. *BioCycle* 26(5):46–49.

U.S. Composting Council. 1996. *Field Guide to Compost Use.* U.S. Composting Council, Amherst, Ohio.

U.S. Environmental Protection Agency (U.S. EPA). 1992. Sampling procedures and analytical methods, p. 41–47. In: *Environmental Regulations and Technology. Control of Pathogens and Vector Attraction in Sewage Sludge.* EPA/626/R-95/013. USEPA, Office of Research and Development, Washington, D.C.

U.S. Environmental Protection Agency (U.S. EPA). 1993. Standards for the use or disposal of sewage sludge. *Federal Register* 58:9248–9415.

Verdonck, O. 1998. Compost specifications. *Acta Horticulturae* 469:169–177.

Whitney, D. 1998. Greenhouse root media, p. 61–64. In: W.C. Dahnke (ed.). *Recommended Chemical Soil Test Procedures for the North Central Region.* North Dakota State University, Fargo. North Dakota Agricultural Experiment Station Bulletin 499 (revised).

Woods End Research Laboratory. 1999a. Compost quality assurance program for the Solvita quality seal. [Online]. Available at http://www.woodsend.org. Verified June 30, 1999. Woods End Research Laboratory, Mt. Vernon, Maine.

Woods End Research Laboratory. 1999b. Guide to Solvita™ Testing for Compost Maturity Index. [Online]. Available at http://www.woodsend.org. Verified Nov. 5, 1999. Woods End Research Laboratory, Mt. Vernon, Maine.

Zucconi, F., M. Forte, A. Monaco, and M. deBertoldi. 1981a. Biological evaluation of compost maturity. *BioCycle* 22(4):27–29.

Zucconi, F., A. Pera, M. Forte, and M. deBertoldi. 1981b. Evaluating toxicity of immature compost. *BioCycle* 22(2):54–57.

Zucconi, F., A. Monaco, M. Forte, and M. deBertoldi. 1985. Phytotoxins during the stabilization of organic matter, p. 73–86. In: J.K.R. Grasser (ed.). *Composting of Agricultural and Other Wastes*. Elsevier Applied Science Publishers, London and New York.

Section II

Utilization of Compost in Horticultural Cropping Systems

CHAPTER 5

Compost Effects on Crop Growth and Yield in Commercial Vegetable Cropping Systems

Nancy E. Roe

CONTENTS

I. INTRODUCTION

Harvested acreage for 25 selected fresh vegetables and melons was 748,677 ha in the United States in 1997 (USDA, 1998). Acreage of 10 processing vegetables added an additional 574,660 ha. The value of production for these 25 fresh market vegetables and 10 processing crops totaled $9.27 billion in 1997 (USDA, 1998). Vegetable production should constitute an ideal use for compost since most crops are grown in annual systems that have a high profit potential. Impediments to the utilization of compost in these intensive production systems include compost costs; transportation costs; lack of adequate application equipment; and lack of clear and consistent demonstration of compost benefits to plant growth, yield, and profits.

1-56670-460-X/01/$0.00+$.50

Research projects in the U.S. and other countries have focused on the use of composts from materials that are readily available in large quantities in the local area. Due to the high bulk density of most composts, local availability is critical for practical and economical usage. In many U.S. vegetable crop production areas, such as south Florida, the proximity of large human populations provides an opportunity for production of composts from municipal wastes. In others, such as parts of North Carolina and the high plains of Texas, large numbers of confined animal feeding operations produce animal manures that can be composted. Feedstocks for composts evaluated on vegetable crops include mixed municipal solid waste (MSW), sewage sludge (biosolids), yard trimmings, wood chips, animal manures, food processing byproducts, and other agricultural wastes. Each of these composts has its own properties and may affect soil characteristics, crop growth and, ultimately, yield in distinct ways. Research projects also generally center on crops that are grown and consumed in the local area or country where the work is being performed. Therefore, research results are difficult to categorize. Nevertheless, worldwide compost quantities are increasing, compost quality is improving, and more commercial vegetable growers are evaluating compost or integrating it into their production systems.

II. COMPOST RESEARCH FOR VEGETABLE CROPPING SYSTEMS

A summary of recent research reporting effects of compost on vegetable crop growth and yield is provided in Table 5.1. The following section provides some details (arranged by crop or crop group) of these studies.

A. Corn

A biosolids/yard trimming compost was applied in 1979 and 1980 on land previously mined for sand and gravel in a study by Hornick (1988). Compost rates were 40, 80, or 160 $Mg \cdot ha^{-1}$ in each year. Control plots were fertilized with 179N-122P-112K ($kg \cdot ha^{-1}$). Sweet corn (*Zea mays* var. *rugosa* Bonaf.) was grown in 1979, 1980, and 1981 as a test crop. In all three years, corn grain yields were not significantly different between compost rates and the control, and there were few differences in grain nutrient concentrations. Residual N from the 80 and 160 $Mg \cdot ha^{-1}$ compost rates was sufficient to keep grain N concentrations similar to those from control plants in the third year.

Hue et al. (1994) conducted a pot study using a highly weathered Ultisol, for which it had been determined that P availability was the main plant nutritional limitation. Rates of yard trimming compost at 75% (by volume) or higher mixed with the soil increased corn growth, but lower rates did not have an effect as compared to corn grown in unamended pots.

Pots filled with mixtures of three tropical soils (an Inceptisol, a Mollisol, and an Oxisol) and an MSW compost at rates from 0 to 25% (by volume) were seeded with corn, and placed in a greenhouse (Paino et al., 1996). Plants were grown for 85 days, followed by two additional corn crops in the same soil mixes. Although there were some differences in the effects of the compost on the three soil types, biomass produced was generally higher in mixtures which contained higher rates of compost.

B. Cruciferous Crops

Municipal waste compost at 0, 7, 14, and 27 Mg·ha^{-1} did not affect head yields of broccoli (*Brassica oleracea* L. Italica group) fertilized with 84 or 168 kg·ha^{-1} of N on a fine sand in a study by Roe et al. (1990).

Low rates of a vegetable waste and manure compost (3 Mg·ha^{-1}) with fertilizer N at 75 kg·ha^{-1} significantly improved broccoli crop response and N use efficiency when compared to a fertilizer-only treatment of 150 kg·ha^{-1} N plus 50 kg·ha^{-1} P (Buchanan and Gliessman, 1991). Increasing applications of compost alone (3, 7.5, and 30 Mg·ha^{-1}) tended to increase broccoli yield and N accumulation, but decreased N use efficiency.

Smith et al. (1992) reported no detrimental effects on cabbage (*Brassica oleracea* L. Capitata group) yields from a biosolids/straw compost used at rates up to 100% of the N requirement. At any given rate of applied N, optimal cabbage yields were obtained when half the N was supplied from an organic source (compost) and half from ammonium nitrate. Compost application improved the efficiency of mineral fertilizer use. The beneficial effects of compost were attributed to favorable effects on soil physical conditions and to the gradual release of essential phytonutrients.

Chinese cabbage (probably *Brassica rapa* L. Chinensis group) yields were increased by the addition of swine waste compost at 25 Mg·ha^{-1}, with or without sawdust, compared to no-compost plots with an acid field soil (pH $\leq$ 5.0), but not with a neutral soil (Kao, 1993). All plots also received fertilizer at a rate of 80N-9P-33K (kg·ha^{-1}). With the acid soil, Zn and Cu concentrations in the leaves from plots with sawdust/swine waste compost were higher than in leaves from no-compost plots.

Maynard (1994) reported that yields of broccoli and cauliflower (*Brassica oleracea* L. Botrytis group) from unfertilized plots amended with a mixed compost (poultry manure, horse manure, spent mushroom compost, and sawdust) at 56 or 112 Mg·ha^{-1} were similar to or greater than yields from plots fertilized with 150 N-66P-125K (kg·ha^{-1}).

C. Cucurbits

Winter (butternut) squash (*Cucurbita moschata* Duch. ex Poir.) seedlings emerged slightly faster from plots mulched with MSW compost than from polyethylene mulched plots, but fruit yields were unaffected (Roe et al., 1993).

A summer squash (*Cucurbita pepo* L.) crop was grown following a tomato (*Lycopersicon esculentum* Mill.) crop in a field where two MSW composts had been applied at 0, 33, or 67 Mg·ha^{-1} and a third MSW compost at 0, 67, and 135 Mg·ha^{-1}, before tomato planting. Total squash yields and mean fruit size were increased by all rates of two of the composts and not affected by the other, compared to plots without compost (Bryan et al., 1994).

Table 5.1 Summary of Recent Research Reporting Effects of Compost on Vegetable Crop Growth and Yields

Crop	Compost	Growth Response[z]	Yield Effects[z]	Reference
Alliaceae				
Onion	BS/AW	NA	+, =	Smith et al., 1992
	BS/WC	NA	+	Bevacqua and Mellano, 1993
Asteraceae				
Lettuce	BS/WC	NA	+	Bevacqua and Mellano, 1993
Brassicaceae				
Broccoli	MSW	NA	=	Roe et al., 1990
	AM, AW	NA	+	Buchanan and Gliessman, 1991
	AM	NA	+, =	Maynard, 1994
Cabbage	BS/AW	NA	+	Smith et al., 1992
Cauliflower	AM	NA	+, =	Maynard, 1994
Chinese cabbage	AM/AW	NA	+, −	Kao, 1993
Chenopodiaceae				
Spinach	BS	NA	+, =	Mellano and Bevacqua, 1992
Cucurbitaceae				
Cucumber	AW	+	+	Kostov et al., 1995
Summer squash	MSW	NA	+, =	Bryan et al., 1994
Winter squash	MSW	NA	=	Roe et al., 1993
Fabaceae				
Cowpea	MSW	NA	+	Bryan and Lance, 1991
Snap bean	MSW	NA	+, =	Ozores-Hampton and Bryan, 1993a
	AM	NA	+, =	Allen and Preer, 1995
	YT	+	+, =	Gray and Tahwid, 1995
Soybean	wood	+	NA	Lawson et al., 1995
Malvaceae				
Okra	MSW	+	+	Bryan and Lance, 1991
Poaceae				
Corn	BS/YT	NA	−	Hornick, 1988
	YT	+	NA	Hue et al., 1994
	MSW	+	NA	Paino et al., 1996
Solanaceae				
Eggplant	MSW	NA	+	Ozores-Hampton and Bryan, 1993b
Pepper	MSW	NA	=	Roe et al., 1992
	MSW	NA	−	Ozores-Hampton and Bryan, 1993b
	MSW	NA	−	Clark et al.. 1994
	MSW	NA	−	Roe et al., 1994
	BS/YT	NA	=	Roe and Stoffella, 1994b
	leaf	NA	=	Maynard, 1996
	BS/YT	+, =	+	Roe et al., 1997
Tomato	MSW	NA	+	Bryan and Lance, 1991
	MSW	NA	+	Manios and Kapetanios, 1992
	MSW	NA	−, =	Bryan et al., 1994
	MSW	NA	+	Clark et al., 1994
	AW	NA	+, =, −	Maynard, 1994
	MSW	NA	+,−	Obreza and Reeder, 1994
	MSW	NA	+, −	Ozores-Hampton et al., 1994

Table 5.1 Summary of Recent Research Reporting Effects of Compost on Vegetable Crop Growth and Yields (Continued)

Crop	Compost	Growth Response[z]	Yield Effects[z]	Reference
	BS/YT	NA	=	Roe and Stoffella, 1994a
	BS	NA	+	Allen and Preer, 1995
	Various	NA	+, –	Alvarez et al., 1995
	MSW	NA	+	Bryan et al., 1995
	MSW	NA	+	Maynard, 1995
	AW	+	+	Stoffella and Graetz, 1997

Note: BS, biosolids; AW, agricultural wastes; WC, wood chips; MSW, municipal solid waste; AM, animal manures; YT, yard trimmings.

[z] NA, +, –, = represent: information not available, increased, decreased, or equal, respectively.

Kostov et al. (1995) reported that greenhouse cucumbers (*Cucumis sativus* L.) grown on a medium containing composting vegetable wastes with the addition of synthetic nutrients produced fruit 10 to 12 days earlier and had a yield 48 to 79% higher than those grown in soil mixed with cattle manure at a 2:1 ratio (dry weight basis). The composting wastes raised soil temperatures, increased CO_2 production and microbial biomass, and released nutrients for plant utilization.

D. Legumes

Recognition of the need for more research into the relationship between soil microbiological populations and organic matter may result in more studies of compost effects on legume nodulation and N fixation. Lawson et al. (1995) reported that soybeans (*Glycine max* L.) grown in acid or saline soil amended with 4% wood waste compost had improved nodulation and shoot growth when compared with those in unamended soil.

Other studies of vegetable legume crop responses to composts have focused on yields. With N added at 84 $kg \cdot ha^{-1}$, 13 and 20 $Mg \cdot ha^{-1}$ of MSW compost gave higher cowpea (*Vigna unguiculata* [L.] Walp.) pod yields than 7 $Mg \cdot ha^{-1}$ of compost or no compost. With 168 $kg \cdot ha^{-1}$ N, yields were higher with 7, 13, and 20 $Mg \cdot ha^{-1}$ compost than with no compost (Bryan and Lance, 1991).

An MSW compost incorporated at 90 and 135 $Mg \cdot ha^{-1}$ into a calcareous limestone soil resulted in snap bean (*Phaseolus vulgaris* L.) yields that were similar to beans grown without compost in the first crop, but quadratic yield increases with compost rate increases (starting from the zero-rate control) in the subsequent crop (Ozores-Hampton and Bryan, 1993a).

Composts from biosolids, horse manure, and yard trimmings were applied for 2 years to identical plots of a silt loam soil at rates of 53 $Mg \cdot ha^{-1}$ (Allen and Preer, 1995). Snap beans from the manure compost plots produced yields equal to those from fertilized control plots in the first year. In the second year, the manure and yard trimmings compost plots produced the highest yields.

Gray and Tawhid (1995) reported that snap bean seedling emergence and plant survival in unmulched plots were increased by the addition of 2.5 cm of leaf compost as a mulch over rows after seeding.

E. Solanaceous Crops

Many of the studies involving compost utilization for solanaceous crop production have been conducted in Florida. The combination of a large vegetable industry on soils low in organic matter, plus high urban populations producing large quantities of organic wastes has supported extensive compost research in Florida.

When 10 $Mg \cdot ha^{-1}$ of MSW compost was applied in trenches in combination with 6.7 to 13.4 $Mg \cdot ha^{-1}$ of MSW compost incorporated into beds on a gravelly soil, tomato yields were higher than with no compost (Bryan and Lance, 1991).

Manios and Kapetanios (1992) studied MSW compost use in greenhouse tomato production. Although all treatments were supplied with equal amounts of fertilizer through irrigation, yields of greenhouse tomatoes grown in soil were highest with the highest MSW compost application rates (10 m^3 compost per 1000 m^2 soil), compared to 5 m^3 compost per 1000 m^2 soil or no-compost. They also reported that compost stored outside and exposed to natural conditions for one winter affected yields similarly to compost that was stored under cover, despite a lower electrical conductivity (EC) in the former compost.

Roe et al. (1992) evaluated MSW compost as a mulch, compared with a standard polyethylene mulch, on bell pepper (*Capsicum annuum* L.) production systems. They reported that biosolids/yard trimmings compost used as a mulch at 112 and 224 $Mg \cdot ha^{-1}$ on bell peppers grown on raised beds increased total fruit yields when compared with no mulch, but yields were similar to or lower than with polyethylene mulches. Municipal solid waste compost used as mulches at 13, 40, or 121 $Mg \cdot ha^{-1}$ decreased bell pepper yields as compared with polyethylene mulches, even though all plots were fertilized with a total of 269N-45P-192K ($kg \cdot ha^{-1}$). However, yields increased linearly with increasing compost mulch rates. In another experiment, total bell pepper fruit yields from plots mulched with MSW compost at 224 $Mg \cdot ha^{-1}$ were less than half of those from polyethylene-mulched plots (Roe et al., 1994).

Ozores-Hampton and Bryan (1993b) reported increased total marketable and large fruit from eggplant (*Solanum melongena* L.) and higher yield of large bell pepper fruit grown in plots amended with MSW compost at 90 and 134 $Mg \cdot ha^{-1}$ than from unamended plots.

In another experiment, one MSW compost was applied at 0, 33, or 67 $Mg \cdot ha^{-1}$ and another at 0, 67, and 135 $Mg \cdot ha^{-1}$, and tomatoes were planted, followed by squash (Bryan et al., 1994). Additional compost was applied at identical rates prior to planting a subsequent tomato crop. In both tomato crops, growth and yields were reduced by one of the composts, but not affected by the other.

In a four-season experiment, MSW compost applied at 67 and 135 $Mg \cdot ha^{-1}$ on drip- irrigated plots, with fertilizer at 215, 309, or 403 $kg \cdot ha^{-1}$ of N, 44 $kg \cdot ha^{-1}$ of P, and 248, 356, or 464 $kg \cdot ha^{-1}$ of K, reduced yields in the initial crop of bell peppers in compost plots. A subsequent tomato crop had more extra large and total marketable fruit, when compared with no-compost plots (Clark et al, 1994). This compost may have been initially immature, since another pepper crop grown on the identical plots resulted in increased yields. Fertilizer applied to compost plots for that crop did not affect yields, but increased yields in no-compost plots. Yields from early and final harvests and extra large fruit in an additional tomato crop also were higher in compost plots than in no-compost plots.

Maynard (1994) reported that tomato and bell pepper fruit yields from plots amended with compost produced from poultry manure with other agricultural wastes were similar to or greater than yields from fertilized plots, except in one crop of tomatoes where they were lower.

Obreza and Reeder (1994) reported that immature MSW composts at 13, 27, 75, and 112 $Mg{\cdot}ha^{-1}$ generally did not change or decreased yields of tomatoes for 2 years, when compared with plants grown without compost and fertilized at the same rate (56N-49P-93K $kg{\cdot}ha^{-1}$ preplant and 172N-57P-230K $kg{\cdot}ha^{-1}$ applied through the drip system).

With N at rates of 240 $kg{\cdot}ha^{-1}$, fruit yields from tomatoes grown in soil amended with one MSW compost at 48 $Mg{\cdot}ha^{-1}$ or another at 24 $Mg{\cdot}ha^{-1}$ were similar to those from plants grown in plots without composts (Ozores-Hampton et al., 1994).

Transplanting tomato and pepper plants into a field containing an uncured (immature) and newly incorporated biosolids/yard trimming compost at 135 $Mg{\cdot}ha^{-1}$ (fresh weight) immediately or up to 4 weeks after compost application did not result in yield differences in pepper or tomato fruit when compared with unamended plots (Roe and Stoffella, 1994a, 1994b).

Tomatoes produced higher yields when grown with amendments of horse manure or biosolids compost at 53 $Mg{\cdot}ha^{-1}$ than with the same rate of yard trimmings compost, biosolids/yard trimmings compost, or fertilizer at 220N-97P-183K ($kg{\cdot}ha^{-1}$) in one year, but in the second year, highest yields were from the fertilized or biosolids compost-amended plots (Allen and Preer, 1995).

Alvarez et al. (1995) reported that three of four commercial composts incorporated into a soil increased growth of tomato plants, while one compost depressed tomato growth. Compost amendments caused only small variations in the total numbers of bacteria, actinomycetes, and fungi in the rhizosphere of tomato plants. However, the addition of some composts increased the incidence of certain rhizobacteria antagonistic to soilborne pathogens such as *Pythium ultimum* and *Rhizoctonia solani*.

Auclair et al. (1995) compared organic growing media for greenhouse tomato production. When tomatoes were grown on peat moss and shrimp compost, fruit contents of Ca, Cu, Fe, P, and Zn increased and fruit ripened later than when tomatoes were grown on composted cattle manure.

Marketable yield of tomatoes grown in calcareous soils was increased by additions of two MSW composts, one at 37 and 74 and the other at 74 and 148 $Mg{\cdot}ha^{-1}$, compared with similarly fertilized plots without compost (Bryan et al., 1995). Rates were selected so that the total N added would be 370 and 740 $kg{\cdot}ha^{-1}$ for the two rates of each of the composts. Fruit size from compost plots was similar in the first year and larger in the second year when compared with fruit from unamended plots.

An MSW compost applied just before planting each spring at 56 and 112 Mg·ha^{-1} with fertilizer at 146N-64P-121K (kg·ha^{-1}) resulted in tomato fruit yield increases in three consecutive years, compared with fertilizer only (Maynard, 1995).

Undecomposed leaves (15.2 cm depth) tilled into plots in spring or fall or leaf compost (112 Mg·ha^{-1}) incorporated in spring for three years with fertilizer at 146N-64P-121K (kg·ha^{-1}) resulted in similar bell pepper yields in the control and compost plots while yields were lowest from both treatments with undecomposed leaves in the first year (Maynard, 1996). In the second year, plants in compost-amended plots produced higher yields than plants in control plots or in plots with a fall application of leaves, but similar yields to plants in plots with a spring application of leaves. In the third year, yields were similar among all treatments.

When biosolids/yard trimming compost at 134 Mg·ha^{-1} or no-compost was combined in a factorial arrangement with 0, 50, and 100% of a grower's standard fertilizer (71N-39P-44K kg·ha^{-1} broadcast and 283N-278K kg·ha^{-1} banded in bed centers), highest bell pepper fruit yields occurred in the plots with compost and 50% fertilizer (Roe et al., 1997).

In other studies, compost made from filtercake, a sugarcane (*Saccharum officinarum* L.) processing waste, was used (Stoffella and Graetz, 1997). Tomatoes were transplanted into pots filled with a 1:1 (v:v) mixture of the compost and a sandy field soil, the field soil only, or the compost only. Plants from pots with compost or compost mixtures had higher shoot weights, thicker stems, and larger shoot to root ratios than plants grown in unamended field soil. In a field experiment, plants from plots with the filtercake compost at 224 Mg·ha^{-1} were larger and produced higher yields than plants grown without compost, regardless of fertilizer rates (Stoffella and Graetz, 1997).

F. Other crops

Okra (*Abelmoschus esculentus* [L.] Moench) grown in pots with MSW compost mixed at 10 to 30% (v:v) with a very gravelly loam soil had increased lateral root development and early fruit yields compared to plants grown in unamended soil (Bryan and Lance, 1991).

Onion (*Allium cepa* L.) yield on a sandy loam soil increased with increasing rate of organic matter application, when the organic matter was biosolids/straw compost, or digested or raw biosolids (Smith et al., 1992)

Biosolids compost at 12 and 25 dry Mg·ha^{-1} increased onion and spinach (*Spinacia oleracea* L.) yields when incorporated to a soil depth of 10 cm, but not to a 30 cm soil depth (Mellano and Bevacqua, 1992). Onion and lettuce (*Lactuca sativa* L.) plants grown in plots of sandy loam soil with biosolids/wood chips compost applied over a 2-year period, at cumulative totals of 37 and 74 Mg·ha^{-1}, produced higher yields than the unamended control (Bevacqua and Mellano, 1993).

III. CONCLUSIONS

Generalizing from numerous projects that examine the use of different composts at varying rates with or without additional fertilizers on various vegetable crops in diverse soils and assorted climates is extremely hazardous. However, if we cannot find enough similarities to develop guidelines for compost utilization, then this research is unproductive from a practical standpoint.

Responses to composts are often more pronounced when crops are grown less intensively or are under an environmental stress. In their review, Gallardo-Lara and Nogales (1987) summarized vegetable and agronomic crop responses to MSW compost as being more positive in poorer soils, and reported that mixtures of synthetic fertilizers and composts are usually more efficient than either alone in meeting crop nutritional requirements. Gray and Tawhid (1995) reported that pod yields of bush snap beans were increased in a dry season, but not in a wetter one, by a leaf compost mulch. Buchanan and Gliessman (1991) reported that broccoli N use efficiency was highest in treatments that combined N from a synthetic source with compost.

Another consideration is that nutrient levels in composts are not always in the correct proportions for plant growth. There is a potential for buildup of some nutrient concentrations in the soil if composts are applied at high enough rates to supply the most limiting nutrients, usually N. Excessive concentrations of plant nutrient elements raise the potential for environmental damage and may threaten the safety of those consuming the vegetables. With increased interest in food safety and nutrition, researchers are beginning to report the concentrations of elements and compounds in plants that have the potential to be beneficial or to cause harm to humans who are consuming the vegetables. Kao (1993) stated that annual applications of sawdust/swine waste compost at high rates (25 or 50 $Mg \cdot ha^{-1}$) to acid soils would eventually raise soil Zn and Cu to toxic levels. In another study, a compost and a vermicompost decreased the nitrate concentration, but increased the K concentration of lettuce leaf tissue, when compared with synthetic fertilizers (Ricci et al., 1995).

Although much evidence points toward soil and environmental improvements with compost use, as well as crop yield increases in many instances, the use of compost must increase profits in order for it to become an accepted practice among vegetable growers. Kostov et al. (1995) reported that it was more economical to use composting vegetable residues for greenhouse cucumber production than a manured soil. Roe and Cornforth (1997) reported that uncomposted dairy manure and dairy manure compost both increased growth, yield, and net income from melons (*Cucumis melo* L.) and broccoli in a low-input growing system, but it was less expensive to use the uncomposted manure. However, food safety concerns prevent the use of uncomposted manures directly on vegetable crops.

Although compost is organic matter, it can contain potentially harmful pollutants, such as heavy metals and human pathogens, which must be prevented from entering the food chain. Proper handling of feedstocks, composting at correct temperatures, and testing can eliminate most of the pathogens (Farrell, 1993). Concentrations of metals in compost can be controlled by proper choice of feedstocks and awareness of soil–plant reactions to additions of composts (Chaney and Ryan, 1993).

At present, most vegetable growers who use composts are smaller, more specialized, and often grow organically, whether by choice or due to lack of resources. To encourage compost use by larger commercial growers, more evidence for the benefits of compost utilization, especially economic benefits, must be developed.

REFERENCES

Allen, J.R. and J.R. Preer. 1995. Use of municipal waste in vegetable crop production. *Caribbean Food Crops Society Proceedings* 30:199–205.

Alvarez, M.A., S. Gagné, and H. Antoun. 1995. Effect of compost on rhizosphere microflora of the tomato and on the incidence of plant growth-promoting rhizobacteria. *Applied Environmental Microbiology* 61(1):194–199.

Auclair, L., J.A. Zee, A. Karam, and E. Rochat. 1995. Nutritive value, organoleptic quality and productivity of greenhouse tomatoes in relation to production method: organic-conventional-hydroponic. *Sciences des Aliments* 15(6):511–528.

Bevacqua, R.F. and V.J. Mellano. 1993. Sewage sludge compost's cumulative effects on crop growth and soil properties. *Compost Science and Utilization* 1(3):34–40.

Bryan, H.H. and C.J. Lance. 1991. Compost trials on vegetables and tropical crops. *BioCycle* 32(3):36–37.

Bryan, H.H., B. Schaffer, and J.H. Crane. 1994. Solid waste compost for improved water conservation and production of vegetable crops (tomatoes/squash)-Homestead site, p. 6–9. In: W.H. Smith (ed.). *Summary Report for the Florida Composting Conference.* Florida Department of Agriculture and Consumer Services, Tallahassee.

Bryan, H.H., B. Schaffer, R.E. Sanford, and M. Codallo. 1995. Growth and yield of tomato in calcareous soil amended with municipal solid waste compost. *Proceedings of the Florida State Horticultural Society* 108:251–253.

Buchanan, M. and S.R. Gliessman. 1991. How compost fertilization affects soil nitrogen and crop yield. *BioCycle* 32(12):72–77.

Chaney, R.L. and J.A. Ryan. 1993. Heavy metals and toxic organic pollutants in MSW-composts: research results on phytoavailability, bioavailability, fate, etc., p. 451–506. In: H.A.J. Hoitink and H.A. Keener (eds.) *Science and Engineering of Composting: Design, Environmental, Micobiological, and Utilization Aspects.* Renaissance Publications, Worthington, Ohio.

Clark, G.A., C.D. Stanley, and D.N. Maynard. 1994. Compost utilization for improved management of vegetable crops on sandy soils-Bradenton site, p. 11–13. In: W.H. Smith (ed.). *Summary report for the Florida Composting Conference.* Florida Department of Agriculture and Consumer Services, Tallahassee.

Farrell, J.B. 1993. Fecal pathogen control during composting, p. 282–300. In: H.A.J. Hoitink and H.A. Keener (eds.). *Science and Engineering of Composting: Design, Environmental, Microbiological, and Utilization Aspects.* Renaissance Publications, Worthington, Ohio.

Gallardo-Lara, F. and R. Nogales. 1987. Effect of the application of town refuse compost on the soil-plant system: a review. *Biological Wastes* 19:35–62.

Gray, E. and A. Tawhid. 1995. Effect of leaf mulch on seedling emergence, plant survival, and production of bush snap beans. *Journal of Sustainable Agriculture* 6(2/3):15–20.

Hornick, S.B. 1988. Use of organic amendments to increase the productivity of sand and gravel spoils: effect on yield and composition of sweet corn. *American Journal of Alternative Agriculture* 3(4):156–162.

Hue, N.V., H. Ikawa, and J.A. Silva. 1994. Increasing plant-available phosphorus in an Ultisol with a yard-waste compost. *Communications in Soil Science and Plant Analysis* 25 (19&20):3291–3303.

Kao, M.M. 1993. The evaluation of sawdust swine waste compost on the soil ecosystem, pollution, and vegetable production. *Water Science and Technology* 27(1):123–131.

Kostov, O., Y. Tzvetkov, N. Kaloianova, and O. Van Cleemput. 1995. Cucumber cultivation on some wastes during their aerobic composting. *Bioresource Technology* 53(3):237–242.

Lawson, I.Y.D., K. Muramatsu, and I. Nioh. 1995. Effect of organic matter on the growth, nodulation, and nitrogen fixation of soybeans grown under acid and saline conditions. *Soil Science and Plant Nutrition* 41(4):721–728.

Manios, V.I. and E. Kapetanios. 1992. Effect of town refuse compost as soil amendment on greenhouse tomato crop. *Acta Horticulturae* 302:193–201.

Maynard, A.A. 1994. Sustained vegetable production for three years using composted animal manures. *Compost Science and Utilization* 2(1):88–96.

Maynard, A.A. 1995. Cumulative effect of annual additions of MSW compost on the yield of field-grown tomatoes. *Compost Science and Utilization* 3(2):47–54.

Maynard, A.A. 1996. Cumulative effect of annual additions of undecomposed leaves and compost on the yield of field-grown peppers. *Compost Science and Utilization* 4(2):81–88.

Mellano, V.J. and R.F. Bevacqua. 1992. Sewage sludge compost as a soil amendment for horticultural crops. *HortScience* 27:697. (Abstract).

Obreza, T.A. and R.K. Reeder. 1994. Municipal solid waste compost use in tomato/watermelon successional cropping. *Soil and Crop Science Society of Florida Proceedings* 53:13–19.

Ozores-Hampton, M. and H.H. Bryan. 1993a. Municipal solid waste (MSW) soil amendments: influence on growth and yield of snap beans. *Proceedings of the Florida State Horticultural Society* 106:208–210.

Ozores-Hampton, M. and H.H. Bryan. 1993b. Effect of amending soil with municipal solid waste (MSW) compost on yield of bell pepper and eggplant. *HortScience* 28:103. (Abstract).

Ozores-Hampton, M., B. Schaffer, and H.H. Bryan. 1994. Nutrient concentrations, growth, and yield of tomato and squash in municipal solid waste-amended soil. *HortScience* 29:785–788.

Paino, V, J.P. Peillex, O. Montlahuc, A. Cambon, and J.P. Bianchini. 1996. Municipal tropical compost: effects on crops and soil properties. *Compost Science and Utilization* 4(2):62–69.

Ricci, M.D.S. F., V.W.D. Casali, A.M. Cardoso, and H.A. Ruiz. 1995. Nutrient contents of two lettuce cultivars fertilized with organic compost. *Pesquisa Agropecuaria Brasileira* 30(8):1035–1039.

Roe, N.E. and G.C. Cornforth. 1997. Yield effects and economic comparison of using fresh or composted dairy manure amendments on double cropped vegetables. *HortScience* 32:462. (Abstract).

Roe, N.E., S.R. Kostewicz, and H.H. Bryan. 1990. Effects of municipal solid waste compost and fertilizer rates on broccoli. *HortScience* 25:1066. (Abstract).

Roe, N.E., H.H. Bryan, P.J. Stoffella, and T.W. Winsberg. 1992. Use of compost as mulch on bell peppers. *Proceedings of the Florida State Horticultural Society* 105:336–338.

Roe, N.E., P.J. Stoffella, and H.H. Bryan. 1993. Utilization of MSW compost and other organic mulches on commercial vegetable crops. *Compost Science and Utilization* 1(3):73–84.

Roe, N.E. and P.J. Stoffella. 1994a. Influence of immature compost on growth and yields of tomato. *Caribbean Food Crops Society Proceedings* 30: 194–198.

Roe, N.E. and P.J. Stoffella. 1994b. Influence of "uncured" compost on growth and yield of bell pepper. *Proceedings of the National Pepper Conference*, Las Cruces, New Mexico, 14–16 August 1994.

Roe, N.E., P.J. Stoffella, and H.H. Bryan. 1994. Growth and yields of bell pepper and winter squash grown with organic and living mulches. *Journal of the American Society for Horticultural Science* 119:1193–1199.

Roe, N.E., P.J. Stoffella, and D. Graetz. 1997. Composts from various municipal solid waste feedstocks affect vegetable crops II. Growth, yields, and fruit quality. *Journal of the American Society for Horticultural Science* 123:433–437.

Smith, S.R., J.E. Hall, and P. Hadley. 1992. Composting sewage sludge wastes in relation to their suitability for use as fertilizer materials for vegetable crop production. *Acta Horticulturae* 302: 203–215.

Stoffella, P.J. and D.A. Graetz. 1997. Sugarcane filtercake compost influence on tomato emergence, seedling growth, and yields, p. 1351–1356. In: M. deBertoldi, P. Sequi, B. Lemmes, and T. Papi (eds.). *The Science of Composting Part 2*. Blackie Academic and Professional, London, United Kingdom.

United States Department of Agriculture (USDA)-National Agricultural Statistics Service. 1998. *Statistical Highlights 1997–98: Vegetable Crops*. http://usda.mannlib.cornell.edu:80/usda/

CHAPTER 6

Compost Utilization in Ornamental and Nursery Crop Production Systems

George E. Fitzpatrick

CONTENTS

I. INTRODUCTION

Growers of ornamental nursery crops are regarded as high-priority potential customers by people who manufacture and market compost products. Nursery crops,

1-56670-460-X/01/$0.00+$.50

in general, have high profit margins, so growers can afford to provide resources to maintain and improve productivity that might be prohibitively expensive for other kinds of crops. Also, a very high proportion of nursery crops are grown in containers, so when a production cycle is completed, and the crop is sold, the growing substrate is sold with it. Therefore, growers of container crops must acquire new potting medium supplies at the beginning of each new production cycle.

II. NURSERY CROP PRODUCTION

In North America, the first commercial plant nurseries were established in the early 18th century, beginning with Prince Nursery, which opened in Flushing, NY in 1737 (Higginbotham, 1990). From this time until the middle of the 20th century, most nursery crop production was in field nurseries. Container production was a very minor component in ornamental crop production until the late 1940s, but in the past 50 years container production has become increasingly dominant (Davidson et al., 1999). Although many compost marketers prefer to sell their products to field nurseries, because of a high potential volume of material needed (Tyler, 1996), the increasing dominance of container production assures an important niche for compost utilization in horticulture. Moreover, hauling costs and application costs represent significant challenges in any compost field application program (Roe, 1998), but acquisition of new growing substrates is a normal activity for container nurseries.

III. DEVELOPMENT OF COMMERCIAL COMPOST PRODUCTION SYSTEMS

Composting, the controlled decomposition of organic matter to a point where the product can be safely and beneficially used to improve crop productivity, is believed to be a practice as old as agriculture itself. The earliest known written reference to composting is believed to be in the clay tablets of the Akkadian Empire, ca. 2700 B.C. (Rodale et al., 1960). Throughout the many centuries since then, composting has remained a farming activity. Growers who wanted to use compost had to make it themselves from organic materials available to them. Moreover, the perceived benefits of compost use were mostly anecdotal and not the results of controlled scientific studies. It was not until the early years of the 20th century that the first scientific studies on compost efficacy were published (Howard and Wad, 1931). The growing realization of the benefits of compost utilization, supported by controlled scientific studies, encouraged the development of commercial organizations that made compost products with the aim of selling these materials to growers. One of the earliest of these firms, Kellogg Supply, Inc. began making and marketing compost products in 1927 in Carson, CA (Kellogg, 1985). Increasing urbanization during the 20th century augmented this trend, because concentrated populations produced concentrated amounts of waste products. Composition of urban waste streams can vary tremendously from one geographical area to another and from one time of the year to another, but they invariably contain mostly (> 60%) organic substances that are suitable for composting (Obeng and Wright, 1987).

Commercial composting organizations earn income by collecting fees from urban waste haulers as well as income from growers through the sale of the finished compost product. Many commercial composting companies were not able to either achieve or maintain profitability because of intense competition from relatively inexpensive landfills and incinerators, and they ceased operation. The composting businesses that survived were faced with the constant challenge of maintaining compost product quality and product uniformity. Uniformity and consistency of physical and chemical parameters in growing substrates is extremely important in the production of container ornamentals (Poole et al., 1981). The highly diverse nature of urban waste streams can cause significant fluctuations in parameters such as bulk density, porosity, pH levels, soluble salt content, and possibly the presence of phytotoxic substances. Commercial compost organizations that cannot monitor and manage these perturbations will not be able to consistently produce a product that can be successfully utilized by container plant growers.

IV. CHALLENGES TO SUCCESSFUL COMPOST USE

A. Nutritional Content

Since compost products are made from organic materials, it is not surprising that they typically contain substantial levels of certain essential nutrients. However, the concentration of these nutrients, particularly nitrogen (N) and potassium (K), is usually not sufficiently high to provide complete nutritional support for horticultural crops (Chaney et al., 1980; Hue and Sobieszczyk, 1999). Growers must realize that, although a portion of a crop's nutritional requirement may be met by compost present in the rooting substrate, optimum growth cannot normally be achieved without supplemental fertilization.

B. Soluble Salt Levels

Both the feedstocks from which compost is made and the process parameters utilized in the composting facility can have substantial influences on soluble salt levels in compost products. For example, biosolids (sewage sludges) are frequently stabilized and conditioned at wastewater treatment facilities prior to being composted. Chemicals frequently used for this treatment include ferric chloride and lime, and the biosolids and the compost products made from them often have elevated soluble salt levels. When a compost made from biosolids treated with ferric chloride and lime was compared with a compost that was made from biosolids that had not been so treated, the conductivity of the finished composts was affected. The former compost had a conductivity of 7.5 $dS \cdot m^{-1}$ while the latter had a level of only 3.9 $dS \cdot m^{-1}$, and the control growing medium had a conductivity level of 3.5 $dS \cdot m^{-1}$ (Fitzpatrick, 1986). In this same study, the compost with the lower salt level supported significantly greater growth than the higher salt level compost, although both compost products supported significantly greater growth than the control medium for spathiphyllum (*Spathiphyllum* sp. 'Mauna Loa') and dwarf schefflera (*Schefflera arboricola* Hayata). Other studies (Conover and Joiner, 1966; Lumis and Johnson, 1982; Sanderson, 1980) have illustrated the negative effects on plant growth of elevated soluble salt levels in certain types of compost products. Despite quality control measures routinely taken by commercial composters, growers are well advised to regularly monitor new batches of compost products as received. Excessively high soluble salt levels in compost materials can be managed by leaching, where leaching would not pose a threat to surface or ground water resources, or by blending the compost with substrates that have lower soluble salt levels. Indeed, many compost based potting media recommendations contain only 20 to 30% compost, as a means of reducing damage that can be caused by high salt levels or other phytotoxic substances that may be present in certain compost products (see Raymond et al., 1998).

C. Compaction

Porosity is one of the more important physical parameters in container media (Poole et al., 1981), because of the need for effective gas exchange in the root zone. Some compost products have been reported to have satisfactory pore space at the beginning of the plant production period but undergo compaction during the production period (Fitzpatrick and Verkade, 1991). Compost materials used as the complete, or stand-alone, rooting substrate are more likely to settle or compact during the production period, thereby reducing the porosity of the medium. This phenomenon is more likely to occur in fresh, immature compost products. This problem can be treated either by allowing the compost product to age further, or by blending the compost with materials that are not likely to undergo compaction during production.

D. Phytotoxicity

Compost products may contain phytotoxic materials that can come from a variety of sources. The organic material from which the compost is made may contain residues of substances that can be toxic to crops grown in the end product compost. For example, forsythia (*Forsythia intermedia* Zab.) and white cedar (*Thuja occidentalis* L.) grown in potting mixes amended with municipal solid waste (MSW) compost suffered boron (B) toxicity due to B that had been present in the MSW feedstock (Lumis and Johnson, 1982). This can be contrasted with the findings of Ticknor et al. (1985), in which the authors reported very low levels of B in the biosolids compost used and in the foliage in photinia (*Photinia* X *fraseri* Dress.), with the suggestion that growers who use this type of compost should consider applying foliar treatments of B to correct this deficiency. Commercial composting organizations regularly monitor the composts they produce, both for their own quality assurance programs, and because of governmental regulatory requirements. However, there is always the possibility of substantial variation in the composition of the incoming organic feedstock, and small volumes of contaminated product can escape detection in random sampling of compost materials.

A more serious cause of phytotoxicity in compost products can come from the composting process itself. Since commercial composting organizations derive income from charging fees to urban waste haulers as well as from the sale of the compost products to growers, there is an obvious and strong economic incentive to minimize the amount of time that the organic material is composting. The technically correct minimum amount of time that an organic substance must undergo composting in order to have a stable end product compost is variable. It depends on several factors, including the size of the compost pile, the pile's aeration status, the pile's moisture status, the carbon to nitrogen (C:N) ratio of the material, heat levels and range during composting, and other factors. References published prior to the widespread commercialization of composting recommend minimum composting periods of approximately 6 months (Howard and Wad, 1931). Although it may be possible to speed up the composting process to some degree, the economic incentives for commercial composters to sell immature compost products are very real. Commercial plant producers who purchase compost products should be sensitive to the seller's incentives and should also be aware that immature compost products can pose serious threats to the health and vigor of plants grown in them. High microbial activity in immature composts can cause biological blockage of N from the crop. The microbes can literally out-compete the plant for the available N, and the plant would exhibit N deficiency symptoms. Also, the microbes that mediate the composting process secrete certain phytotoxic chemicals, such as short-chain fatty acids like acetic acid, propionic acid, and butyric acid, during the early stages of the composting process. Deformity or death of plant parts caused by the ephemeral production of these chemicals at certain points early in the composting process can be a real threat when growers attempt to use immature composts as growing media (Jimenez and Garcia, 1989). When liners of hibiscus (*Hibiscus rosa-sinensis* L.) were planted in containers with growing media amended with uncomposted biosolids, plants exhibited phytotoxicity symptoms within 5 days after planting, and the symptoms increased in intensity as the biosolids concentration in the growing medium increased (Figure 6.1). Plants growing in the control medium (Figure 6.2A) did not exhibit any phytotoxicity symptoms, while plants in media amended with uncomposted biosolids (Figure 6.2B) exhibited chlorosis, consistent with N blockage, and leaf distortion, consistent with the presence of short-chain fatty acids (Fitzpatrick, unpublished data).

There are numerous tests that can be conducted to determine whether a compost product is sufficiently mature (Jimenez and Garcia, 1989), but most of these tests require equipment and facilities that are not directly available to the typical nursery crop grower. One type of test that can be conducted by most nursery growers is the bioassay procedure. A seed flat is filled with the compost material being considered, and a second flat filled with a control growing medium that is known to be stable. Seeds of a species with rapid germination and rapid growth, such as radish (*Raphanus sativus* L.), are sown in the flats and the germination and growth characteristics of plants in both flats are observed for a 1 to 2 week period. If significant levels of phytotoxic substances are present in the compost material under consideration, visual symptoms should be apparent during this time.

Figure 6.1 Hibiscus (*Hibiscus rosa-sinensis* L.) liners 5 days after planting in 25 cm diameter nursery containers filled with a growing medium amended with uncomposted biosolids. The row on the far left is the control medium, with no biosolids, and no phytotoxicity symptoms are apparent on these plants. The row on the far right contains the medium with the highest biosolids concentration, 30% of the total growing medium, and plants in this substrate have the most severe symptoms. The five middle rows contain biosolids concentrations of 5, 10, 15, 20, and 25% (left to right), and phytotoxicity symptoms appear to be increasing as the biosolids concentration increases.

V. IMPORTANT FACTORS IN A CONTAINER GROWING MEDIUM

Although there is no perfect growing medium for all ornamental crops under all growing conditions, numerous authors have described general recommendations. For example, Poole et al. (1981) recommend for container grown foliage crops the following general parameters: bulk density — 0.30 $g{\cdot}cm^{-3}$ (dry), 0.60 to 1.20 $g{\cdot}cm^{-3}$ (wet); pore space — 5 to 30%; water-holding capacity — 20 to 60%; pH — 5.5 to 6.5; soluble salts — 400 to 1,000 $mg{\cdot}L^{-1}$; cation exchange capacity — 10 to 100 meq per 100 cm^3.

Frequently, commercially made compost products have pH levels higher than those listed above; ranges of pH 6.7 to 7.7 are common (Conover and Joiner, 1966; Fitzpatrick, 1989; Fitzpatrick and Verkade, 1991). High pH values can result from the chemical qualities of the feedstocks, or from materials added to the feedstocks. For example, composts made from biosolids frequently have high pH values because of chemical stabilizers, such as lime, added before composting. Unless milled, (see Fitzpatrick, 1989), pore space and water-holding capacities of commercially made compost products are usually within the acceptable ranges. Soluble salt levels, cation exchange capacity, and bulk density may all be significantly influenced by the composition of the parent material or by preprocessing, so growers of ornamental crops should monitor these parameters regularly.

Figure 6.2 Hibiscus (*Hibiscus rosa-sinensis* L.) liners 5 days after planting in 25 cm diameter nursery containers filled with a growing medium amended with uncomposted biosolids; (A) plants growing in the control medium, with no uncomposted biosolids incorporated, do not exhibit any phytotoxicity symptoms, while (B) plants in medium amended with uncomposted biosolids exhibit chlorosis and leaf distortion.

VI. USING COMPOST PRODUCTS BENEFICIALLY IN NURSERY CROP PRODUCTION

There are several published general reviews illustrating compost use in nursery crop production, including Fitzpatrick and McConnell (1998), Fitzpatrick et al. (1998), Sanderson (1980), and Shiralipour et al. (1992). Generally, compost is used in nursery crop production as a less expensive substitute for peat and other organic components of the growing medium. Also, some compost products have been demonstrated to accelerate growth in some species, thereby decreasing the production period for these crops. Some compost products also have been demonstrated to have a suppressive effect on some plant pathogens (see chapter 12 by Hoitink et al. in this book).

A. Field Nursery Production

Although most published work on compost utilization for ornamental crop production focuses on container plant culture, some publications have detailed the uses of compost products as amendments in field nursery soil. In one of the earlier references on compost use in ornamental crop production, DeGroot (1956) reported enhanced growth in gloxinia (*Gloxinia* X *hybrida* Hort.) grown in five rates (1, 2, 3, 4, and 5 kg·m^{-2}) of MSW compost applied to planting beds, and enhanced growth in begonia (*Begonia* X *tuberhybrida* Voss) at five rates (0.5, 1, 1.5, 2, and 2.5 kg·m^{-2}). Gouin and Walker (1977) reported greater stem length in tulip poplar (*Liriodendron tulipifera* L.) and dogwood (*Cornus florida* L.), and significantly less winter dieback in tulip poplar, when planting beds were treated with three rates (2.5, 5, and 10 cm thickness) of compost made from one part digested biosolids and three parts wood chips compared to unamended beds. Gouin (1977), in a separate study, reported similar or reduced growth of Norway spruce [*Picea abies* (L.) Karst] and white pine (*Pinus strobus* L.) in planting beds treated with compost of the same type and at the same rates as in Gouin and Walker (1977) compared to a control topdressed with a slow-release fertilizer and mulched with aged sawdust. Table 6.1 provides a summary of these published reports.

Table 6.1 Response of Ornamental Crops to Compost Products Used in Field Nursery Production

Crop	Compost Type[z]	Rate	Growth Response[y]	Reference
Gloxinia	MSW	1, 2, 3, 4, and 5 kg·m^{-2}	+	DeGroot, 1956
Begonia	MSW	1, 2, 3, 4, and 5 kg·m^{-2}	+	DeGroot, 1956
Tulip poplar	B/WC	2.5, 5, and 10 cm thickness	+	Gouin and Walker, 1977
Dogwood	B/WC	2.5, 5, and 10 cm thickness	+	Gouin and Walker, 1977
Norway spruce	B/WC	2.5, 5, and 10 cm thickness	=, –	Gouin, 1977
White pine	B/WC	2.5, 5, and 10 cm thickness	=	Gouin, 1977

[z] MSW = municipal solid waste compost; B/WC = biosolids/wood chip co-compost.
[y] +, –, = represent: positive, negative, or neutral, respectively (usually relative to a control).

B. Container Production

One of the earliest reports on container production of ornamentals with compost, describing research conducted between 1948 and 1954, was published in Belgium in 1956 (DeGroot, 1956). In a series of research studies, DeGroot found that MSW compost ("compost de ville") would not support growth in azaleas (*Azalea indica* L.), if the compost rate was greater than 10% of the rooting medium. DeGroot indicated that elevated pH levels in the compost product probably were responsible for the growth inhibition. In the same series of experiments, favorable results were observed with begonia when grown in mixes containing 30% MSW compost. He reported reduced growth ("moyenne" or "mediocre") in begonia when compost concentrations were higher than 40%. In a much larger report published 5 years later, DeGroot (1961) observed favorable results in growing 74 species of ornamental plants and unfavorable results in 6 species, when the growing medium contained 25 to 35% compost. Numerous studies published subsequently by many other authors have elaborated on compost use in containerized ornamental plant production, and also advised against using too high a concentration of compost product in the growing medium blend.

1. Temperate Woody Ornamentals

In a study of three species of woody ornamentals, Sanderson and Martin (1974) reported enhanced growth in Chinese holly (*Ilex cornuta* Lindl. & Paxt.) and white cedar when grown in media containing 33% MSW compost, relative to an untreated control. They reported that viburnum (*Viburnum* X *burkwoodii* Hort. Burkw. & Skipw.), when grown in the 33% MSW compost rate, did not differ significantly in growth from plants grown in the control medium. Working with a different species of viburnum, *V. suspensum* Lindl., Fitzpatrick and Verkade (1991) reported that plants grown in both 40% and 100% MSW compost grew at rates that were not significantly different than the control. They speculated that certain plant species, like viburnum, may be physiologically ambivalent to the composition of the growing medium and may be able to adapt to a wide range of rooting conditions.

The issue of effects of compost rate in the growing medium is clearly illustrated in a recent study of four species of temperate woody ornamentals (Raymond et al., 1998). In this study, all four species (deutzia, *Deutzia gracilis* L.; silverleaf dogwood, *Cornus alba* L. 'Elegantissima'; red-osier dogwood, *C. sericea* L.; and ninebark, *Physocarpus opulifolius* [L.] Maxim.) grew at rates significantly higher than the controls when the growing medium contained 25% waxed corrugated cardboard (WCC) compost. When tested in media containing 50% WCC compost, only the silverleaf dogwood grew at rates higher than the control; deutzia and red-osier dogwood grew at rates comparable to the control and ninebark grew at rates significantly lower than the control. The authors characterized the WCC compost as immature; it is therefore possible that some of the growth suppression observed in some species may have been due to the ephemeral phytotoxicity associated with immature compost products. Research findings of compost efficacy on selected species of temperate woody ornamentals are summarized in Table 6.2.

2. Subtropical and Tropical Ornamentals

In a study of three subtropical ornamental species (jasmine, *Jasminum volubile* Jacq.; ligustrum, *Ligustrum japonicum* Thunb. var. *rotundifolium* Blume; and dwarf oleander, *Nerium oleander* L.), Fitzpatrick (1981) reported enhanced growth in ligustrum and dwarf oleander and no difference in growth of jasmine when grown in a mix consisting of 80% biosolids compost, compared to the control. In a different study, dwarf oleander grown in 100% MSW compost and 100% paper mill sludge compost grew at rates that were significantly greater than the control (Fitzpatrick, 1989). This suggests that dwarf oleander may be a particularly responsive species to even slight differences in the rooting environment, quite the opposite of ambivalent species like viburnum. In the same study, Fitzpatrick (1989) reported that growth of orange-jessamine (*Murraya paniculata* [L.] Jack) in 100% MSW compost and 100% paper mill sludge compost was not significantly different than the control. The composts used in this study were well aged, so there would be little concern

Table 6.2 Response of Temperate Woody Ornamental Crops to Compost Products in the Rooting Substrate

Crop	Compost Type[z]	Compost in Rooting Medium (%)	Growth Response[y]	Reference
Chinese holly	MSW	33	+	Sanderson and Martin, 1974
White cedar	MSW	33	+	Sanderson and Martin, 1974
Burkwood viburnum	MSW	33	=	Sanderson and Martin, 1974
Viburnum suspensum	MSW	40, 100	=	Fitzpatrick and Verkade, 1991
Silverleaf dogwood	WCC	25	+	Raymond et al., 1998
Silverleaf dogwood	WCC	50	+	Raymond et al., 1998
Red-osier dogwood	WCC	25	+	Raymond et al., 1998
Red-osier dogwood	WCC	50	=	Raymond et al., 1998
Deutzia	WCC	25	+	Raymond et al., 1998
Deutzia	WCC	50	=	Raymond et al., 1998
Ninebark	WCC	25	+	Raymond et al., 1998
Ninebark	WCC	50	–	Raymond et al., 1998

[z] MSW = municipal solid waste compost; WCC = waxed corrugated cardboard compost.
[y] +, –, = represent positive, negative, or neutral, respectively (usually relative to a control).

about possible phytotoxicity attributable to compost immaturity. In a study of the interactive effects of sewage effluent irrigation and growing media consisting of 80% biosolids compost and 20% sifted incinerator ash, Fitzpatrick (1985) reported growth rates in four species of tropical trees (West Indian mahogany, *Swietenia mahagoni* [L.] Jacq.; pink tabebuia, *Tabebuia pallia* [Lindl.] Miers; pigeon-plum, *Cocoloba diversifolia* Jacq.; and key lime, *Citrus aurantiifolia* [Christm.] Swingle) were comparable to the control, while one species (schefflera, *Brassaia actinophylla* Endl.), grew faster in the control than in the compost-incinerator ash-effluent treatment. Table 6.3 provides a summary of reports on compost efficacy on selected subtropical and tropical ornamental species.

3. Floriculture and Foliage Crops

Poole (1969) reported reduced rooting for cordatum (*Philodendron scandens* C. Koch & H. Sello. subsp. *oxycardium* [Schott] Bunt.) and golden pothos (*Scindapsus aureus* [Linden & Andre´] Engl.) when a 100% MSW compost was used as a rooting medium, compared to 3 commercial rooting media. Both the pH and soluble salt levels of the compost were significantly elevated compared to levels in the three

Table 6.3 Response of Subtropical and Tropical Ornamental Crops to Compost Products in the Rooting Substrate

Crop	Compost Type[z]	Compost in Rooting Medium (%)	Growth Response[y]	Reference
Jasmine	B	80	=	Fitzpatrick, 1981
Ligustrum	B	80	+	Fitzpatrick, 1981
Dwarf oleander	B	80	+	Fitzpatrick, 1981
Dwarf oleander	MSW	100	+	Fitzpatrick, 1989
Dwarf oleander	PM	100	+	Fitzpatrick, 1989
Orange-jessamine	MSW	100	=	Fitzpatrick, 1989
Orange-jessamine	PM	100	=	Fitzpatrick, 1989
West Indian mahogany	B	80	=	Fitzpatrick, 1985
Pink tabebuia	B	80	=	Fitzpatrick, 1985
Pigeon-plum	B	80	=	Fitzpatrick, 1985
Key lime	B	80	=	Fitzpatrick, 1985
Schefflera	B	80	+	Fitzpatrick, 1985

[z] B = biosolids compost; MSW = municipal solid waste compost compost; PM = paper mill sludge compost.
[y] +, –, = represent positive, negative, or neutral, respectively (usually relative to a control).

control media. Fitzpatrick (1986) reported that two foliage plant species (dwarf schefflera and 'Mauna Loa' spathiphyllum) grew faster in two different types of biosolids compost used as 100% of the growing medium, as compared to plants grown in a control medium. One of the composts was made from biosolids that had been treated with ferric chloride and lime prior to composting. The second compost was made from biosolids that had not been so treated. Although both compost products, which had been aged for ca. 10 months prior to being used as growing media, produced larger plants than the control, the compost made from the chemically treated biosolids produced smaller plants than the compost made from the untreated biosolids. Chrysanthemum (*Chrysanthemum* X *morifolium* Ramat. 'Yellow Delaware' and 'Oregon') exhibited a general increase in number of flowers per pot and a decrease in time required for flowering as the concentration of MSW compost in the growing medium was increased up to and including 100% compost as the complete medium (Conover and Joiner, 1966). Gogue and Sanderson (1975) reported marginal leaf injury in *C.* X *morifolium* grown in MSW compost, and having found elevated levels of B in both the compost and plant tissue, suggested B toxicity as an explanation for this observation. Shanks and Gouin (1984) observed that chrysanthemums grew well in a wide range of media types but did particularly well in media containing vermiculite, whether or not compost was present in the mix. Pansy (*Viola tricolor* L. 'Super Swiss Mix'), and snapdragon (*Antirrhinum majus* L. 'Floral Carpet Red') exhibited enhanced growth in media amended with biosolids compost, as compared to the control medium. Both species had greater fresh weight as compost rate was increased up to 50% of the medium concentration, and snapdragon had more flower buds than the control when grown at the 50% compost rate (Hemphill et al., 1984). Wootton et al. (1981) reported enhanced growth in marigold (*Tagetes erecta* L. 'Golden Jubilee'), zinnia (*Zinnia elegans* Jacq. 'Fire Cracker'), and petunia

(*Petunia hybrida* Hort. 'Sugar Plum') grown in biosolids compost that had been screened through a 2.38 mm sieve (No. 8 sieve). They reported no significant medium compaction for the 2 to 3 month period needed to produce annuals and observed no phytotoxicity symptoms.

Although most reported studies on floricultural uses of compost examined just one type of compost, Klock and Fitzpatrick (1997) reported on the effects of three different composts: biosolids–yard trash (SYT), refuse-derived fuel residuals–biosolids–yard trash (RYT), and MSW used as growing media for impatiens (*Impatiens wallerana* Hook. 'Accent Red'). Examining compost rates up to and including 100% of the growing medium, the authors reported that shoot dry mass of plants grown in SYT compost increased as the percentage of compost in the medium increased, while mass of plants grown in MSW compost decreased as percentage of compost in the medium increased. There were no significant differences in plant mass attributable to rate of RYT compost in the growing medium. Comparable results in average number of flowers per plant and plant size were also reported. Reasons for the disparity between these compost types included (1) higher levels of soluble salts in the MSW compost compared to the other two, and (2) less maturity in the MSW compost, with a C:N ratio of 29, as compared to C:N ratio of 17 for the SYT compost and 15 for the RYT compost. Composts with a C:N ratio <20 are generally considered mature (Jimenez and Garcia, 1989). Floriculture and foliage crop findings are summarized in Table 6.4.

VII. THE FUTURE OF COMPOST USE IN ORNAMENTAL PLANT PRODUCTION

Over the last several decades, numerous authors have made predictions about the future of compost utilization in horticultural crop production. There is a wide diversity of opinion on this subject, but certain threads of commonality are apparent, such as the belief that there is great potential for increased use of compost. Some of the predictions for the great potential of compost are several decades old, so it is certainly appropriate to consider why there has not been greater exploitation of compost products by nursery crop growers and other horticultural producers.

One of the major barriers to greater utilization of compost products is the economic instability of the composting industry. Many commercial compost producers have made major changes in the specific types and amounts of products they have manufactured, such as changing the types of organic materials they compost, the mix ratios of feedstocks, various preprocessing procedures, active composting periods, and postprocessing procedures. In many cases, such management decisions were made for valid business reasons, but with little regard to the influence such changes could have on the efficacy of the compost product. Moreover, numerous composting companies have gone out of business during the past decade or two. Growers who have tried compost products and decided to continue using them have often been unable to acquire the same product, or even something close to the same product from the manufacturer, because of changes in the compost product's physical

Table 6.4 Response of Floriculture and Foliage Crops to Compost Products in the Rooting Substrate

Crop	Compost Type[z]	Compost in Rooting Medium (%)	Growth Response[y]	Reference
Cordatum	MSW	100	–	Poole, 1969
Golden pothos	MSW	100	–	Poole, 1969
Dwarf schefflera	B	100	+	Fitzpatrick, 1986
'Mauna Loa' spathiphyllum	B	100	+	Fitzpatrick, 1986
Chrysanthemum (2 cultivars)	MSW	100	+	Conover and Joiner, 1966
Chrysanthemum	MSW	100	–	Gogue and Sanderson, 1975
'Super Swiss Mix' pansy	B	50	+	Hemphill et al., 1984
'Floral Carpet Red' snapdragon	B	50	+	Hemphill et al., 1984
'Golden Jubilee' marigold	B	100	+	Wootton et al., 1981
'Fire Cracker' zinnia	B	100	+	Wootton et al., 1981
'Sugar Plum' petunia	B	100	+	Wootton et al., 1981
'Accent Red' impatiens	B/YT	100	+	Klock and Fitzpatrick, 1997
'Accent Red' impatiens	R/YT	100	=	Klock and Fitzpatrick, 1997
'Accent Red' impatiens	MSW	100	–	Klock and Fitzpatrick, 1997

[z] MSW = municipal solid waste compost; B = biosolids compost; B/YT = biosolids/yard trimmings co-compost; R/YT = refuse-derived fuel residuals/biosolids/yard trimmings co-compost.
[y] +, –, = represent positive, negative, or neutral, respectively (usually relative to a control).

and chemical parameters or because the manufacturer was no longer in the composting business.

There are relatively few published studies that illustrate how changes in feedstock composition or process parameters can influence efficacy of the compost product. Some studies (Fitzpatrick, 1986; Fitzpatrick, 1989; Fitzpatrick and Carter, 1983; Fitzpatrick et al., 1993; Klock and Fitzpatrick, 1997) have provided insight on how compost product quality may be influenced by the feedstocks from which the compost is made, the ways these materials are processed prior to composting, the amount of time these materials are allowed to compost, and the ways these materials are processed after composting. These and other studies clearly showed that such factors can cause major changes (such as pH and soluble salt elevation, introduction of phytotoxic materials, and other perturbations) in the compost product's ability to provide a suitable rooting environment for nursery crops.

The U.S. Composting Council, recognizing the need for greater standardization of testing and characterization procedures for composting and compost products, has developed the publication *Test Methods for the Examination of Composting and Compost* (Leege and Thompson, 1997). This document, although currently in draft form, is available for purchase from its publisher. When completed, it would allow compost users and their advisors more specific analytical tools to compare compost products manufactured by different companies and different types of compost products made by the same organization. As more information elucidates the impacts of specific processes undertaken prior to, during, and after the composting period, growers will be able to make more precise and reliable decisions on how compost products can be most effectively used to increase crop productivity.

REFERENCES

Chaney, R.L., J.B. Munns, and H.M. Cathey. 1980. Effectiveness of digested sewage sludge compost in supplying nutrients for soilless potting media. *Journal of the American Society for Horticultural Science* 105(4):485–492.

Conover, C.A. and J.N. Joiner. 1966. Garbage compost as a potential soil component in production of *Chrysanthemum morifolium* 'Yellow Delaware' and 'Oregon.' *Proceedings of the Florida State Horticultural Society* 79:424–426.

Davidson, H., R. Mecklenberg, and C. Peterson. 1999. *Nursery Management, Administration and Culture*, 4th edition. Prentice–Hall, Upper Saddle River, New Jersey.

DeGroot, R. 1956. L'utilisation du compost de ville en horticulture. *Revue de l'agriculture (Bruxelles)* 9:165–171.

DeGroot, R. 1961. L'utilisation du compost urbain dans la culture des plantes ornamentales de la region Gantoise. *Revue de l'agriculture (Bruxelles)* 14:517–523.

Fitzpatrick, G. 1981. Evaluation of potting mixes derived from urban waste products. *Proceedings of the Florida State Horticultural Society* 94:95–97.

Fitzpatrick, G. 1985. Container production of tropical trees using sewage effluent, incinerator ash and sludge compost. *Journal of Environmental Horticulture* 3:123–125.

Fitzpatrick, G. E. 1986. Sludge processing effects on compost quality. *BioCycle* 27(9):32–35.

Fitzpatrick, G.E. 1989. Solid waste composts as growing media. *BioCycle* 30(9):62–64.

Fitzpatrick, G. and N.S. Carter. 1983. Assessment of sewage sludge compost mixtures as container growing media. *Proceedings of the Florida State Horticultural Society* 96:257–259.

Fitzpatrick, G.E. and D.B. McConnell. 1998. Compost uses for the landscape and nursery industries, p. 43–45. In: D. Tonnessen (ed.). *Compost Use in Florida*. Florida Center for Solid and Hazardous Waste Management, Florida Department of Environmental Protection, Tallahassee.

Fitzpatrick, G.E. and S.D. Verkade. 1991. Substrate influence on compost efficacy as a nursery growing medium. *Proceedings of the Florida State Horticultural Society* 104:308–310.

Fitzpatrick, G.E., P.A. Davis, and M.L. Lamberts. 1993. Effects of processing technologies on efficacy of sludge-yard trash cocompost. *Compost Science and Utilization* 1(2):73–78.

Fitzpatrick, G.E., E.R. Duke, and K.A. Klock-Moore. 1998. Use of compost products for ornamental crop production: research and grower experiences. *HortScience* 33(6):941–944.

Gogue, C.J. and K.C. Sanderson. 1975. Municipal compost as a medium amendment for chrysanthemum culture. *Journal of the American Society for Horticultural Science* 100:213–216.

Gouin, F.R. 1977. Conifer tree seedling response to nursery soil amended with composted sewage sludge. *HortScience* 12:341–342.

Gouin, F.R. and J.M. Walker. 1977. Deciduous tree seedling response to nursery soil amended with composted sewage sludge. *HortScience* 12:45–47.

Hemphill, D.D., Jr., R.L. Ticknor, and D.J. Flower. 1984. Growth response of annual transplants and physical and chemical properties of growing media as influenced by composted sewage sludge amended with organic and inorganic materials. *Journal of Environmental Horticulture* 2(4):112–116.

Higginbotham, J.S. 1990. Four centuries of planting and progress: a history of the U.S. nursery industry. *American Nurseryman* 171(12):36–59.

Howard, A. and Y.D. Wad. 1931. *The Waste Products of Agriculture, Their Utilization as Humus*. Oxford University Press, London, United Kingdom.

Hue, N.V. and B.A. Sobieszczyk. 1999. Nutritional value of some biowastes and soil amendments. *Compost Science and Utilization* 7(1):34–41.

Jimenez, E.I. and V.P. Garcia. 1989. Evaluation of city refuse compost maturity:a review. *Biological Wastes* 27:115–142.

Kellogg, C. 1985. Marketing to targeted users. *BioCycle* 26(5):44.

Klock, K.A. and G.E. Fitzpatrick. 1997. Growth of *Impatiens* 'Accent Red' in three compost products. *Compost Science and Utilization* 5(4):26–30.

Leege, P.B. and W.H. Thompson (ed.). 1997. *Test Methods for the Examination of Composting and Compost*, 1st edition. The United States Composting Council, Bethesda, Maryland.

Lumis, G.P. and A.G. Johnson. 1982. Boron toxicity and growth suppression of *Forsythia* and *Thuja* grown in mixes amended with municipal waste compost. *HortScience* 17:821–822.

Obeng, L.A. and F.W. Wright. 1987. *The Cocomposting of Domestic Solid and Human Wastes*. World Bank Technical Report No. 57, The World Bank, Washington, D.C.

Poole, R.T. 1969. Rooting response of four ornamental species propagated in various media. *Proceedings of the Florida State Horticultural Society* 82:393–397.

Poole, R.T., C.A. Conover, and J.N. Joiner. 1981. Soils and potting mixtures, p. 179–202. In: J.N. Joiner (ed.). *Foliage Plant Production*. Prentice–Hall, Englewood Cliffs, New Jersey.

Raymond, D.A., C. Chong, and R.D. Voroney. 1998. Response of four container grown woody ornamentals to immature composted media derived from waxed corrugated cardboard. *Compost Science and Utilization* 6(2):67–74.

Rodale, J.I., R. Rodale, J. Olds, M.C. Goldman, M. Franz, and J. Minnich. 1960. *The Complete Book of Composting*. Rodale Books, Emmaus, Pennsylvania.

Roe, N.E. 1998. Compost utilization for vegetable and fruit crops. *HortScience* 33(6):934–937.

Sanderson, K.C. 1980. Use of a sewage-refuse compost in the production of ornamental plants. *HortScience* 15:173–178.

Sanderson, K.C. and W.C. Martin, Jr. 1974. Performance of woody ornamentals in municipal compost medium under nine fertilizer regimes. *HortScience* 9(3):242–243.

Shanks, J.B. and F.R. Gouin. 1984. Compost suitability for greenhouse ornamental plants. *BioCycle* 25(1):42–45.

Shiralipour, A., D.B. McConnell, and W.H. Smith. 1992. Uses and benefits of municipal solid waste composts: a review and assessment. *Biomass and Bioenergy* 3(3–4):267–279.

Ticknor, R.L., D.D. Hemphill Jr., and D.J. Flower. 1985. Growth response of *Photinia* and *Thuja* and nutrient concentrations in tissues and potting medium as influenced by composted sewage sludge, peat, bark and sawdust in potting media. *Journal of Environmental Horticulture* 3(4):176–180.

Tyler, R.W. 1996. *Winning the Organics Game*. American Society for Horticultural Science Press, Alexandria, Virginia.

Wootton, R.D., F.R. Gouin, and F.C. Stark. 1981. Composted, digested sludge as a medium for growing flowering annuals. *Journal of the American Society for Horticultural Science* 106:46–49.

CHAPTER 7

Compost Utilization in Landscapes

Ron Alexander

CONTENTS

1-56670-460-X/01/$0.00+$.50

I. INTRODUCTION

Perhaps the greatest area of compost utilization in the U.S. is in landscape applications. Although compost is used in large quantities by many other market segments, most compost marketers would agree that landscapers are the largest purchasers of compost in the U.S. Not only is professional landscape usage significant, but according to *Organic Gardening* magazine's "Gardening in America II" survey (Organic Gardening, 1995), 27 million Americans use compost for gardening and landscape activities. No doubt, the trend of compost usage in landscapes will continue to expand because of popularity in homeowner gardening and professional landscaping applications.

II. WHY DO LANDSCAPERS USE COMPOST?

The primary reason compost is used in landscape applications is because it works, and it works economically. Compost is the only amendment that can effectively improve soil characteristics physically, chemically, and biologically (Alexander, 1996). Unlike other organic soil amendments, compost has the ability to positively affect soil quality in a variety of ways. As the public's understanding of factors affecting soil quality continues to grow, and with the knowledge that most of the soils used in landscape situations are less than ideal, more emphasis will be paid to soil improvement and compost usage will continue to expand. Although many end users will equate the benefit of compost use to lush green plant growth, caused by plant-available nitrogen (N), the most important benefits of using compost are long term and related to its rich content of organic matter (Alexander, 1995). Compost is also used because its unique attributes make it an extremely versatile product, beneficial in many landscape applications such as a soil amendment, media component, mulch, and turf topdressing. In most instances, compost is also less expensive than high-quality peat products and competes favorably against decomposed bark products. Composts are also considered renewable resources, which can be produced without adverse effect on the environment. Today, many end users favor the use of a more environmentally sustainable product, which is one reason why certain landscapers will not use peat moss or topsoil that has been harvested from farm fields. Actually, compost production and utilization has been shown to be a great asset towards a cleaner environment (Composting Council Research & Educational Foundation, 1997). Compost's ability to bind heavy metals, making them less bioavailable, and to degrade certain organic pollutants, such as petroleum and pesticide residues, makes it quite beneficial to the environment (Composting Council Research & Education Foundation, 1997). Also, like the bark mulch industry, the compost industry utilizes an organic feedstock, once considered worthless (actually a liability) and manufactures it into a high-quality product.

Another reason why compost has become so popular in landscape applications is because of its ease of usage. When using a high-quality compost product, often the need for immediate supplemental fertilization is eliminated, or at least minimized. The same can be said with pH adjustment in many areas of the country. Of course, the addition of other plant and soil supplements will be based on soil and compost characteristics and on plant requirements. For this reason, the purchase of a consistently high-quality compost product is required. A description of a high-quality compost is given in Table 7.1. After a year or more of operation, compost producers should be able to accurately estimate the quantitative value of important product characteristics, such as pH, soluble salt content, nutrient content, etc. and provide them to their customers even though seasonal variations may occur. Therefore, tracking data related to compost pH, soluble salt content, nutrient content, particle size, stability/maturity, pasteurization (for weed seed and plant and human pathogen destruction) and perhaps, water-holding capacity are of primary importance for plant growth and management of the growing system. Data related to product bulk density, moisture, organic matter, and inert (foreign matter) content are important to other customers. Data related to the amount of trace elements, heavy metals, pesticide residues, and polychlorinated biphenyls (PCBs) in the compost will also be important to track, depending upon the feedstock of the compost product, the product's intended use, and state and federal regulations.

III. WHY DON'T LANDSCAPERS USE COMPOST?

The most prevalent reasons why landscapers don't use compost are because they have a misunderstanding of what compost actually is; compost is not readily available to them in a convenient form (could be bulk or bagged, depending on the customer's requirements); only poor-quality compost is available to them; they had an unsatisfactory experience with compost in the past; or because they simply do not believe in using any soil amendments. Some landscapers also claim to be satisfied with the products they are currently using and thus are reluctant to change. A small percentage still possess stigmas against compost produced out of specific feedstocks (e.g., biosolids, municipal solid waste [MSW]), although very few will argue that the products do not work. In almost all cases where landscapers are not currently using compost, they would use it under specific circumstances if given the proper education and guidance. Although landscapers may require proof that compost works for them, educational efforts sponsored by national trade associations and industry trade journals have illustrated many successful uses of compost in landscape applications. Once high-quality compost becomes more readily available, landscape use is likely to increase.

Table 7.1 Typical Characteristics of Municipal Feedstock-Based Composts[z]

Parameter	Typical Range	Preferred Range for Various Applications Under Average Field Conditions
pH	5.0–8.5	6.0–7.5
Soluble salts (dS·m^{-1})	1–10	5 or below
Nutrient content (%) (dry weight basis)	N 0.5–2.5	N 1 or above
	P 0.2–2.0	P 1 or above
	K 0.3–1.5	
Water holding capacity (%) (dry weight basis)	75–200	100 or above
Bulk density (kg·m^{-3})	415–712	475–593
Bulk density (lbs per yd^3)	700–1200	800–1000
Moisture content (%)	30–60	40–50
Organic matter content (%)	30–70	50–60
Particle size	—	Pass through 1.3 cm (1/2 inch) screen or smaller
Trace elements/heavy metals	—	Meet U.S. EPA Part 503 Regulations
Growth screening	—	Must pass seed germination, plant growth assays
Stability	—	Stable to highly stable

[z] Municipal feedstock-based composts are primarily derived from yard trimmings, biosolids, municipal solid waste, or food byproducts, or a combination of one or more of these feedstocks.

Adapted from Alexander, 1996.

IV. USING COMPOST AS A PLANTING MEDIUM AMENDMENT IN THE LANDSCAPE

Success in establishing landscapes is dependent upon knowing the soil that exists on site, the species of plants to be planted, and the organic amendment to use. All three of these elements are of equal importance (Gouin, 1997). Compost and other organic amendments have a wide variation in the percent organic matter, pH, nutrient content, soluble salts, etc. It is essential to know the biological, chemical, and physical characteristics of the organic amendments available. This allows selection of the amendment that best improves the soil and meets plant requirements (Gouin, 1997).

A. Soil Structure Considerations

Compost can greatly enhance the physical structure of soil. In fine-textured (clay, clay loam) soils, the addition of compost will reduce soil bulk density, improve friability (workability and porosity), and increase its gas and water permeability, thus reducing erosion (Boyle et al., 1989). When used in sufficient quantities, the addition of compost has both an immediate and long-term positive impact on soil structure. Compost resists compaction in fine-textured soils and increases the water-holding capacity and improves soil aggregation in coarse-textured (sandy) soils (Boyle et al., 1989). The soil-binding properties of compost are due to its humus content. The constituents of the humus act as a soil "glue," holding soil particles together, making them more resistant to erosion and improving the soil's ability to hold moisture (Alexander, 1996). These soil building properties are particularly important to landscapers since many landscapes fail because of the poor management of the plants in structurally deficient soils. Many landscapers now plant trees and shrubs with their root balls only partially buried, because the plants would literally drown if the root ball was totally buried in the fine-textured soils. Conversely, many landscapes fail because of drought conditions and lack of adequate watering.

Compost incorporation significantly prolonged the period between irrigation and the occurrence of turfgrass wilting in Florida research (Cisar and Snyder, 1995). Therefore, the addition of compost may provide for greater drought resistance and more efficient water utilization, thereby reducing the frequency and intensity of irrigation required. Improved water retention from the addition of compost to sandy soils as well as improved moisture dispersion under plastic-mulched beds also has been reported (Obreza, 1995). The compost-amended soil allowed water to more readily move laterally from its point of application (microirrigation tubing).

Although it is difficult to use an excessive amount of compost on sandy soils (as long as soluble salts are not excessive), the excessive use of compost on clay soils may be problematic. When incorporating compost at a 20% inclusion rate or higher, some clay soils may hold excess moisture. This can make the soil slow to drain and difficult to work even when the soil is slightly wet. In turf and other permanent planting areas, this would not be a concern. However, it should be considered in areas where on-going mechanical cultivation is practiced (e.g., annual flower beds) (Gouin, 1997).

B. Modification of pH

The addition of compost to soil may modify the pH of the blended soil. Depending on the pH of the compost and of the native soil, compost addition may raise or lower the soil/compost blend's pH. Therefore, the addition of a neutral or slightly alkaline compost to an acidic soil will increase soil pH if added in appropriate quantities. In specific conditions, compost has been found to affect soil pH even when applied at quantities as low as 22.4 to 44.8 $Mg{\cdot}ha^{-1}$ (10 to 20 tons per acre) (Hortenstine and Rothwell, 1973). The incorporation of compost also has the ability to buffer or stabilize soil pH. This property will allow some landscapers to avoid the initial addition of pH adjustment agents where compost is utilized, as well as potentially reduce the on-going addition of these supplements (Alexander, 1996). Because compost may have an effect on the pH of soil within the treated area, the soil pH should be assessed before any amendments (lime or sulfur) are applied. Ideally, a soil test should be conducted first, in order to verify the pH requirements of the soil itself. Then, knowing the soil requirements and the characteristics of the compost, the appropriate pH adjusting agents can be applied.

Aside from turf and ornamental grasses, most ornamental plants perform best at a pH of 7.0 or below. Although the addition of large amounts of organic amendments to soils allows one to grow plants over a wider range of pH's, in time the roots of plants will extend far beyond the initial planting area and the amount of organic matter in the soil will decrease. Additional pH adjustment recommendations can be found in Table 7.2.

Table 7.2 pH Adjustments Based on Existing Soil Conditions and Plants to be Established

Soil pH is Less than 5.0 & Establishing Non-Acid Loving Plants

If existing soil pH is below 5.0 and the soil has less than 6% organic matter and only plants that grow best in mildly acid soils are to be planted, add limestone in addition to compost unless compost made from lime dewatered biosolids is available. If limed compost is available, there is generally sufficient lime in the compost to adjust the pH to the desired level.

Soil pH is Less than 5.0 & Establishing Acid Loving Plants

If existing soil pH is below 5.0 and the soil has less than 6% organic matter, select a compost that has a pH at or below neutral (pH 7.0) and does not contain any liming agents (e.g., limestone, hydrated lime, ash, etc.). Although the compost will raise the pH of the soil to above the desired range, the increased organic matter content will compensate for the difference.

Soil pH is Greater than 5.0 & Establishing Non-Acid Loving Plants

If existing soil pH is above 5.0 and non-acid loving plants are being grown, compost containing liming agents should not be used except in areas where turf and ornamental grasses are to be established. Only compost without liming agents should be used for amending soils in ornamental plantings of ericaceous crops and plants that prefer mildly acid soils. Ornamental grasses and turf species are more tolerant to high pHs than are most broadleaf species.

Soil pH is Greater than 5.0 & Establishing Acid Loving Plants

Most acid loving plants perform best when planted in soils having an abundant supply of organic matter. However, despite the pH buffering capacity of organic matter, it is important to maintain a pH as close to ideal as possible. Under such soil pH conditions, it is often better to use peat moss or pine fines, and not compost as a soil amendment, and supply nutrients using chemical fertilizers. Using a 1:1 blend (v/v) of peat mosses or pine fines and compost (unlimed) can also be beneficial. Since most kinds of peat moss (Canadian, Sphagnum) have a pH near 3.5, there is often sufficient acidity in the peat moss to neutralize the higher pH of the compost.

Adapted from Gouin, 1997.

C. Fertility Effects

Composts are a source of plant nutrients and also have a profound effect on availability of plant nutrients. The addition of compost can also add soluble salts.

1. Improved Cation Exchange Capacity

Amending soils with compost will increase their cation exchange capacity (CEC) (Hortenstine and Rothwell, 1973), enabling them to more effectively retain nutrients.

Amending soils with compost will also allow crops to more effectively utilize nutrients, while reducing nutrient loss by leaching (Brady, 1974). Thus, the fertility of soils is often tied to their organic matter content. Improving the CEC of sandy soils by adding compost can greatly improve the retention of plant nutrients in the root zone (Alexander, 1996). This could allow landscapers to reduce fertilizer application rates, and lessen concerns about nutrient leaching (e.g., N and phosphorus [P]).

2. Source of Plant Nutrients

Compost products contain a considerable variety of macro- and micronutrients. Although often seen as a good source of N, P, and potassium (K), compost also contains sulfur (S), calcium (Ca), and magnesium (Mg), as well as micronutrients essential for plant growth. Because compost contains relatively stable sources of organic matter, these nutrients are supplied in a slow-release form. Compost is usually applied at much higher rates than inorganic fertilizer; thus it can have a significant cumulative effect on nutrient loading and availability. The addition of compost can affect both fertilizer and pH adjustment (lime/sulfur) addition (Alexander, 1996). Initial plant nutrient requirements can sometimes be satisfied when compost is used at the recommended rate. When additional fertilization is required, rates should be adjusted to account for elements and salts provided by the compost.

Supplemental fertilization will be necessary on an on-going basis. The nutrient requirements of the plant species, the type and quantity of fertilizer used, the nutrient content of the compost, and the availability of those nutrients in the soil will affect rates and frequency of supplemental fertilization. Typically, fertilization will not be necessary during the first 6 to 12 months following crop establishment. Composts containing relatively low nutrient levels may, however, need supplemental fertilization in the short term. Using specific types of compost may reduce fertilizer requirements for several years, depending upon climatic conditions.

Compost made from biosolids often has a higher N and P concentration than compost made from animal manures and yard trimmings. Composts made from animal manures and yard trimmings generally contain elevated levels of K and lower levels of P. Information on nutrient contents of compost can provide guidance in compost selection and reduce chances of creating nutrition related concerns in the future. Although not a typical occurrence, compost that contains extremely high levels of Ca has the potential of binding P and essential trace elements in both the compost and soil, thus preventing their uptake by plants (Gouin, 1997).

The overall best compost to use can also be further determined through soil test results. If soils are low in P, using a compost made from biosolids can reduce or eliminate the need to add commercial phosphate fertilizers. If the soils are deficient in K but rich in P, then using a compost from yard trimmings and/or animal manures in place of biosolids is preferred. For amending soils possessing high levels of Ca, one should avoid using a compost that contains additional liming agents (Gouin, 1997).

Typically, the practice of incorporating fertilizer into the planting bed before planting may be eliminated when stable composts are used at appropriate rates. However, yard debris and MSW composts are more likely to require supplemental fertilization, whereas stable biosolids are not. If unstable compost is used, stunted plant growth and other symptoms of N deprivation may be observed. If so, fertilization (primarily N) will need to be applied soon after plant establishment.

3. Addition of Soluble Salts

Most commercial composts contain a significant amount of nutrients in the form of fertilizer salts. These fertilizer salts are also referred to as soluble salts. Since excessive amounts of soluble salts can stunt or kill plants, caution should be taken when using compost in the culture of salt sensitive plant species. For composts that contain higher levels of soluble salts (over 5 $dS \cdot m^{-1}$), one should not exceed a 20% inclusion rate in a soil mix where salt sensitive species are to be established. Greater amounts of compost can be used with composts containing low to moderate levels of soluble salts. Although salt-related injury is not common, thorough watering at the time of planting will significantly reduce potential risk (Gouin, 1997). Repeat applications of compost in the same planting bed may also increase soluble salt levels which may be damaging to more sensitive crops. Compost should be applied every other year in planting beds, or at half the rate at which it was applied the previous year, unless salt levels are being monitored or relatively salt-tolerant crops are being grown (Alexander, 1995).

D. Improved Soil Biology/Microbiology

The activity of soil organisms, essential in productive soils, is largely based on the presence of organic matter. Microorganisms play an important role in organic matter decomposition, which in turn leads to humus formation and nutrient availability. Microorganisms can also promote root activity as specific fungi work symbiotically with plant roots, assisting them in the extraction of nutrients from soils. Sufficient levels of organic matter also encourages the growth of earthworms, which through tunneling, increase water infiltration and aeration (Alexander, 1996). Landscapers are now starting to understand the critical role that soil organisms play in the health and success of their landscapes.

E. Reduced Incidence of Soil-Borne Diseases

Incidence of soil-borne diseases on many plants may be influenced by the level and type of organic matter and microorganisms present in soils. An increased population of certain microorganisms may suppress specific plant pathogens such as *Pythium* and *Fusarium*, as well as nematodes (Nelson, 1992). Because many plant species are susceptible to soil-borne diseases, the benefit of compost usage, especially in the period following planting, can be paramount as far as plant survival is concerned.

F. Comparing Compost to Other Planting Media and Soil Amendments

Comparing compost to other planting media and soil amendments is not an easy task due to the variability of different compost products and the need to compare the effectiveness of these products in varying applications. Within this section is a discussion of various horticultural products that are used in conjunction with, or instead of, compost. A comparison of the physical and chemical characteristics of a typical compost to other planting media and soil amendments can be found in Table 7.3.

Table 7.3 Comparison of Compost to Other Planting Media and Soil Amendments

	Compost[z]	Organic Soil[y]	Native Peat[x]	Canadian Peat[w]
Organic matter (%)	46	12	74	97
pH	7.4	7.5	5.2	4.2
Soluble salts ($dS \cdot m^{-1}$)	2.23	0.64	0.31	0.07
Bulk density ($kg \cdot m^{-3}$)	515	1125	228	112
Bulk density (lbs per ft^3)	32.2	70.2	14.3	7.0
Moisture-holding capacity (%)	227	53	428	1307
Cation exchange capacity (meq per 100g)	17.3	13.6	4.0	3.1

[z] Represents a biosolids/yard trimmings compost.
[y] Represents an organic Florida muck soil.
[x] Represents a Florida reed sedge peat.
[w] Represents a Canadian sphagnum peat moss.
Adapted from Alexander, 1996.

Peat moss is derived from *Sphagnum* that grows in bogs and becomes covered with water when it dies. Because of the cold, wet climate in which *Sphagnum* grows, peat moss accumulates to great depths, undergoing partial anaerobic decomposition. Over the years, peat moss changes both physically and chemically due to harvesting methods and its location in the bog. Coarse chunky peat with a pH above 5.0 is seldom available. Peat moss which is marketed today usually is a finer material that has a pH between 3.3 to 3.5. This finer peat moss shrinks rapidly and requires two, and sometimes three, times more limestone to neutralize its acid concentration than with peat harvested in previous years (Gouin, 1989). Although peat moss initially starts with a high CEC, it decreases with time, thus reducing its ability to hold nutrients as the aging process continues.

Sedge peat or *native peat* generally consists mainly of sedges and grasses that grow in bogs. When these grasses and sedges die, their tops sink into the water and undergo partial anaerobic decomposition. Since these plants are high in cellulose and contain little lignin, they decompose more rapidly than peat moss and contain few fibers (Gouin, 1989). Although sedge peat and native peat can be used as a substitute for peat moss, they are generally not as satisfactory in certain nursery applications. Also, they are highly variable from bog to bog and can be equally as acidic as peat moss. The CEC of sedge peat and native peat is similar to that of peat moss.

Softwood bark has become a major source of organic matter for the ornamental horticultural industry. Products such as pine (*Pinus*), fir (*Abies*), hemlock (*Tsuga*), redwood (*Sequoia*), and cypress (*Cupressus*) barks are used throughout specific regions of North America. Because softwood barks are low in cellulose and high in lignins, they can be used either fresh or composted and do not decompose rapidly. Cypress and redwood sawdusts are also low in cellulose and can be used in much the same way. However, only coniferous barks with less than 10% cellulose can be used fresh. Coniferous bark with 10% or more cellulose must be composted. For optimum growth, when used as a soil amendment or growing media component, the bark products should be milled to particle sizes no larger than about 1.5 cm (0.5 in.) diameter (Gouin, 1989). Unlike peat moss, sedge peat, or native peat, the CEC of bark improves with age. However, not all barks are the same and their availability is diminishing in certain regions. The landscaping industry also uses coarsely ground coniferous bark products as decorative mulches.

Hardwood bark, sawdust, shavings, or wood chips should never be used as a soil amendment unless they have been thoroughly composted. These materials are high in cellulose and low in lignins; therefore, they shrink rapidly and will rob plants of N. The competition for N may not be effectively offset by supplying additional N in a fertilizer program. The use of these materials in field applications should be limited to areas where planting will not occur for several months. Using a fine-textured and well-aged, or composted, hardwood bark will minimize negative effects.

Topsoil is defined as "the surface or upper part of the soil profile." Landscapers who use topsoil often define it as a naturally produced medium consisting of sand, silt, clay, organic matter, trace amounts of nutrients, and other inerts capable of supporting plant growth. However, in many parts of the U.S., many of the soils purchased as topsoil and used for horticultural applications are not true topsoils but rather are mineral subsoils obtained from below the true topsoil layer. These subsoils are often devoid of organic matter and essential plant nutrients and do not possess the physical structure required for optimum plant growth. These materials are typically processed (screened or shredded) to remove debris before marketing. In some areas, sand and muck-type materials are sold as topsoils. Neither of these materials possesses properties essential for optimum plant growth.

V. COMPOST UTILIZATION IN VARIOUS LANDSCAPE SITUATIONS

The versatility of high-quality compost products allows them to be utilized in a variety of landscape applications. However, to use them effectively, it is essential to match a compost that possesses particular characteristics to its best specific application. One compost product is not the best product for all landscape applications. Once the appropriate compost product is selected, it is necessary to understand how best to apply it, as well as its effects on the overall growing system.

A. General Guidelines for Compost Application

The use of compost can influence the short- and long-term characteristics of the soil or medium in which you are planting. As discussed earlier, compost can modify the physical, chemical, and biological characteristics of a growing medium. Thus, it is important to understand that typical cultural practices may need to be modified where compost is used.

1. Compost Descriptions

Garden beds, landscape planters and turf establishment — Compost used in these applications should be stable to highly stable and must pass growth screening tests so as not to cause a depletion of available N or seedling/plant injury. Composts containing various amounts of organic matter may be used and the product's moisture content should be between 35 and 55% in order to improve handling. Compost with a pH of 5.5 to 8.0 may be used; however, it is important that the pH of the amended soil meets the pH requirements of the plant species to be established. Composts produced from lime dewatered biosolids should not be used in garden beds or planters where acid-loving species are to be established. Compost may favorably affect the soil's pH eliminating the need to lime the soil prior to plant establishment.

The soluble salt content of the compost may vary as long as the concentration within the amended soil is below the soluble salt tolerance level of the plant species to be established. The soluble salt content of the amended soil should not exceed 2.5 $dS \cdot m^{-1}$ (based on a saturated paste extract method) where ornamental transplants or seedling plants are to be established, and approximately half that level (1.25 $dS \cdot m^{-1}$) where seeds are to be planted. Young seedlings may be more sensitive to soluble salts than transplants (Alexander, 1995). A good rule of thumb is if the compost has a soluble salt content above 5.0 $dS \cdot m^{-1}$, then no greater than a 20% inclusion rate of compost should be used (Gouin, 1997). The soluble salt concentration of the amended soil should not exceed 4.0 $dS \cdot m^{-1}$ for turf. Turf established by seed may be more sensitive to soluble salts than turf established by sod or sprigs.

To help determine initial and on-going fertilizer requirements, the content and availability of the nutrients contained in the compost needs to be identified. The quantity of N in the compost and the form in which it exists is of particular importance. Only compost with low ammonium nitrogen (NH_4-N) levels should be used unless the amended soil is aged before planting or very low quantities of compost are used. High NH_4-N levels can inhibit seed germination and cause the death of young seedlings in most crops. Typically, the practice of incorporating fertilizer into the seedbed before turf establishment can be eliminated when compost is used at appropriate rates. Performing a soil test on the upgraded soil will aid in determining subsequent fertilizer application rates (Alexander, 1995).

Composts of various particle size may be used. However, compost screened through a 1.3 cm (0.5 in.) screen or smaller is preferred. Compost particle size will also be dependent upon the compost feedstock and texture classification of the soil being amended. Coarser compost particles (wood based) located on the soil surface can impede seed-to-soil contact, reducing grass seed germination rate. Compost should also meet all applicable state and federal health and safety standards. Sharp objects and fragments (i.e., glass and metal) should be avoided, man-made inerts should be minimized, and the product should be free of weeds (Alexander, 1995).

Mulching and erosion control — The type of compost used as a decorative and/or functional mulch may vary widely since mulches have numerous functions: agronomic (e.g. reduction of soil temperature fluctuation, reduction of moisture stress), grounds management (e.g., weed control), and aesthetic. For some, the aesthetic characteristics of the mulch are the most significant factors; those factors are the most subjective. An aesthetically acceptable mulch is consistent in appearance, contains no foreign matter, and has minimal odors. In some parts of the U.S., dark mulches are preferred, while light colored (cypress) or red mulches are preferred elsewhere.

The particle size of the product may vary since some end users are accustomed to fine mulches, and others are accustomed to coarse, woody mulches. Coarser mulches may be more effective in reducing weed growth, preventing wind erosion, and allowing the infiltration of surface-applied water (Stewart and Pacific, 1993). The soluble salt content of the compost mulch may also vary; however, soluble salts leached from the mulch can cause detrimental effects to more salt sensitive crops, such as geraniums (*Pelargonium* sp.). The stability of the compost used may vary widely, as everything from fine, well-composted materials to uncomposted wood chips are used as mulch. Composts containing high salt levels, high NH_4 levels, or other phytotoxins should not be used if crops are to be established on the mulched site soon after application (Alexander, 1995).

To improve the weed control capabilities of the compost mulch, various plastic products, landscape fabric, newspaper, or even herbicides may be applied before the mulch is applied. Because a stable compost mulch will readily supply mineral elements, many plants will not require additional nutrition during the first 12 months following mulch application. This is particularly true with composts rich in N. Additional mineral elements, particularly N, may be necessary where less stable or carbonaceous compost products are used (Alexander, 1995).

Compost used for erosion control should be slightly coarse to coarse. The particle size may vary widely. If the compost contains some particles 1.3 cm (1/2 in.) in size or greater, these coarser particles will overlap on the soil surface, creating a stable mat which resists water and wind erosion. A maximum compost particle size has not been established; however, very coarse products may be more difficult to apply and problematic if the slope is to be landscaped or seeded. Fine compost may tend to be more affected by wind erosion, but may be able to more effectively absorb moisture. Caution should be employed when using nutrient-rich compost products in or adjacent to environmentally sensitive areas or where water quality concerns exist. Only yard debris compost, compost made from uncontaminated wood waste-based materials, or well-stabilized biosolids compost are recommended at this time.

If biosolids composts are used, they should contain N in a primarily water-insoluble form, low in nitrates (Alexander, 1995).

Planting backfill mix amendment — Stable composts with relatively high plant nutrient levels are desired, as such composts may eliminate the need to add fertilizer during planting. If fertilizers are to be applied at planting, only slow-release forms are recommended. Where salt-sensitive ornamental crops are planted, up to half the recommended fertilizer rate should be applied. The soluble salt content (concentration) of compost used in backfill mixes may vary. However, the preferred soluble salt content of the finished backfill mix should be 3 $dS{\cdot}m^{-1}$ or less. Maximum salt tolerance levels are soil and species dependent. Soluble salts should not be a significant problem with most woody ornamentals. However, care should be given when bare root, ericaceous, and other salt-sensitive crops are planted. The quality of the compost used in this application depends upon the specific crop being planted, the age and sensitivity of the plant, and the type of native soil in which it is blended. In this application, however, the higher quality the compost, the better (Alexander, 1995).

Soil blend component — Compost used for blending must be moderately to highly stable and must pass growth screening tests. If the blended topsoil is to be aged before end use, a moderately stable compost may be used. The use of coarser composts may necessitate the need to screen the finished topsoil blend. The soluble salt content of the compost used in the soil blend may vary; however, the soluble salt content of the finished blend should not surpass the maximum salt tolerance levels of the crop(s) to be planted in the soil blend. Although not typically a concern, composts which contain large quantities of very fine particles (passing through 1 mm sieve) may reduce the porosity of the soil blend (Alexander, 1995).

More specific data are needed that compare the effects of particular compost products on landscape plants. Although greenhouse and nursery research, as well as field experience, allows for general recommendations to be made, additional research is necessary in this area of study.

2. Compost Application Methods

The method by which compost is applied is typically based on the compost's characteristics, how the compost is being used, the size of the project, and field conditions. For small planting or mulching projects, compost may be obtained in bags and spread by hand using a rake. For larger projects, compost may be obtained in bulk, transported to the site in a dump truck or wheelbarrow and spread manually, or by using tractor-drawn equipment. Where slopes are being mulched for decorative purposes or erosion control, or compost is being applied to sites that are difficult to access, blower type units have been used to propel the compost up to 60 m (200 ft). Pneumatic blower trucks even exist which can apply product through a wide hose (over 90 m [300 ft] in length) and apply material around plants and other objects. A more common method to apply compost for various uses is with a manure spreader or topdressing unit (Figure 7.1). A manure spreader uses rotating flails (paddles) to project the compost into the air, whereas a topdressing unit uses a rotating, cylindrical brush to project the compost down towards the soil surface. Both units may be

Figure 7.1 Application of a thin layer of compost using a topdressing (brush type) spreader.

calibrated to apply lower rates (0.6 to 1.3 cm; 1/4 to 1/2 in. layer) or higher rates (2.5 cm; 1 in. layer) of compost; however, the application of higher rates is slow and may require more than one pass over the site. Side discharge manure spreaders have been used to apply compost inside planting rows and tractor-trailers have been fitted with flails to allow large volumes to be spread. Equipment has even been developed to apply compost to a depth of 1.3 to 2.5 cm (1/2 to 1 in.) over a raised nursery bed. Often, when rates of 2.5 cm (1 in.) or more are applied, piles of compost are strategically placed throughout the site and a grading blade, York rake, or front-end loader/bulldozer blade is used to spread the compost. With experience and care, accurate application rates are achievable (Alexander, 1996).

Continued innovations in compost application equipment will increase compost usage. Thus, equipment is now available for purchase or rental to allow users to more efficiently apply compost. Users should remember that the moisture content and particle size of the compost will affect its spreadability. Standard "box spreaders" and agricultural or commercial fertilizer/lime spreaders often have difficulty spreading coarse or wet compost (Alexander, 1996).

A guide to estimating the volume of compost required to cover a specific area is given in Table 7.4.

B. Application Instructions for Specific Uses of Compost in the Landscape

One of the most useful attributes which high-quality compost products possess is their versatility. The ability to use compost products in a variety of end uses is

Table 7.4 Compost Use Estimators

Cubic Yards (yd^3) of Compost Required to Cover 1,000 ft^2			Cubic Meters (m^3) of Compost Required to Cover $100 m^2$		
1/4-in. layer	⇒	0.75 yd^3	0.7 cm layer	⇒	0.6 m^3
1/2-in. layer	⇒	1.5 yd^3	1.4 cm layer	⇒	1.2 m^3
1-in. layer	⇒	3.0 yd^3	2.7 cm layer	⇒	2.5 m^3
$1^1/_2$ in. layer	⇒	4.5 yd^3	4.1 cm layer	⇒	3.7 m^3
2-in. layer	⇒	6.0 yd^3	5.5 cm layer	⇒	4.9 m^3
$2^1/_2$ in. layer	⇒	7.5 yd^3	6.8 cm layer	⇒	6.2 m^3
3-in. layer	⇒	9.0 yd^3	8.2 cm layer	⇒	7.4 m^3

Adapted from Alexander, 1996.

one of the primary reasons why the use of compost in landscape applications, both professional and non-professional, has grown so dramatically over the past 10 years. However, to have success *in the field,* the proper compost products must be used and the product itself must be used correctly.

Following is a detailed discussion of six common areas of compost use in landscaping.

1. Garden Beds and Landscape Planters

Probably the most popular use for compost today is in garden bed establishment and renovation. In this application, the product's numerous attributes have allowed for glowing successes from coast to coast. Two important factors that contribute to the use of compost in planting beds are poor soil conditions on construction sites once grading is completed, and the necessity of landscapers to be successful in their planting endeavors, the first time. As a normal practice, builders will scrape soil from a construction site in order to bring it to the correct grade. They will either stockpile the soil, and reapply it later, or sell it to a topsoil dealer. Often, when it is reapplied, the actual topsoil is mixed with subsoil, reducing its quality, or the topsoil is buried under poor-quality soil when reapplied. In either case, the need to improve soils around residential and commercial structures exists. When a homeowner or landscaper invests in landscaping a site, they expect that immediate and positive results will occur, and persist. Therefore, using composts as a soil amendment to help ensure their success is seen as a good investment.

Application instructions — The compost application rate will vary depending upon soil conditions, compost characteristics, and plant species to be established. Compost has been successfully applied at a rate of approximately 1.7 cm (Maynard, 1998) to 7.5 cm (Beeson, 1995) (2/3-in. to a 3-in. layer), then incorporated to an approximate depth of 15 to 20 cm (6 to 8 in.), resulting in an inclusion rate of 10 to 50% by volume. Performing a soil test will assist in determining proper compost application rates. Typical application rates are between 2.5 to 5.0 cm (1- to 2-in.) layer, 20 to 30% by volume (Smith and Treaster, 1991). Lower inclusion rates may be necessary for salt-sensitive crops such as geraniums (*Pelargonium* sp.) or where composts with higher salt levels are used. Once the compost inclusion rate is chosen, a blend of soil and compost may be produced and tested prior to planting. This will identify the soil characteristics, including soluble salt and organic matter content, as well as identify the appropriate rate of fertilization and pH adjustment necessary for optimum plant growth. Compost should be broadcast uniformly and incorporated with a shovel or rototiller until the compost/soil mix is homogeneous. The treated area can be smoothed if necessary before planting. The amended area should then be irrigated, if necessary, to settle the soil, to provide moisture to the plant(s), and to help leach salts out of the root zone (Alexander, 1995).

If desired, materials used to adjust soil pH (e.g., lime or sulfur [S]) may be added to the soil prior to incorporation of the compost, as may any additional nutrients. However, where possible, it would be more beneficial to apply these materials after compost incorporation and soil testing. Once planting is completed, the planting area should be fertilized if necessary with a starter fertilizer and thoroughly watered.

Although compost is typically applied "as is" (unblended) and incorporated into planting beds, in several states, garden blend soils containing 20 to 40% compost are sold to establish or renovate garden beds. Soils modified for ornamental planting mixes should be designed to contain at least 5% organic matter. By using compost as the organic matter source, landscapers get added benefits, such as various micro- and macronutrients, a stabilized pH and a healthy supply of microbes. Often, these garden planter mixes contain a 25 to 33% compost inclusion rate. At these inclusion rates, many annual and perennial plants need no additional fertilization. The compost used in these landscape mixes must meet the requirements of the crops being established. For instance, ericaceous plants (such as those in the genus *Rhododendron*) and other acid-loving crops should not be planted using composts that contain appreciable amounts of lime. Crops which are salt sensitive, such as conifers, should not be planted with compost products which have a high soluble salt content. In general, composts used in planter mixes should possess the same basic characteristics of those described earlier for garden beds (Alexander, 1995).

Planting berms, which are used as borders in some landscapes and as landscape focal points in others, can be constructed by blending existing soils with compost at a 25 to 30% compost inclusion rate. In this type of mix, the soil is used for long-term stability of the raised berm. However, in rooftop planter mixes, sand or a sandy loam soil should be used for stability. Sandy soils can usually provide enough ballast for shrubs and small-size trees, but they are much lighter than clay and silt based soils. In rooftop mixes, the weight of this mix must be kept to a functional minimum. A standard rooftop mix for shrubs and ground covers should contain 30 to 40% sand or sandy soil, 30% compost, 10 to 20% pine fines, and 10% of a light weight aggregate (Alexander, 1999).

Where trees are to be planted in rooftop mixes, a good standard mix should contain 40% sandy loam soil, 20% sand, 30% compost, and 10% pine fines. Pine fines are used in these outdoor mixes because they provide excellent long-term CEC. Similar to the rooftop planting mixes, large planter mixes (outdoor containers) should contain 60% sand, 10% pine fines, and 30% compost (Alexander, 1999).

2. Mulching

Many types of products are used successfully as aesthetic and functional mulches. Usage is often based on customer preference, desired functionality, and regional trends. For this reason, common mulch products include decorative stone/rocks, wood chips, tire chips, ground yard debris, various types of tree bark, and compost. Although the use of compost as mulch is often met with some skepticism, it is being used with much success. Composts which contain coarser wood particles, preferably uniform in size, are typically desired.

Application instructions — Compost used as mulch is typically transported to the application site using a wheelbarrow and then applied around existing plant materials using a shovel or rake. The product can be smoothed using a rake or by hand. In large beds, the compost may be transported and positioned using a dump truck and then evenly applied using a rake, or may be transported to the site and applied using a pneumatic blower unit. In some instances, the mulch is applied and then planting holes are dug through the mulch layer and into existing soil. Once properly planted in the hole, the compost mulch should be distributed around the plant base. Compost should be applied at a depth of 2.5 to 7.5 cm (1 to 3 in.) beneath trees, shrubs, and other plant materials in garden beds. Avoid placing mulch against the tree trunk or main leader of the shrub, to prevent potential disease and insect damage. Biosolids composts used as mulch typically should not be applied at rates greater than 5.0 cm (2-in.) deep (Smith and Treaster, 1991), whereas many yard debris composts can be applied in up to a 7.5 cm (3-in.) layer (Ewing and Allen, 1994). Salt sensitive species may react negatively to application rates greater than 2.5 cm (1-in.) of certain composts. For individual trees and shrubs, the product should be applied at rates described earlier, from the tree's stem/trunk to its drip line, or further if desired (Alexander, 1995).

Apply the compost evenly in the garden bed or around the trees and shrubs, creating a solid mat of compost mulch. For singular trees and shrubs, a rim may be formed at the outside of the mulch layer in order to capture and hold water. Once applied, the mulch may be watered-in to help keep it in place and to help leach salts. If the compost is high in soluble salts, reduced amounts should be applied and the mulch should be well watered. Composts with higher soluble salt contents should be used with caution on herbaceous and salt-sensitive plants (Alexander, 1995).

Similar to other mulches, compost should not be over-applied, especially when immature and unstable composts are used. Caution should be used when applying composts that have a high pH where acid-loving species are planted. For certain applications, adding S to areas where these crops are grown may be necessary.

It may be necessary to rake the compost mulch layer occasionally to help maintain its uniform appearance. When compost is used in annual beds and when perennial beds are being prepared for replanting, the old compost mulch layer should be incorporated into the existing bed. This old layer of mulch will actually become a soil amendment to help prepare the area for replanting. In perennial beds and around trees and shrubs, where the compost has not been incorporated, use a rake or a shovel to break up the existing layer of compost mulch, ensuring that a crusted layer has not formed before reapplying new mulch (Alexander, 1995).

Where soluble salt levels of the soil are problematic, field experience has shown that repeat applications should not exceed 2.5 cm (1-in.) if biosolids composts are used. Greater rates may be possible with yard debris compost; however, some yard debris products may possess an elevated soluble salt level, particularly in areas of the country where road salts are applied for snow and ice management. If washouts occur where the compost mulch has been applied, a rake can be used to repair and smooth these areas. Reapplication of mulch will likely be necessary on a yearly basis for aesthetic purposes and weed control (Alexander, 1995).

3. Planting Backfill Mixes

Although the technique of amending the soil placed around a newly planted tree or shrub has been used extensively throughout the horticultural industry, the concept has been met with much controversy. Conflicting research exists regarding the benefits of improving backfill soils with soil amendments (Birdel et al., 1983; Smalley and Wood, 1995). However, many landscapers claim that the use of compost in backfill mixes has reduced the amount of plants they have to replace (because of death) on their landscaping jobs. This is plausible since compost provides nutrition, improves the moisture-holding capacity of the soil, and assists in the control of soil-borne diseases (Gouin, 1997; Nelson, 1992).

Application instructions — The inclusion rate of compost in the backfill mix may vary based on the species to be grown and the characteristics of the soil to be blended. An inclusion rate of 25% (Watson et al., 1993) to 50% (Smalley and Wood, 1995) compost by volume, blended with the native soil, has been widely used. However, the preferred and most popular inclusion rate is at approximately 33% compost by volume. Where trees or shrubs are to be planted, adequate drainage in the area is of the utmost importance. Plastic drainage lines, gravel filled holes, and other methods can be used to assure proper site drainage below or around the planting holes. Prepare balled and burlapped (B&B), containerized, or bare root plants for planting in accordance to industry standard methodologies before planting (Alexander, 1996).

The planting hole should be dug slightly shallower than the height of the root ball and two to four times its width (Watson and Kupkowski, 1991). In dense soils or poorly drained sites, the planting hole can be dug only two thirds the depth of the root ball. The soil removed from the planting hole should be stockpiled near the hole and mixed at an appropriate ratio of two parts soil to one part compost. The soil and compost should be blended by hand or with a shovel until uniform (Figure 7.2). The tree or shrub should be placed in the planting hole and the blended backfill mix should be added around the root ball, firming it occasionally to remove air pockets and assure a firm footing. Once firmed in place, larger trees should be anchored or supported using one of a variety of techniques, guy wires or propping. Supports should be removed 1 to 3 months after planting, depending on tree size and site conditions. Once the planting hole is filled and firmed with the backfill mix, a soil berm should be constructed around the edge of the plant root ball to help retain moisture. The plant should then be watered well and mulched (Alexander, 1996).

Figure 7.2 Blending compost with existing soil.

4. Turfgrass Establishment and Topdressing

The use of compost in the establishment and renovation of turfgrass has become popular in a variety of situations, including residential and commercial lawns, athletic fields, golf courses, and even utility turf. Benefits of compost use include faster turf establishment, improved turf density and color, increased root growth, and a reduced requirement for fertilizer and irrigation (Landschoot, 1996). Whether using compost to establish turf by seed, sod, or sprig, excellent results should be obtained.

Application instructions — Compost should be applied at a 2.5 cm (Landschoot and McNitt, 1994) to 5.0 cm (Angle et al., 1981) (1 to 2 in.) depth, then incorporated to an approximate depth of 12.5 to 17.5 cm (5 to 7 in.), resulting in an inclusion rate of 20 to 30% by volume. The compost application rate will vary depending upon soil conditions, compost characteristics, and turf species to be established. Compost application rates should be altered depending upon the potential tillage depth. Compost may be applied with a manure spreader, grading blade, front-end loader, raking device, or other equipment. Once applied, the compost should be incorporated using a rototiller, rotovator, or disc until the compost is uniformly mixed. If compost is suspected to have an elevated soluble salt concentration, the amended soil should be irrigated to leach the salt out of the root zone prior to planting. Once incorporated, a proper seed bed should be established by raking or dragging, and rolling the soil surface until smooth. Seed may be applied using a hydroseeder or cultipack seeder, or it may be broadcast over the soil surface, then lightly incorporated using a drag mat or leaf rake. Depending upon the species, sprigs may be incorporated along with the compost or be spread on the prepared soil surface and knifed into the soil with a disc or specialty implement, followed by rolling. Sod may be applied directly onto the soil surface either manually or by using specialized machinery. Once planting is completed, the planting area should be fertilized with a starter fertilizer, as necessary, and watered on an on-going basis to assure establishment (Alexander, 1995).

Topdressing has long been a reliable turf maintenance practice in the golf course industry. The practice entails applying a thin uniform layer of topdressing material over an established and usually declining turf area. Topdressing is performed for many reasons, including promoting seed germination, increasing the organic matter content of the soil and leveling the surface of turf areas. Topdressing is usually done in conjuction with aeration and reseeding. Core aeration should be completed using hollow or spoon tines, 1.3 to 2.5 cm (1/2 to 1 in.) in diameter. The tines will remove plugs of soil, or cores, from the soil surface. After aeration, the topdressing is applied and through mechanical dragging, the holes are refilled with the topdressing material. Typically, a 0.3 to 1.3 cm (1/8 to 1/2 in.) layer of compost is applied during the topdressing procedure (Figure 7.3). When the topdressing procedure is performed along with aeration, many other benefits are obtained, including improved water percolation, improved air exchange, thatch degradation, increased water-holding capacity of soil, and reduced soil compaction (Alexander, 1991).

Figure 7.3 Profile of aerated turf, topdressed with compost.

Commonly used topdressings are sand, sand-based mixes, and compost. Topdressing with finely screened 0.9 cm (3/8-in. screen), or smaller, nutrient rich, stable compost products are preferred. Topdressing is often used as a maintenance practice on turf areas that are overused, or on the decline. When topdressing is applied in conjunction with seeding, seed germination will be improved (Alexander, 1991).

Topdressing with compost has become popular because few reasonably priced top-dressing products are available for maintaining large turf areas.

Further information on the use of compost on athletic fields has been given by Alexander and Tyler (1992).

5. Topsoil Blending

Billions of tons of topsoil are lost every year because of environmental and geologic conditions and phenomena, and poor soil management practices. Farm soils are commonly harvested and sold to the landscape industry, as are soils harvested from construction sites. Even though these soil harvesting practices continue, it has become increasingly difficult for landscapers and gardeners to purchase high-quality topsoil for use in their landscape projects. This has lead to a dramatic expansion of the manufactured or blended topsoils industry. Topsoil blenders typically blend lower quality soils with compost, to produce a blended product that more closely compares to a high-quality topsoil. In several parts of the U.S., topsoil blenders are the largest users of commercially produced compost.

Application instructions — The inclusion rate of compost in the blends may vary based on the types of crops being grown, the characteristics of soil to be blended, the specific application of the topsoil blend, and the specific needs of the end user. For instance, compost may be used at higher rates specifically to modify the physical and chemical characteristics of soil, or may be added in more specific quantities to produce a topsoil blend which meets a specific organic matter level.

The addition of other products such as lime or S to modify pH, or the addition of bark, sand, or other topsoils to adjust physical parameters of the finished topsoil blend, may also be desirable. The necessity of adding these amendments will depend on the qualities of the compost, the requirements of the specific plant species being grown, and customer preference. An inclusion rate of 20 to 30% (Landschoot and McNitt, 1994) compost by volume is recommended, depending upon the quality and physical property of the soil to be amended and the organic matter content of the compost. However, rates of 10 to 50% are also commonly used. An inclusion rate of 20% may be sufficient where organic matter-rich composts are used, while a 30% inclusion rate would be recommended in sandier soils and when using composts with a lower organic matter content. Higher rates of compost inclusion may yield superior results if blended with extremely poor-quality soils (Tester and Parr, 1983), certain subsoils, or soil-like aggregate byproducts. The pH of the blended topsoil should be adjusted to meet crop requirements (Alexander, 1995).

The compost should be blended with the topsoil and any other amendment until a homogeneous mix is achieved. Blending can be done by using front-end loaders, rotating drum-type mixers, augers, or soil shredders. The ingredients can be effectively blended using a front end loader by layering the ingredients, then rolling the pile by lifting and dumping the mix forward with the loader bucket. Various amendments and additives may also be added during the mixing process to develop specialized blends ideal for specific crops or sites. The finished blend can be screened to a specific size to meet customer requirements (Alexander, 1995).

6. Erosion Control

In recent years, erosion and sediment control has become a major issue in the construction and landscape industries because of regulations enacted on federal, state, and local levels. These regulations have been enacted to conserve topsoil and reduce surface water pollution. The goal is to stop, or at least reduce, the displacement of soil particles.

Application instructions — The compost application rate may vary depending upon severity of slope, as well as soil or compost characteristics. Applications of a 7.5 cm (Stewart and Pacific, 1993) to 10 cm (Michaud, 1995) (3 to 4 in.) layer of compost on the soil surface will effectively control soil erosion on a slope of up to 45% (Michaud, 1995) for a period of 1 to 3 years. Prior to the application of compost, the exposed soil should be tracked (compacted) with a tracked bulldozer, rather than smoothed, if possible. The product can be efficiently applied through a mechanical slinger or blower-type apparatus which can apply the product both up and down a sloped area (Figure 7.4). However, applying compost with this type of apparatus may generate dust (Alexander, 1995).

Figure 7.4 Compost applied pneumatically to a sloped area.

Compost can also be applied and graded by bulldozing it up or down a sloped area. It is also feasible to apply compost by dumping it down a slope in bulk quantities, then spreading it with a tractor-pulled grading blade or manually using large rakes. On excessively unstable soils (wet), compost should be applied using a slinger or blower-type apparatus, or manually. An excavator or backhoe may also be used. Dry compost should not be applied in windy conditions. Once spread, moisture should be applied over the layer of compost for compacting purposes.

Figure 7.5 A compost berm used to replace silt fencing at a contruction site.

When possible, the compost layer should be tracked, especially on heavier soils, so water does not move freely between the compost–soil interface. Tracking will incorporate the compost into the soil surface to some degree. In order to prevent rill formation, compost should be applied to cover the entire exposed soil surface and the layer should extend approximately 1 m (3 ft) over the top of the slope (Stewart and Pacific, 1993) or mesh into existing vegetation. Best results will be achieved if a sediment fence is used at the base of the slope in conjunction with the compost. If used, the sediment fence fabric should be laid on the soil surface with the lip facing the slope itself. A 0.5 m ($1^1/_2$ ft) high by 1 m (3 ft) wide berm of compost should then be applied to the base of the sediment fence and over the fence fabric lip. This will act as a prefilter for the sediment fence. By applying the compost over the fabric lip, digging a trench to bury the fabric will be avoided and so will associated costs. Alternatively, a compost berm (mound) alone may be placed at the base of the slope in lieu of the sediment fence (Figure 7.5). The berm may be up to 0.6 m (2 ft) high by 1.2 m (4 ft) wide (Stewart and Pacific, 1993) depending upon the severity of the slope. As an alternative to silt fencing or constructing a berm, a toe could be constructed by excavating a shallow ditch at the base of the slope and backfilling it with crushed stone or gravel (Alexander, 1995).

If actively vegetating the slope is preferred, it may be completed as desired. If the compost product is carbonaceous and does not appear to be stable, then the product should be field stabilized (aged) before seeding. Seeded sites should be watered if possible. If not actively vegetating, in most conditions, natural vegetation will intrude over time. Seeding should not occur until field aging has allowed salts to leach, volatile organic acids to decompose, and NH_4 to be converted to nitrate or hydrolized into ammonia. If washouts occur on the slope, spot applications of compost may be performed. However, long-term maintenance should be minimal as vegetation should cover the area in time, aiding in long-term erosion control. If MSW composts are used which contain man-made inerts (e.g., plastics), they may tend to float over time, which may create an aesthetic problem (Alexander, 1995).

VI. CONCLUDING REMARKS

The versatility of high-quality compost, as well as its unique characteristics, make it an ideal amendment to landscape soils and planting media. However, it is the positive field results that have allowed compost to be used so extensively in areas where it is produced and marketed in bulk. Only through a well-developed bulk and bagged distribution program can large and small landscapers obtain compost in the quantities and the price range necessary to allow its wide-scale use. These types of compost distribution programs are becoming more and more common throughout the U.S. and the world.

REFERENCES

Alexander, R. 1991. Sludge compost: can it make athletic fields more playable? *Lawn & Landscape Maintenance Magazine* 12(7):46–52.

Alexander, R. 1995. *Suggested Compost Parameters and Compost Use Guidelines.* The Composting Council, Alexandria, Virginia.

Alexander, R. 1996. *Field Guide to Compost Use.* The Composting Council, Alexandria, Virginia.

Alexander, R. 1999. Blending improves marketability of compost II. *Composting News* 7(11):1–10.

Alexander, R. and R. Tyler. 1992. Using compost successfully. *Lawn & Landscape Maintenance Magazine* 13(11):23–34.

Angle, J.S., D.C. Wolf, and J.R. Hall III. 1981. Turfgrass growth aided by sludge compost. *Biocycle* 22(6):40–43.

Beeson, R., Jr. 1995. Personal communication.

Birdel, R., C. Whitcomb, and B.L. Appleton. 1983. Planting techniques for tree spade dug trees. *Journal of Arboriculture* 9:282–284.

Boyle, M., W.T. Frankenberger, Jr., and L.H. Stolzy. 1989. The influence of organic matter on soil aggregation and water infiltration. *Journal of Production Agriculture* 2:290–299.

Brady, N.C. 1974. *The Nature and Properties of Soils.* 8th edition. Macmillan Publishing Company, New York, p. 99.

Cisar, J.L. and G.H. Snyder. 1995. Amending turfgrass sand soils to improve water retention and reduce agrochemical leaching, p. 137–160. In: W.H. Smith (ed.). *Florida Water Conservation/Compost Utilization Program Final Report, March 1995.* University of Florida, Gainesville.

Composting Council Research & Educational Foundation. 1997. *The Soil & Water Connection.* Composting Council Research & Educational Foundation, Alexandria, Virginia. March.

Ewing, K. and S.M. Allen. 1994. *Growth Study of a Yard Waste Compost.* Center for Urban Horticulture, University of Washington, Seattle.

Gouin, F.R. 1989. Peat Moss and Peat Substitute. Department of Horticulture, University of Maryland, College Park. Bulletin HE 134–85.

Gouin, F.R. 1997. Selecting organic soil amendments for landscapes, p. 2–5. In: *Landscape Architect Specifications for Compost Utilization*. Clean Washington Center (CWC), Seattle, Washington.

Hortenstine, C.C. and D.F. Rothwell. 1973. Pelletized municipal refuse compost as a soil amendment and nutrient source for sorghum. *Journal of Environmental Quality* 2:343–345.

Landschoot, P. 1996. *Using Compost to Improve Turf Performance*. The Pennsylvania State University, Bulletin 5M496ps5733.

Landschoot, P. and A. McNitt. 1994. Improving turf with compost. *BioCycle* 35(10):54–57.

Maynard, A.A. 1998. Utilization of MSW compost in nursery stock production. *Compost Science and Utilization* 6(4):38–44.

Michaud, M. 1995. *Recycled Materials Used as Erosion Control Mulches*. Kennebec County Soil and Water Conservation District. Maine Waste Management Agency, Augusta, Maine.

Nelson, E.B. 1992. *Biological Control of Turfgrass Diseases*. Cooperative Extension Service, Cornell University, Ithaca, New York. Information Bulletin 220.

Obreza, T. 1995. Solid waste compost for improved water conservation and production of vegetable crops, p. 15–31. In: W.H. Smith (ed.). *Florida Water Conservation/Compost Utilization Program Final Report, March 1995*. University of Florida, Gainesville.

Organic Gardening. 1995. *Gardening in America II*. Rodale Press, Inc., Emmaus, Pennsylvania.

Smalley, T.J. and C.B. Wood. 1995. Effect of backfill amendment on growth of red maple. *Journal of Arboriculture* 21:247–250.

Smith, E.M. and S.A. Treaster. 1991. Application of composted municipal sludge in the landscape, p. 19–21. In: *Ornamental Plants: A Summary of Research, 1991*. The Ohio State University, Ohio Agricultural Research and Development Center, Wooster. Special Circular 137.

Stewart, W. and W.H. Pacific. 1993. *Final Report: Demonstration Project Using Yard Debris Compost for Erosion Control*. Metropolitan Service District, Portland, Oregon.

Tester, C.F. and J.F. Parr. 1983. Intensive vegetable production using compost. *BioCycle* 24(1):34–36.

Watson, G.W. and G. Kupkowski. 1991. Soil moisture uptake by green ash trees after transplanting. *Journal of Environmental Horticulture* 9:227–230.

Watson, G.W., G. Kupkowski, and K.G. von der Heide-Spravka. 1993. Influence of backfill soil amendments on establishment of container-grown shrubs. *HortTechnology* 3:188–189.

CHAPTER 8

Compost Utilization in Fruit Production Systems

Flavio Pinamonti and Luciano Sicher

CONTENTS

I. INTRODUCTION

Fruit production is widespread in the tropical, subtropical, and temperate regions of the world. Fruit cultures have developed into highly specialized and intensive production systems that generally exploit soil to its maximum productivity. Recently, the limited use of cow manures or soil organic amendments, lack of crop rotations, frequent use of clean cultivation, minimal introduction of sod or cover crops, insufficient fallow time, increase in the traffic of orchard machinery, and intensive inorganic fertilization and herbicide programs have accelerated soil exploitation. These factors have been identified as some of the major constraints of an intensive fruit monoculture (Bauer et al., 1992; Delas et al., 1982; Giulivo, 1986; Van Huyssteen,

1-56670-460-X/01/$0.00+$.50

1988) resulting in less than optimal root system function and productivity of cultivated fruit species.

Soil management practices affect the relationships between plant and soil. Adequate gas exchange, water-retention capacity, organic matter content, biological activity, lack of soil compaction, and fertility are fundamental elements for a optimum water and nutrition exchange balance between soils and plants. Optimum plant and soil relationships result in production of cultivated fruit species with high agronomic yields (Giulivo, 1989; Marangoni and Scudellari, 1989).

Cultivated fruit species are grown in an array of soil–climatic environments, resulting in differential water, nutritional, and cultural requirements. Therefore, a soil management strategy should be developed for each environment to insure optimum use of available resources and optimum crop productivity. At the same time, growers must conserve specific soil and cultural characteristics that may add qualitative value to the crop and its products (Bauer et al., 1992; McCarthy et al., 1992; Seguin, 1981).

Generally, organic matter content in soils is minimal and is receding to alarming levels. Additions of organic amendments (composts) can reverse losses in soil fertility and replace the lack of traditional cow manure applications in fruit production systems (Delas et al., 1982).

A variety of composts have been used as soil amendments or mulches in fruit production systems (Table 8.1). The purpose of this chapter is to provide a review of compost as an integral component of fruit crop production and management systems.

II. COMPOST, SOIL FERTILITY, AND SUSTAINABLE FRUIT PRODUCTION

The concept of sustainable agriculture is "the long-term use of resources without degradation" and is a major topic of study at agricultural institutions all over the world. Environmental hazards associated with modern agriculture technique in developed countries have resulted in endangerment of wild species and pollution of ground and surface water (Dick, 1992; Paoletti et al., 1992). Environmental groups and governments have demanded reduction of inorganic fertilizer usage in agricultural systems to minimize nutrient leaching into groundwater or pollution from surface water runoff. Recent integration between agricultural policy and environmental policy is an example of how political measures (agro-environmental measures) can promote sustainable farming systems (Agenda 2000, 1998; Garaguso and Marchisio, 1993; Reg. CEE 2078/92, 1992). Similarly, principles and technical guidelines of integrated crop production systems in Europe are focused on the preservation and promotion of long-term soil fertility through sustainable agriculture (integrated and organic farming practices) (Boller et al., 1998a; El Titi et al., 1993).

Fertility is a function of balanced physical soil characteristics, chemical reactions, and biological activities. Therefore, sustainable agriculture models should incorporate a humic balance that maintains an adequate soil organic matter content. Allison (1973) stated that "soil organic matter has, since the dawn of history been the key to soil fertility and productivity," in part because organic matter constitutes the part of soil that can store available energy and can be eventually used by soil organisms (Sequi, 1976).

Bauer et al. (1992) proposed a limit of 2 to 3% organic matter content in the Ap horizon for cultivated vineyard soils of cold-temperate regions. Soil organic matter content can be increased and maintained by management of manure applications. In virgin soils, a balance among vital chemical and biological processes maintains adequate levels of organic matter content and existing fertility. Intensive agricultural activity tends to break this equilibrium, exposing soil to losses of organic matter (Prasad and Power, 1997). To restore an equilibrium between losses and restitution of soil organic content, agronomic practices such as frequency and depth of clean cultivation, irrigation techniques, crop rotation, recycling of plant residue, and applications of manures or compost should be harmonized.

Generally, fruit production systems depend exclusively on inorganic fertilizers to maintain elevated soil fertility since availability of cow manure is limited. Therefore, new sources of organic material that supplement the balanced turnover of organic matter should be investigated. Compost utilization is a method of recovering soil organic matter content (Press et al., 1996; Sequi et al., 1996).

Decomposition of organic matter is influenced by soil types and soil management (Table 8.2). The need for organic matter is evident in all fruit crop production regions, especially where there is insufficient use of sod culture or cover crops. Permanent sod in vineyards is usually used in cold-temperate regions with soil organic matter content of 1.5 to 2% in Ap horizon and 250 to 300 mm of well-distributed precipitation in the vegetative season (Bauer et al., 1992). If soil organic matter is limiting (<1.5%), then permanent sod cannot be established during the dry season. However, after several years of green manure, mulching with organic materials, or additions of organic soil amendments, permanent sod can be established in vineyards.

In Italy, more than 50% of the cultivated soils have an organic matter content of < 2% (Tellarini, 1994; Zamborlini et al., 1990) and are considered unfavorable for sustainable agricultural production. In Switzerland, every grape (*Vitis vinifera* L.) grower participating in the integrated production (IP) program must adhere to a humus content of >1% (if <1%, then it is mandatory to provide and execute a rehabilitation plan); observe legal maximum heavy metal content in compost (provide analytical data of compost batch); retain brushwood and return all organic waste of vinification to the vineyard; maintain sod on at least 50% of the ground (every second row), including during winter; and use corrective measures where soil erosion is visible (Boller et al., 1998b). Failure in one area would result in the disqualification of financial governmental support for the entire farm.

The need for organic matter has been reported for Mediterranean soils (Felipò, 1996) and for many agricultural soils of the cold-temperate regions (Bauer et al., 1992; Delas et al., 1982). In tropical soils, following deforestation, soil organic matter content declines dramatically (Jenkinson, 1988). In Italy, soils have an annual mean deficit of 500 $kg{\cdot}ha^{-1}$ organic matter, which is primarily associated with intensive agronomic practices (Bartolini, 1982; Tellarini, 1994). Data on organic matter deficits vary due to differences in climate, soil typology, plant species, soil management techniques, and crop rotation.

Table 8.1 Results of Some Compost Utilization Studies on Various Fruit Crops

Fruit Crop	Compost Type	Primary Results	Reference
Vitis vinifera (grape)	MSW[z]	Reduced soil erosion. Increased soil organic matter content.	Bosse, 1967
Vitis vinifera	MSW	Increased soil organic matter. Did not reduce must quality.	Walter, 1980
Vitis vinifera	MSW	Increased soil organic matter. Increased fruit yield.	Enkelmann and Völkel, 1982
Vitis vinifera	MSW	Reduced water runoff and soil losses.	Carsoulle et al., 1986
Vitis vinifera	MSW	Improved soil chemical and physical properties. Did not reduce must quality.	Scienza et al., 1987
Vitis vinifera	MSW	Reduced water runoff and soil losses.	Ballif and Herre, 1988
Vitis vinifera	MSW	Increased soil organic matter. Reduced water runoff and soil erosion.	Delas, 1989
Vitis vinifera	MSW and Biosolid/ bark	Improved soil chemical and physical properties. Increased growth of young vines and fruit yield. Did not reduce must quality.	Pinamonti et al., 1991
Vitis vinifera	Bark	Increased fruit yield, sugar content, and total acid concentration.	Wang et al., 1991
Vitis vinifera	MSW	Improved soil physical properties. Reduced soil erosion. Increased herbicide effectiveness.	Moncomble and Descotes, 1992
Vitis vinifera	MSW	Increased soil pH and heavy metal adsorption in sandy, acidic, and degraded soils.	Delas, 1993
Vitis vinifera	MSW	Improved nutritive status of vines. Did not reduce fruit yield and must quality.	Balanyà Martí et al., 1994
Vitis vinifera	MSW	Reduced water runoff and soil erosion.	Ballif et al., 1995
Vitis vinifera	MSW and Bark	Reduced soil erosion. Did not reduce vine performance. Did not change qualitative characteristics of enologic products.	Sauvage, 1995
Vitis vinifera	Bark	Improved soil physical properties. Controlled weed development. Stimulated root growth of young vines. Improved fruit color. Increased fruit weight and sugar content.	Fujiwara, 1996
Vitis vinifera	Biosolid/ bark	Improved soil chemical and physical properties. Improved vineyard establishment.	Pinamonti et al., 1996
Vitis vinifera	MSW and Biosolid/ bark	Improved soil chemical and physical properties. Reduced chemical weed control. Allowed substitution of chemical fertilizers with no loss in vine vigor, yield, or quality of musts.	Pinamonti, 1998a
Malus xdomestica (apple) and *Vitis vinifera*	MSW and Biosolid/ bark	Improved soil chemical and physical properties. Increased growth of young apple trees and young vines. Obtained earlier maximum yields.	Pinamonti and Zorzi, 1996

Table 8.1 Results of Some Compost Utilization Studies on Various Fruit Crops (Continued)

Fruit Crop	Compost Type	Primary Results	Reference
Malus xdomestica and *Vitis vinifera*	MSW and Biosolid/bark	Improved soil chemical and physical properties. Maintained a vegetative production balance with no reduction in product quality.	Pinamonti, 1998b
Malus xdomestica	MSW	Reduced diurnal fluctuations in soil temperature. Increased earthworm activity.	Hartley and Rahman, 1994
Malus xdomestica	MSW	Increased soil bioactivity. Increased cellulose degradation activity.	Hartley et al., 1996
Malus xdomestica	Manure	Increased soil nitrate and phosphorous levels.	Walsh et al., 1996
Citrus sinensis (orange)	MSW	Improved soil physical properties. Increased soil biological activity.	Canet et al., 1998
Citrus spp. seedlings	MSW	Reduced infection by Phytophthora nicotianae.	Widmer et al., 1998
Prunus persica (peach)	MSW	Increased growth of young trees. Increased fruit yield.	Strabbioli and Angeloni, 1987
Olea europaea (olive)	MSW	Improved soil chemical and physical properties. Improved nutritive status and growth of trees. Increased fruit production with no reduction in fruit quality.	Aguilar Torres et al., 1996
Macadamia integrifolia (macadamia)	Macadamia husk-manure	Improved soil chemical properties. Reduced soil nitrate content. Maintained yield of nut-in-shell or salable kernel. Did not decrease kernel quality.	Bittenbender et al., 1998

[z] MSW = municipal solid waste.

Humification balance and biological equilibrium of soils represent fundamental elements for an integrated control of rhizo-biosphere disorders. Fruit tree root absorption disorders (nutritional deficiencies, fruit tree replant problems) and root diseases may be a consequence of biological soil degeneration related to anomalous nonhumifying pathways of organic matter degradation in soils (Zucconi and De Bertoldi, 1987; Zucconi et al., 1981). Reduction in humification can be associated with the accumulation of monogenic organic residues (i.e., a monocultural succession cropping) in the soil and decreased microbial variety due to the biocidal action of inorganic fertilizers, pesticides, and fumigants (Zucconi, 1993; Zucconi and Monaco, 1987).

Fruit tree replanting success may depend on associations among species and on crop rotations. However, if these methods are not practical (i.e., small specialized fruit-growing farms), then substantial amounts of soil organic amendments should be applied to enhance replanting success (Zucconi, 1988; Zucconi and Monaco 1987).

Table 8.2 Factors Influencing Decomposition of Soil Organic Matter in Vineyards Established in Cold-Temperate Regions

Slow Decomposition	Fast Decomposition
Clay soils	Sandy soils and stony soils, skeleton rich (poor fine soil)
Minimal cultivation	Frequent cultivation
Sod culture	Persistent solar radiation on cultivated soils
Organic mulches (compost, straw, barks, others)	Without mulches or without sod culture
Brushwood left on the soil surface and all organic waste from vinification returned to the vineyard soil surface	Brushwood is not left on the vineyard soil surface
Estimated annual dry organic matter decomposition is 3000–4000 $kg \cdot ha^{-1}$	Estimated annual dry organic matter decomposition is 7000 $kg \cdot ha^{-1}$

From Bauer, K. et al., 1992. Bodenpflege, p. 2–19. In: K. Bauer et al. (eds.). *Ökologisch Orientierte Bodenpflege und Düngung in Qualitätsweinbau.* Bundesministerium für Land- unf Forstwirtschaft Publication, Vienna, Austria. With permission.

III. COMPOST UTILIZATION IN FRUIT PRODUCTION SYSTEMS

Compost utilization should be compatible with practical agricultural production systems. Compost must be used so as to assure optimum productivity and fruit quality of cultivated fruit crops while minimizing environmental concerns. Investigations have focused on compost as a beneficial soil amendment that can be integrated into various fruit production systems (Table 8.1).

Compost agricultural quality and safety characteristics have been identified and implemented through state and country regulations (Accotto et al., 1996; U.S. EPA, 1994). Compost utilized as a soil amendment in fruit production systems should be free of pathogens, viable weed seeds, phytotoxins, and foul odors. There should be minimal amounts of glass and plastic, and heavy metal concentrations should be below government limits. The compost should have a particle size adequate for the desired use; be stable and mature; have a balanced percentage of nitrogen (N), phosphorus (P), and potassium (K); and be cost effective.

Pinamonti (1998b) analyzed chemical and physical characteristics of 100 samples of compost and other soil organic amendments produced in Italy (Table 8.3). Composts were drier, had higher ash content, and had lower organic matter and nutrient content than cow manure. At identical application rates, composts resulted in proportionately higher organic matter content than cow manures. Composts are generally lighter, and finer, have homogeneous particle size, and are easier to handle and transport than manures. Poultry manure, with a high nutrient content, is considered as a fertilizer rather than a soil organic amendment in Italy (Perelli, 1994) whereas peat, with a high organic matter content and low nutritional content is considered as a soil organic amendment and not a fertilizer. However, high costs reduce the use of peat. Compost utilization in various orchard management practices is summarized in Table 8.4.

Table 8.3 Analytical Characteristics of Various Soil Organic Amendments and Composts

Item	Cow Manure	Poultry Manure	Peat	Yard Trimmings Compost	Biosolid/Bark Compost	Municipal Solid Waste Compost
Number of samples	5	12	44	11	58	30
Moisture (%)	74	25	61	49	45	30
pH	8.41	8.47	4.42	7.64	7.36	7.90
E.C.[z] ($\mu S \cdot cm^{-1}$)	2510	6800	235	1050	1260	3800
Ash (%)	27.1	34.4	2.9	54.6	35.7	50.2
Organic matter (%)	66	57	88	44	55	40
C:N ratio	19.0	11.9	54.6	14.7	18.1	18.9
N (%)	2.01	3.13	0.93	10.13	1.76	1.27
P (%)	0.87	1.98	0.06	0.25	0.74	0.31
K (%)	1.48	2.59	0.08	0.37	0.52	0.54

[z] E.C.= electrical conductivity.

Adapted from Pinamonti, 1998b.

Table 8.4 Compost Utilization Practices in Fruit Production Systems

Utilization Practice	Compost Rate and Application Technique	Benefits
Corrective dressing (before planting)	$50–100\ t \cdot ha^{-1}$	Restoration of organic matter
	Surface-applied and mixed with topsoil by conventional tillage	Incorporation and build up of less mobile nutrients (P and K)
Maintenance dressing	$40–60\ t \cdot ha^{-1}$ every 2–3 years	Restoration of organic matter
	Surface-applied or placed on the under-row area, with or without mixing with the topsoil by conventional tillage	Restoration of nutrients removed with the fruits Maintenance of an adequate vegetative production balance
Mulching	$30–100\ t \cdot ha^{-1}$ every 2–3 years Placed on the under-row area (fruit tree strips)	Weed control Improvement of water balance and reduction of large fluctuations in soil temperature Restoration of nutrients removed with the fruits
Application into planting holes	$5–20\ t \cdot ha^{-1}$ depending on planting arrangement	Improved rooting
	Direct application near the root system at the time of planting	Improvement of edaphic environment of the root system

Data on chemical and physical characteristics of composts are not sufficient to evaluate their horticultural value and impact on the environment. Therefore, experiments should be conducted to evaluate compost quality, its biological and economical benefit in fruit production systems, and any potential adverse environmental impacts.

A. Compost for Corrective Dressing (Preplant Applications)

A corrective dressing in fruit production systems is the application of inorganic fertilizers and/or soil organic amendments in quantities that provide a sufficient reserve of nutrients and/or organic matter for several years (Robinson, 1992). Generally, soil organic amendments and inorganic fertilizers are cultivated into the topsoil. Soil surface-applied organic amendments and fertilizers are less effective within a short-term period in improving topsoil fertility.

Fruit orchards, especially those of small sizes (<1 ha), have been established as traditional monocultural continuous cultivations. Replanting the same species often results in poor fruit tree establishment. Fruit tree replant problems are frequently attributed to low levels of humic reserves (Delas et al., 1983). Soil organic amendments reduce fruit tree replant problems by providing plant nutrients and by stimulating growth and development of the root system (Richards, 1983).

In corrective dressing, compost with large particle sizes (sieve diameter 20 to 30 mm) can be incorporated into the surface soil if it has humified organic matter with sufficient phyto-nutritive properties (Pinamonti, 1998b). Compost application rates can be determined by traditional methods adopted for manure. For example, based on the analytical characteristics shown in Table 8.3, the dry organic matter content contained in 100 t of cow manure is identical to that in 57 t of biosolid/bark (BB) compost, 61 t of municipal solid waste (MSW) compost, and 77 t of yard trimmings compost. Compost can provide a continues supply of nutrients and improve efficiency of inorganic fertilizers (Bittenbender et al., 1998; Ferreira and Cruz, 1992; Sikora, 1996). Abad Berjón et al. (1997) reported that compost applications reduced optimal levels of inorganic N fertilization, thereby reducing risks of nitrate eutrophication.

In Europe, compost was used in fruit production systems during the 1960s and 1970s. Many multiple annual trials on vineyard soils demonstrated that 50 to 100 $t{\cdot}ha^{-1}$ of MSW compost increased soil organic matter content from 0.5 to 1.0% (Bosse, 1967; Enkelmann and Völkel, 1982; Walter, 1980).

In Italy, eight trials were conducted in apple (*Malus* x*domestica* Borkh.) orchards and vineyards comparing biosolid/bark and MSW composts (50 $t{\cdot}h^{-1}$) to cow manure (60 $t{\cdot}h^{-1}$) as preplant corrective dressings (Pinamonti and Zorzi, 1996). After 3 years of application, composts had increased soil organic matter content, nutrients, porosity, and percentage of available water, and had improved structural stability of the aggregates as compared with manure (Table 8.5).

Fully stable soil organic amendments are known for their slow mineralization (Epstein, 1997). Soil organic matter and nutrient content increased following preplanting compost application (Bevacqua and Mellano, 1993; Cortellini et al., 1996; Paino et al., 1996). Soil structure has been reported to be positively influenced by compost utilization (Ballif et al., 1995; Nogales et al., 1996). Increases in soil porosity, aggregate structural stability, water retention, and water infiltration velocity, along with a notable reduction of soil erosion, have been reported with composts.

Compost may not satisfy the complete N, P, and K requirements of cultivated fruit trees, but should be appropriately integrated with inorganic fertilizers. Compost–inorganic fertilizer combinations may be a viable agronomic alternative in fruit

Table 8.5 Effects of Compost and Manure Applications on Soil Chemical and Physical Characteristics after 3 Years

Amendment and Rate[z]	pH	Organic Matter (%)	C:N Ratio	P (mg·kg^{-1})	K (mg·kg^{-1})	Available Water (%)	Porosity (% vol)	Water Stability Index (%)
Cow manure (60 t·ha^{-1})	7.75	3.04 b[y]	8.57	27.7 b	115 b	12.6 b	42.1 b	33.4 b
BB compost (50 t·ha^{-1})	7.74	3.78 a	8.95	32.1 a	147 a	13.7 a	43.1 a	38.9 a
MSW compost (50 t·ha^{-1})	7.78	3.98 a	8.88	30.4 ab	145 a	13.7 a	42.9 a	28.1 c

Note: These are mean values observed in 8 trials.

[z] BB = biosolid/bark; MSW = municipal solid waste.

[y] Means followed by the same letter are not statistically different (Duncan multiple range test, $P \leq 0.05$).

From Pinamonti, F. and G. Zorzi, 1996. Experiences of compost use in agriculture and in land reclamation projects, p. 520. In: M. De Bertoldi et al. (eds.). *The Science of Composting*. Blackie Academic and Professional, London. With permission.

production systems on poor soils (Businelli and Gigliotti, 1994; Ferreira and Cruz, 1992). Compost alone cannot supply the total N requirement of cultivated fruit plants during the establishment years, even on fertile soils. However, compost was reported as a supplemental nutritional source that maintained optimum quantitative and qualitative production levels (Aguilar Torres et al., 1996; Balanyà Martí et al., 1994; Bittenbender et al., 1998).

Corrective dressing in fruit production systems is used to restore soil fertility. Compost contribution to corrective dressing is directed at improving soil physical, chemical, and biological properties. Poor soils are often used in fruit production areas due to the monocultural succession crop production systems. Delas (1993) reported that compost (150 t·ha^{-1}) applied on sandy, acidic, and degraded soils resulted in improved soil pH and increased heavy metal adsorption. Compost used as corrective dressing in fruit production systems should be at rates of 60 to 80% of those used for cow manures.

B. Compost Use in Maintenance Dressing

Maintenance dressing is regular (annual or every few years) applications of sufficient nutrients to supply the ongoing needs of the plant (Robinson, 1992). During the vegetative cycle (20 years) of most cultivated fruit plants, maintenance dressing is used to maintain adequate soil organic matter content, to maintain optimum nutritional status of the plants, and to restore nutrients removed with harvested fruits. To achieve these results, soil organic amendments are usually applied once in 3 to 5 years, whereas inorganic fertilizers are applied annually. Soil organic amendments obtained from mixes of cow manure, poultry manure, waste from skin and hide workshops, olive (*Olea europaea* L.) husks etc. are processed into packets or pellets. Soil organic amendments are generally surface-applied and placed on the under-tree row area with or without soil incorporation. The use of soil organic amendments in maintenance dressings is especially important in fruit production areas that are hot, or in more temperate areas with limited precipitation, because introduction of sod culture or cover crops is often difficult. When soils are exposed, losses of organic matter can be high.

During the vegetative cycle of cultivated fruit crops, use of manures containing N mainly in a form that is quickly available to plants can be problematic. Excessive release of N from application of manure, particularly in slurry forms, may result in imbalances between vegetative and reproductive growth, culminating in low-quality production. Additions of compost with large particle size (sieve diameter 20 to 30 mm) may be an alternative to these manures, thereby avoiding excess N availability. Delas (1989) reported that compost should be applied at 40 to 60 $t{\cdot}ha^{-1}$ every 2 to 3 years to maintain sufficient organic matter in vineyard soils. Compost, unlike manure, contains more unmineralized N and therefore does not result in excessive vegetative vigor and reduction in product quality (Pinamonti, 1998b). Vineyards and orchards soils that received 40 to 60 $t{\cdot}ha^{-1}$ of compost every 2 to 3 years had good crop production and sufficient plant nutritional balance (Bittenbender et al., 1998; Canet et al., 1998; Strabbioli and Angeloni, 1987; Wang et al., 1991). Compost applied to grape, apple, peach (*Prunus persica* [L.] Batsch.), orange (*Citrus sinensis* L.), and macadamia (*Macadamia integrifolia* Maiden et Betch.) trees resulted in improved plant nutritional status and fewer physiological disorders (bunchstem necrosis, iron-induced chlorosis, bitter pit of apple, and premature leaf drop) than the exclusive use of inorganic fertilizers. Although soil organic amendments supply nutrients, they also improve the physical and hydrological conditions of the soil, thereby reducing the likelihood of a nutritional imbalance in fruit crops (Kimura and Fujiwara, 1992; Pinamonti, 1998b).

Roots of the cultivated fruit species may be improved by compost applications. Orchards can be rejuvenated by providing new areas for root exploration. This requires aeration and ample underground drainage (chisel-plows, draining-plows). An alternative technique to improve aeration and drainage is to incorporate soil organic amendments into trenches located between plant rows (Zucconi, 1988). The process can be repeated in the following years until it covers a large part of the orchard. Aguilar Torres et al. (1996) reported that MSW compost (45 $t{\cdot}ha^{-1}$) increased the soil organic matter and nutrient contents of an olive grove located in Spain. MSW compost increased water infiltration by 56%, increased available soil water by 62%, improved nutritive status of the olive trees, and increased olive production by 68% with no reduction in fruit quality. In viticulture systems, compost can supply nutrients in sufficient quantities to vines without reducing qualitative fruit characteristics (Balanyà Martí et al., 1994; Scienza et al., 1987).

C. Compost Use as Mulches

In orchards, mulches are more commonly applied in strips beneath fruit trees but can be applied in alleys between tree rows. Mulches can provide weed control, reduce surface water evaporation, increase water infiltration (hence less erosion), and improve soil physical conditions (Lanini et al., 1988; McCarthy et al., 1992). Mulches can be very effective in moderating and stabilizing soil temperatures.

Favorable surface soil conditions under mulch without tillage can encourage root branching and root growth in surface layers (Richards, 1983; Van Huyssteen, 1988). Clippings (from alley swards during mowing), straw, bark, or plastic film can serve as effective mulches in fruit production systems. However, organic mulches are not commonly used for soil management in orchards. Clean cultivation and chemical weeding are more common practices used in fruit production systems because they provide effective and economical weed control. Concerns have been raised about the environmental and agronomic impacts of clean cultivation and herbicide use, such as soil structure decay, erosion, interaction with soil biological activity (microorganisms, earthworms), flora replacement, and pollution of groundwater (Boubals, 1991; Mantinger, 1990). Plastic mulch also is reported to be effective in weed control and of benefit to the vegetative and reproductive development of fruit crops (Pool et al., 1990; Van Der Westhuizen, 1980).

Compost, if used as a mulch, should be applied along the row of fruit trees, 5 cm thick and 50 cm wide. Composts with a minimum level of maturation, as long as sufficiently stabilized, with large particle sizes (sieve diameter 20 to 30 mm), and without viable weed seeds can be effective as mulches. Composts as mulches last for 2 to 3 years, after which a repeated application is needed. Composts or mulches are of benefit in maintenance dressing of fruit production systems because they can restore nutrients removed by fruit trees, particularly during the reproductive stages, without the use of inorganic fertilizers (Pinamonti et al., 1991).

Compost rates necessary to obtain effective mulching vary according to the planting arrangement. For example, a strip of mulch 50 cm wide and 5 cm thick with rows spaced at 5, 4, 3, and 2 m, requires compost (bulk density 0.5 $kg \cdot L^{-1}$) rates of 25, 31, 42, and 62 $t \cdot ha^{-1}$, respectively.

Erosion has been aggravated in vineyards located on hilly slopes of France due to parcel extension with removal of natural obstacles (green covered headland, dry walls, hedges, rows of trees), increase in the traffic of vineyard machinery, and more herbicide usage (Delas, 1989). Organic mulching is a suitable technique to control soil surface erosion. Besides physically covering the soil, organic mulches can improve water-retention capacity, increase effective water infiltration, and provide reliable water runoff control (Delas, 1989). MSW compost has been effective in reducing soil erosion in Champagne vineyards with a no-tillage system for 20 years (Delas, 1989).

Composts utilized as mulches can be applied at rates varying from 90 to 120 $t \cdot ha^{-1}$ in the first year and then at rates varying from 40 to 60 $t \cdot ha^{-1}$ every 3 to 4 years. The effectiveness of composts as mulches in fruit production systems has been reported (Ballif and Herre, 1988; Carsoulle et al., 1986). Water runoff and soil losses were reduced from clay-calcareous soil with a slope of 34% with MSW compost applications (Table 8.6) (Ballif and Herre, 1988). Similar results were noted in grapevine plots with a rotating rainfall simulating system (Carsoulle et al., 1986). When applied to eroded soils, compost can restore both organic matter content and soil structure (Kashmanian et al., 1990). However, compost mulch does not change the qualitative characteristics of enologic products (Sauvage, 1995).

In Italy, fruit and viticulture crops are grown on about 2 million hectares and often are cultivated on hilly areas, with low fertility and low soil organic matter

Table 8.6 Effects of Soil Mulching by MSW Compost on Water Runoff and Soil Losses[z]

Treatment	Control (Untilled and Bare Soil)	MSW Compost Mulch (150 t·ha^{-1})
Water runoff (mm)	11.8	1.2
Water runoff resulting from storm rainfalls (mm)	10.0–22.0	2.2–1.6
Soil losses (kg·ha^{-1})	1382	5.3

[z] MSW = municipal solid waste.

From Ballif, J.L. and C. Herre, 1988. Contribution a l'étude de ruissellment de sols viticoles in Champagne. Effects d'une couverture de compost urbain. *Comptes Rendus de l'Académie d'Agriculture de France* 74:108. With permission.

content. The under row strips are periodically cultivated or herbicides are applied to control weeds (Fregoni and Miravalle, 1991).

In new vineyards, mulching with compost was compared with black polyethylene mulching and a control (mechanical tillage and chemical weeding) (Pinamonti, 1998a; Pinamonti et al., 1996; Pinamonti et al., 1991). Soil moisture content of compost mulched plots was consistently higher than in the control or polyethylene mulched plots (Figure 8.1). Compost mulches improved water permeability and water storage and reduced water evaporation. Polyethylene mulch prevented water infiltration, but reduced surface evaporation. Compost mulches reduced daily and seasonal low and high temperature fluctuations, whereas the polyethylene resulted in extreme soil temperatures fluctuations (Figures 8.2 and 8.3).

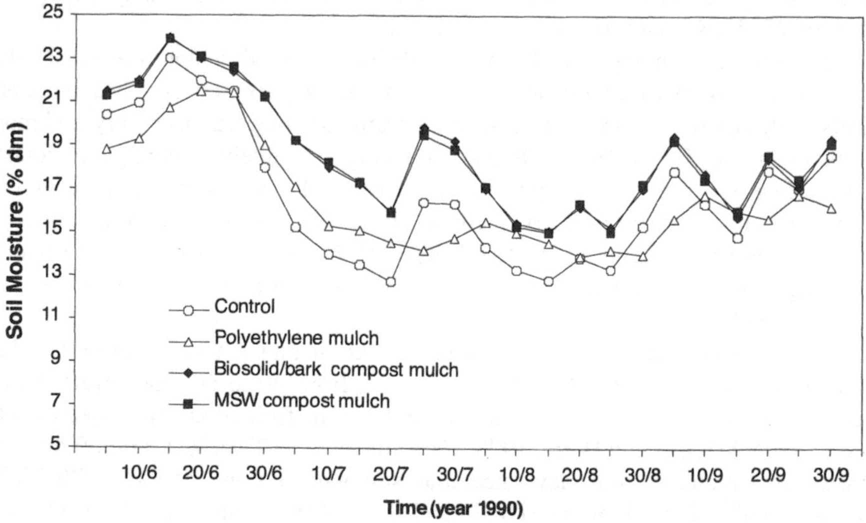

Figure 8.1 Soil moisture content (20 cm depth) in the grapevine row as influenced by mulches in a vineyard. (From Pinamonti, F., 1998a. Compost mulch effects on soil fertility, nutritional status and performance of grapevine. *Nutrient Cycling in Agroecosystems* 51:242. With permission.)

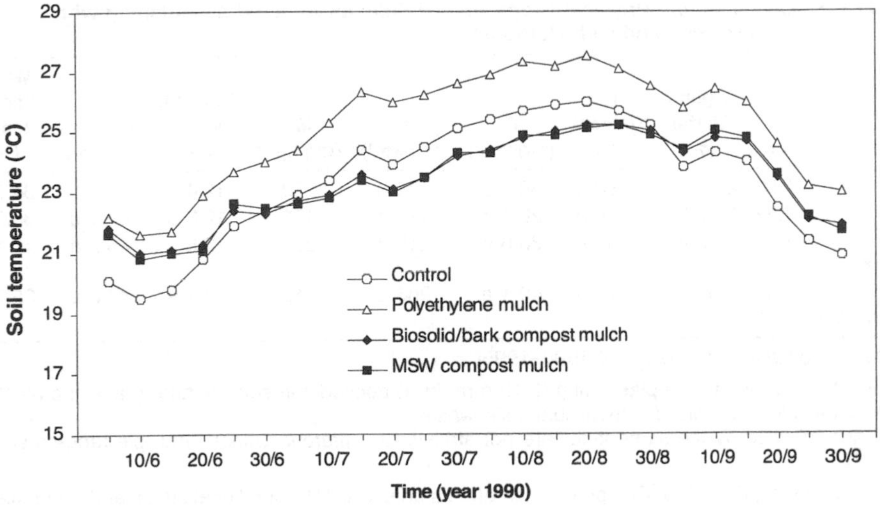

Figure 8.2 Soil temperature (20 cm depth) in the grapevine row as influenced by mulches in a vineyard. (From Pinamonti, F., 1998a. Compost mulch effects on soil fertility, nutritional status and performance of grapevine. *Nutrient Cycling in Agroecosystems* 51:242. With permission.)

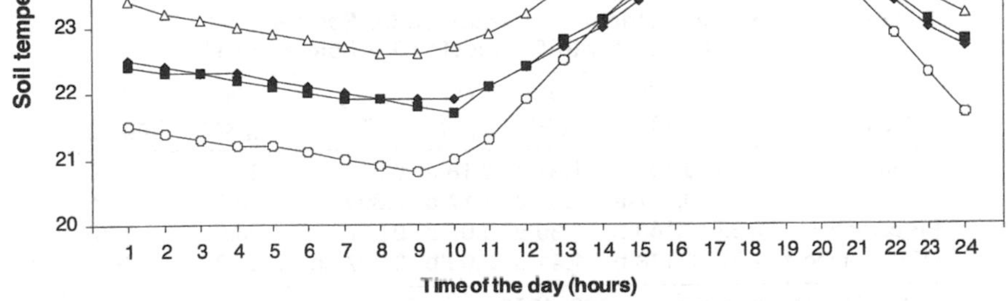

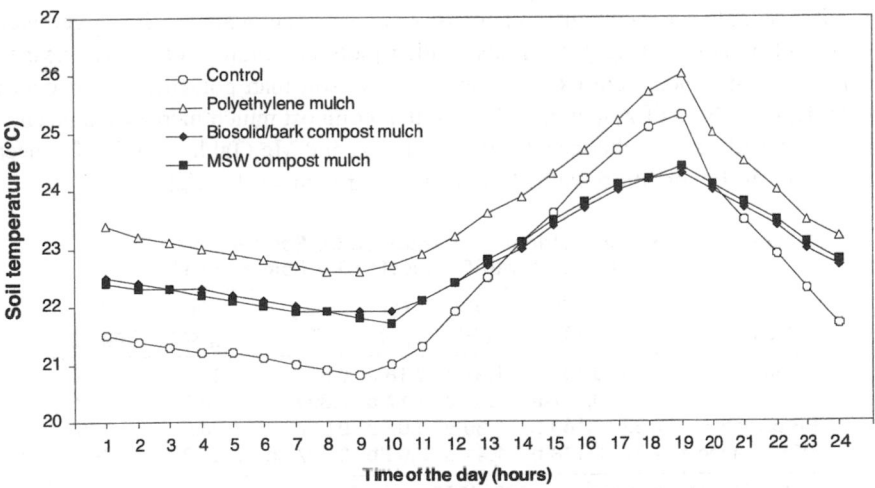

Figure 8.3 Soil temperature measured on 18 August 1990 (20 cm depth) in the grapevine row as influenced by mulches in a vineyard. (From Pinamonti, F., 1998a. Compost mulch effects on soil fertility, nutritional status and performance of grapevine. *Nutrient Cycling in Agroecosystems* 51:243. With permission.)

Table 8.7 Chemical and Physical Properties of Soils as Affected by Mulching with Polyethylene and with Compost

Mulches[z]	pH	Organic Matter (%)	N Kjeldahl (%)	P (mg·kg⁻¹)	K (mg·kg⁻¹)	Mg (mg·kg⁻¹)	Available Water (%)	Porosity (% Vol)	Water Stability Index (%)
Control	7.58	2.67 b[y]	0.160 b	30.7 b	176 b	203	12.4 b	43.5 b	34.7 b
PE	7.62	2.63 b	0.155 b	28.9 b	177 b	191	12.3 b	43.1 b	33.3 b
BB compost	7.60	3.12 a	0.181 a	38.6 a	215 a	207	13.5 a	45.2 a	41.1 a
MSW compost	7.61	3.08 a	0.177 a	40.1 a	206 a	212	13.1 a	44.8 a	35.7 b

Note: The values are means of 1990–1995.

[z] PE = black polyethylene plastic film (0.12 mm thick) applied immediately after grapevine planting; BB = biosolid/bark; MSW = municipal solid waste.

[y] Means followed by the same letter are not statistically different (Duncan multiple range test, $P \leq 0.05$).

From Pinamonti, F., 1998a. Compost mulch effects on soil fertility, nutritional status and performance of grapevine. *Nutrient Cycling in Agroecosystems* 51:243. With permission.

Weed growth was reduced by compost mulch to a similar level of the control (mechanical tillage and chemical weeding). The weed control provided by the compost mulch diminished during the first growing season, but the compost mulch still allowed a reduction in the number of herbicide applications. The polyethylene film controlled weeds up to the sixth year.

Compost mulch increased organic matter content, total N, available P, exchangeable K, available water content, and porosity of the soil compared to the polyethylene or control plots (Table 8.7). Compost mulch plots had more root growth near the soil surface and more root exploration of the topsoil than polyethylene or control plots. Leaf analysis of grapevines showed that compost mulch increased concentrations of K and decreased concentrations of P, Ca, and Mg (Table 8.8). Leaf concentrations of N, Fe, or Mn remained unchanged throughout the trial.

Table 8.8 Grape Leaf Nutrient Concentration for Soil Mulched with Polyethylene and with Composts (Dry Weight Basis)

Mulches[z]	N (%)	P (%)	K (%)	Ca (%)	Mg (%)	Fe (mg·kg⁻¹)	Mn (mg·kg⁻¹)
Control	2.59	0.157 a[y]	1.31 b	2.18 a	0.313 a	89.1	175
PE	2.61	0.157 a	1.30 b	2.17 a	0.309 a	90.3	178
BB compost	2.58	0.143 b	1.39 a	2.07 b	0.287 b	96.2	175
MSW compost	2.57	0.148 b	1.41 a	2.09 b	0.297 ab	89.8	162

Note: Values are means of 1990–1995.

[z] PE = black polyethylene plastic film (0.12 mm thick) applied immediately after grapevine planting; BB = biosolid/bark; MSW = municipal solid waste.

[y] Means followed by the same letter are not statistically different (Duncan multiple range test, $P \leq 0.05$).

From Pinamonti, F., 1998a. Compost mulch effects on soil fertility, nutritional status and performance of grapevine. *Nutrient Cycling in Agroecosystems* 51:244. With permission.

Table 8.9 Influence of Mulching with Polyethylene and with Compost on Pruning Weight from Grapevines and Grape Yields

	Pruning Weight (g/vine)						Yield of Grapes (kg/vine)			
Mulches[z]	1990	1991	1992	1993	1994	1995	1992	1993	1994	1995
Control	10.9 c[y]	111 b	186 ab	421 ab	248	265	0.94 b	3.33 ab	1.93	2.14 b
PE	36.8 a	136 a	194 a	450 a	241	258	1.09 a	4.09 a	2.16	2.58 a
BB compost	26.2 b	78 c	165 c	363 c	256	271	0.91 b	3.11 b	2.06	2.19 b
MSW compost	23.6 b	98 bc	175 bc	402 bc	241	256	1.13 a	2.84 b	2.33	2.09 b

[z] PE = black polyethylene plastic film (0.12 mm thick) applied immediately after grapevine planting; BB = biosolid/bark; MSW = municipal solid waste.

[y] Means followed by the same letter are not statistically different (Duncan multiple range test, $P \leq 0.05$).

From Pinamonti, F., 1998a. Compost mulch effects on soil fertility, nutritional status and performance of grapevine. *Nutrient Cycling in Agroecosystems* 51:244. With permission.

The number of dead vines on compost mulch plots at the end of the first vegetative season was 1 to 2% lower than in the control (3.9%) or polyethylene mulch (8.5%) plots. Grapevines from compost mulch plots had improved general performance and plant growth during the first years. Pruning weights of vines in compost-mulched plots were 120 to 140% higher than in control plots after the first year. However, these differences decreased with time, despite additional compost applications at the beginning of the fourth year (Table 8.9).

Compost mulch did not cause differences in qualitative characteristics of the musts compared to the control, except for slight changes in the acidic balance. Pinamonti and Zorzi (1996) and Pinamonti (1998b) reported similar results with apple trees. Generally, more favorable effects of mulching with compost occur on vineyards than apple orchards due to the lower organic matter content in vineyard soils than in apple orchard soils. Other studies confirm that compost mulch improves the development of young fruit trees or young vines in the establishment years and results in earlier maximum yields (Fujiwara, 1996; Pinamonti and Zorzi, 1996).

Improved soil moisture content has been reported with the use of compost mulch. Pinamonti (1998a) reported that the soil moisture content in compost mulch plots was higher than in control plots. This is attributed to the permeability capacity of the compost layer that allows water storage in the soil with minimal water losses by evaporation. The compost mulch layer has an insulation effect that reduces diurnal fluctuations in soil temperature (Hartley and Rahman, 1994).

Fujiwara (1996) reported that weed growth within rows of vineyards was restricted for a few months (one vegetative season) after compost mulch was applied. However, herbicides are eventually required even on the mulched strip (Moncomble and Descotes, 1992).

Leaf samples from grapevine and apple plots with compost mulch had higher P and reduced K, Ca, and Mg contents than control plots. The nutritional changes were partially attributed to changed soil conditions (moisture and temperature) and a diversified root structure that grew less in depth due to mulching (Pinamonti et al., 1996; Walsh et al., 1996). Compost mulch in viticulture, together with a system of partial fertilization and surface tillage, stimulated root growth, improved fruit color, and increased fruit weight and sugar content (Fujiwara, 1996).

Compost mulch was reported to increase total CO_2 emission in the field and cellulose degradation activity (Hartley et al., 1996), and to improve earthworm activity (Hartley and Rahman, 1994). Agassi et al. (1998) reported that 85% of rainwater percolated into the soil in compost mulched plots as compared with 42% in the control plots. Compost mulch stimulated both enzymatic activity (dehydrogenase and fluorescein diacetate) and soil microbial activity.

Compost mulch has beneficial effects on soil fertility (higher content of organic matter and improvement of physical properties); therefore it is a viable alternative to inorganic fertilizers used for maintenance dressing in fruit production systems. The nutrients contained in compost are released slowly (Sikora, 1996), thereby preventing physiological nutrient imbalances in vines and fruit trees or reduction in product quality. Application of composts as mulches can improve water balance and thermal conditions of soils resulting in more vegetative development of young fruit trees or young vines in the establishment phase. However, over an extended time, compost as mulch did not have significant effects on the nutritive, vegetative, or reproductive status of fruit trees or grapevines. The agronomic beneficial contribution from composts as mulches occurs during the establishment years of young orchards. However, compost applications improve soil fertility, reduce erosion, and can be a partial substitute for inorganic fertilizers (Sikora, 1996).

D. Compost Application into Planting Holes

Grafted trees with bare roots are generally used to establish orchards. When saplings or young trees are planted, soils in planting holes should be managed to optimize root development and vegetative growth. Planting holes are generally filled with fresh soil but may have an application of 5 to 10 L of peat per hole. Orchards that are renovated with the same species (monocultural succession cropping) utilize these same soil management practices for plant establishment, primarily to reduce potential fruit tree replant problems (Zucconi and Monaco, 1987).

Organic amendments (composts) applied in planting holes should be highly stabilized, have low concentrations of soluble salts (electrical conductivity less than 2000 $\mu S \cdot cm^{-1}$), be free of phytotoxic compounds, and have optimum particle size (screened through 10 mm diameter sieve). These compost attributes are especially important since the material has intimate contact with the root system. Compost should also have a moisture content between 35 to 50% (Epstein, 1997). Compost with a high moisture content is expensive to transport and difficult to handle. Compost that is very dry is dusty, does not imbibe water effectively, and can induce water stress to the tree upon planting (Pinamonti, 1998a).

In Italy, biosolid/bark compost, MSW compost, and peat were evaluated as amendments in planting holes (5 L per hole, 5 to 20 $t \cdot ha^{-1}$) for six apple cultivars (Table 8.10) (Pinamonti and Zorzi, 1996). After 4 years, apple shoot radius and yields were similar between MSW compost and untreated control plots (Table 8.10). However, biosolid/bark compost and peat amendments increased both growth and yield of apples compared to the control (Table 8.10). Perhaps the high pH and soluble

Table 8.10 Influence of Compost in Planting Holes on Vigor and Yield of Apple Trees

	Shoot Radius (cm)				Yield (kg/tree)	
Amendment and Rate[z]	**1989**	**1990**	**1991**	**1992**	**1991**	**1992**
Control	1.76 b[y]	2.73 b	3.54 b	4.33	3.63 b	8.69 b
Peat (5 L/tree)	1.88 a	2.83 a	3.66 a	4.36	4.36 a	9.73 a
BB compost (5 L/tree)	1.86 a	2.85 a	3.68 a	4.41	4.18 a	9.97 a
MSW compost (5 L/tree)	1.78 b	2.71 b	3.52 b	4.29	3.60 b	8.09 c

Note: Values are means of six trials.

[z] BB = biosolids/bark; MSW = municipal solid waste.

[y] Means followed by the same letter are not statistically different (Duncan multiple range test, $P \leq 0.05$).

From Pinamonti, F. and G. Zorzi, 1996. Experiences of compost use in agriculture and in land reclamation projects, p. 520. In: M. De Bertoldi et al. (eds.). *The Science of Composting*, Part 1. Blackie Academic and Professional, London. With permission.

salt content of the MSW compost contributed to the lack of a beneficial response. Improved growth responses were reported with annual applications of compost in mound-layered trees for the production of apple rootstocks in comparison to peat and rice (*Oryza sativa* L.) chaff (Pinamonti, 1998b). Compost applications in citrus (*Citrus* spp.) planting holes also can reduce the presence of *Phytophthora nicotianae* both in the nursery and in the field (Widmer et al., 1998).

IV. CONCLUSION

Compost utilization in fruit production systems as a soil organic amendment increases organic matter content, supplies nutrients, and improves soil physical properties. The use of compost alone is not sufficient to meet crop nutritional requirements. Therefore, compost integrated with reduced rates of inorganic fertilizers may be an effective alternative to improve infertile soils used for fruit culture. In fact, mixed compost/inorganic fertilizer applications are not only complementary but also synergistic since soil organic amendments provide a greater efficiency of inorganic fertilizers and irrigation water for plant nutrient availability. In fruit production systems, compost provides a slow and gradual release of nutrients that does not induce excessive vegetative growth or reduce fruit quality.

Mulching with compost in orchards improves soil water balance and soil physical properties, reduces erosion, and allows decreases in inorganic fertilizer rates.

Compost applied both in corrective dressing (before planting) and in planting holes can reduce fruit tree replant problems associated with monocultural succession cropping, and thus can serve as an alternative to fruit crop rotation.

Compost has been used as an alternative and/or complement to traditional cow manure application in fruit production systems. Numerous trials have shown the potential advantages following application of compost as a soil organic amendment (Table 8.11). However, some precautions are necessary when compost is utilized in fruit production systems. Key factors to be considered are compost compatibility with plants (phytotoxicity) and effects on soil fertility. Compost should be free of

Table 8.11 Potential Benefits of Compost Utilization in Fruit Production Systems

Soil Conditions	Compost Benefits[z]	Fruit Crop Plant Growth	Compost Benefits[z]
Organic matter	+	Development of young plants in the establishment years	+
Structural stability of the aggregates	+	Root growth near the soil surface and root exploration of the topsoil	+
Porosity, aeration and drainage	+	Irrigation requirements during production	–
Compaction by orchard machinery	–	Plant nutritional status	+
Infiltration and permeability	+	Vegetative production balance	=
Erosion and runoff	–	Product quantity	=
Water retention and available water to fruit crop plants	+	Product quality	=
Water losses by surface water evaporation	–	Fruit tree replant problems	–
Fluctuation in soil temperature	–	Nutritional deficiencies	–
Effectiveness of inorganic fertilizers and irrigation water	+	Physiological disorders	–
Bioavailabilty of nutrients	+	Number of young fruit crop plants replaced during orchard establishment	–
Eutrophication and nutrients losses	–		
Biological and enzymatic activities	+		
Suppression of soilborne plant pathogens	+		
Number of herbicide applications	–		

[z] + increase/improvement, – reduction, = maintenance.

pathogens, viable weed seeds, phytotoxins, and foul odors. There should be minimal amounts of glass and plastic, and heavy metal concentrations should be below government limits. The compost should have a particle size adequate for the desired use; be stable and mature; and have a balanced percentage of nutrients. Stabilization of compost plays a critical role because it increases humus and mineral concentration and improves compatibility with the root system of the fruit crop.

The potential for using compost for the applications discussed in this chapter is promising. The utilization of compost can be promoted via adequate marketing strategies based on quality control measures to assure the compost product quality (label of compost facility; ecological label of the European Community — Ecolabel; international norms as UNI EN ISO 9002). Economic aspects are important, especially because compost rates in fruit production systems are somewhat elevated (at least 40 to 50 $t{\cdot}ha^{-1}$). On the one hand, neither compost pelletization nor compost packaging may be justifiable (the benefits may not be worth the additional costs). On the other hand, government subsidies and cost sharing programs could help to defray purchase, transportation, and application costs. The objective of institutional subsidies is to promote on-farm use of compost as a conservation practice of sustainable agriculture. Encouraging farmers to properly use compost can reduce erosion, preserve soil fertility, and improve water quality. Compost is a renewable organic matter resource. In the near future, local availability of compost will increase due to a proliferation of compost production facilities.

REFERENCES

Abad Berjón, M., M.D. Climent Morató, P. Aragón Revuelta, and A. Camarero-Simón. 1997. The influence of solid urban waste compost and nitrogen-mineral fertilizer on growth and productivity in potatoes. *Communications in Soil Science and Plant Analysis* 28(17&18):1653–1661.

Accotto, E., F. Azzariti, M. Glisoni, A. Trombetta, and T. Vercellino. 1996. La legislazione europea, nazionale e regionale su produzione ed uso del compost. *Acqua Aria* 3:281–287.

Agassi, M., A. Hadas, Y. Benyamini, G.J. Levy, L. Kautsky, L. Avrahamov, and H. Zhevelvev. 1998. Mulching effects of composted MSW on water percolation and compost degradation rate. *Compost Science & Utilization* 6(3):34–41.

Agenda 2000. 1998. *Agenda 2000. Quale avvenire per l'agricoltura europea?* Commissione europea, Direzione generale Agricoltura (DG VI), Brussels, Belgium, p. 3.

Aguilar Torres, F.J., P. González Fernández, and M. Pastor Muñoz-Cobo. 1996. Miglioramenti della fertilità del terreno dell'oliveto con l'applicazione periodica di compost di residui solidi urbani. Confronto con il sistema di non lavorazione con suolo nudo. *Olivae* 64:40–45.

Allison, F.E. 1973. *Soil Organic Matter and its Role in Crop Production.* Elsevier, Amsterdam, p. 210.

Balanyà Martí, T., J. Saña Vilaseca, M.L. González Tardiu, and M. De La Peña Kruter. 1994. Utilización de compost de residuos sólidos urbanos en un viñedo del Penedés. *Viticultura Enologia Profesional* 31:20–25.

Ballif, J.L. and C. Herre. 1988. Contribution à l'étude du ruissellment de sols viticoles en Champagne. Effects d'une couverture de compost urbain. *Comptes Rendus de l'Académie d'Agriculture de France* 74:105–110.

Ballif, J.L., C. Herre, and D. Gobert. 1995. Les eaux de ruissellement et d'infiltration d'un sol viticole Champenois. Résultats de couvertures de compost urbains et d'écorces broyées 1985–1994 (France). *Progrès Agricole et Viticole* 112, 24:534–544.

Bartolini, R. 1982. La sostanza organica nel terreno è sempre meno. *Terra e Vita* 10:183–186.

Bauer, K., H. Amann, A. Fardossi, and W. Wunderer. 1992. Bodenpflege, p. 2–19. In: K. Bauer, H. Amann, A. Fardossi, and W. Wunderer (eds.). *Ökologisch Orientierte Bodenpflege und Düngung im Qualitätsweinbau.* Bundesministerium für Land–unf Forstwirtschaft Publication, Vienna, Austria.

Bevacqua, R.F. and V.J. Mellano. 1993. Sewage sludge compost's cumulative effects on crop growth and soil properties. *Compost Science & Utilization* 1(3):34–37.

Bittenbender, H.C., N.V. Hue, K. Fleming, and H. Brown. 1998. Sustainability of organic fertilization of macadamia with macadamia husk-manure compost. *Communications in Soil Science and Plant Analysis* 29(3&4):409–419.

Boller, E.F., J. Avilla, J.P. Gendrier, E. Jörg, and C. Malavolta. 1998a. Integrated plant protection in the context of a sustainable agriculture. *Bulletin of the International Organization for Biological and Integrated Control of Noxious Animals and Plants/West Palaeartic Regional Section*, 21(1):19–22.

Boller, E.F., E. Jörg, J. Avilla, C. Malavolta, and J.P. Gendrier. 1998b. Practicing integrated production:methods and constraints. *Bulletin of the International Organization for Biological and Integrated Control of Noxious Animals and Plants/West Palaeartic Regional Section*, 21(1):29–41.

Bosse, I. 1967. Ein Versuch zur Bekaempfung der Bodenerosion in Hanglagen des Weinbaues durch Muellkompost. *Weinberg und Keller* 15:385–397.

Boubals, D. 1991. Technique d'entretien du sol et environment, p. 3–10. In: Association Nationale pour la Protection des Plantes (ANPP). *Proceedings of the 3rd International Symposium on No-Tillage and Other Soil Management Techniques in Vines*, Montpellier, France 18–20 Nov. 1991. ANPP Annales 3, Paris.

Businelli, M. and G. Gigliotti. 1994. Applicazione del compost da residui solidi urbani in agricoltura, p. 111–115. In: Osservatorio Agroambientale Cesena (OAC). *Proceedings of Giornate di Studio Compost:Dai Rifiuti una Risorsa per l'Agricoltura*, Forlì, Italy 14–15 Apr. 1994. OAC Publication.

Canet R., F. Pomares, M. Estela, and F. Tarazona. 1998. Efecto de diferentes enmiendas orgánicas en las propiedades del suelo de un huerto de cítricos. *Agrochimica* 42(1–2):41–49.

Carsoulle, J., J.P. Canler, and J.J Gril. 1986. Influence de quelques techniques culturales sur le ruissellement et l'erosion en vignoble de coteaux (Beaujolais), p. 345–352. In: Association Nationale pour la Protection des Plantes (ANPP). *Proceedings of the 2nd International Symposium on No-Tillage in Vines*, Montpellier, France 26–28 Nov. 1986. ANPP Annales 1, Paris.

Cortellini L., G. Toderi, G. Baldoni, and A. Nassisi. 1996. Effects on the content of organic matter, nitrogen, phosphorous and heavy metals in soil and plants after application of compost and sewage sludge, p. 457–468. In: M. De Bertoldi, P. Sequi, B. Lemmes, and T. Papi (eds.). *The Science of Composting*. Proceedings of the International Symposium on The Science of Composting, Bologna, Italy 30 May–2 Jun. 1995. Blackie Academic & Professional, London.

Delas, J. 1989. Utilization of city refuse compost in viticulture, p. 282–293. In:Istituto Agrario di San Michele all'Adige (IASMA). *Proceedings of the International Symposium on Compost Production and Use*, S. Michele all'Adige, Trento, Italy, 20–23 Jun. 1989. IASMA Publication.

Delas, J. 1993. L'impiego del compost in viticoltura, p. 2–9. In: Osservatorio Agroambientale Cesena (OAC). *Proceedings of the 2nd Convegno Sull'agricoltura Sostenibile*, Cesena, Italy 6–7 May 1993. OAC Publication.

Delas, J., P. Rivet, and J.P. Soyer. 1982. La matière organique dans les sols viticoles du Médoc. Résultats d'une enquête. *Connaissance Vigne Vin* 16(4):219–232.

Delas, J., C. Molot, and J.P. Soyer. 1983. Apport de matières fertilisantes avant plantation de vigne, Résultats d'un essai en sol de Graves du Bordelais. *Connaissance Vigne Vin* 17(1):31–42.

Dick, R.P. 1992. A review: long-term effects of agricultural systems on soil biochemical and microbial parameters, p. 25–36. In: M.G. Paoletti and D. Pimentel (eds.). *Biotic Diversity in Agroecosystems*. Symposium on Agroecology and Conservation Issues in Tropical and Temperate Regions, Padova, Italy 26–29 Sept. 1990. Elsevier, Amsterdam.

El Titi, A., E.F. Boller, and J.P. Gendrier. 1993. Integrated Production. Principles and Technnical Guidelines. *Bulletin of the International Organization for Biological and Integrated Control of Noxious Animals and Plants/West Palaeartic Regional Section*, 16(1):13–38.

Enkelmann, R. and R. Völkel. 1982. Einsatz von muellklaerschlammkompost und muellkompost im weinbau. *Landwirtscaft Forschung* 35(1–2):77–89.

Epstein, E. 1997. *The Science of Composting*. Technomic Publishing Company, Lancaster, Pennsylvania.

Felipò, M.T. 1996. Compost as a source of organic matter in Mediterranean soils, p. 402–412. In: M. De Bertoldi, P. Sequi, B. Lemmes, and T. Papi (eds.). *The Science of Composting*. Proceedings of the International Symposium on The Science of Composting, Bologna, Italy 30 May–2 Jun. 1995. Blackie Academic & Professional, London.

Ferreira, M.E. and M.C.P. Cruz. 1992. Estudo do efeito de vermicomposto sobre absorção de nutrientes e produção de metéria seca pelo milho e propriedades do solo. *Científica São Paulo* 20(1):217–227.

Fregoni, M. and R. Miravalle. 1991. Lcs techniques de culture appliquees dans le domaine de l'entretien du sol des vignobles soumis a la viticulture biologique, p. 299–306. In: Association Nationale pour la Protection des Plantes (ANPP). *Proceedings of the 3rd International Symposium on No-Tillage and Other Soil Management Techniques in Vines*, Montpellier, France 18–20 Nov. 1991. ANPP Annales 3, Paris.

Fujiwara, T. 1996. Improvement of the physical properties of heavy clayey soils in newly developed vineyard and its effect on the yield and quality of grapes. *Bulletin of the Hiroshima Prefectural Agriculture Research Center* 63:2–54.

Garaguso, G.C. and S. Marchisio. 1993. *Rio 1992: Vertice per la Terra*. FrancoAngeli, Milano, Italy, p. 460.

Giulivo, C. 1986. Concimazione del frutteto e tecnica agronomica. *Rivista di Frutticoltura* 9–10:23–37.

Giulivo, C. 1989. Tecnica colturale del terreno, fertilizzazione e irrigazione del melo. *Rivista di Frutticoltura* 6:9–17.

Hartley, M.J. and A. Rahman. 1994. Use of mulches and herbicides in an apple orchard, p. 320-324. In: A.J. Popay (ed.). *Proceedings of the 47th New Zealand Plant Protection Conference*, Waitangi, New Zealand 9–11 Aug. 1994. Ncw Zealand Plant Protection Society Publication, Rotorua, New Zealand.

Hartley, M.J., J.B. Reid, A. Rahman, and J.A. Springett. 1996. Effect of organic mulches and a residual herbicide on soil bioactivity in an apple orchard. *New Zealand Journal of Crop and Horticultural Science* 24:183–190.

Jenkinson, D.S. 1988. Soil organic matter and its dynamics, p. 564–607. In: A. Wild (ed.). *Russel's Soil Conditions and Plant Growth*. Longman Scientific & Technical, New York.

Kashmanian, R.M., H.C. Gregory, and S.A. Dressing. 1990. Where will all the compost go? *BioCycle* 31(10):38–39, 80–83.

Kimura, H. and T. Fujiwara. 1992. Studies on the improvement and maintenance of soil physical properties on a hardened soil in a orchard field. *Bulletin of the Hiroshima Prefectural Agriculture Research Center* 55:65–71.

Lanini, W.T., C.E. Elmore, and J.M. Shribbs. 1988. Experience acquise aux U.S.A. pour le paillage du sol des vergers. *Le Fruit Belge* 423:228–249.

Mantinger, H. 1990. Pratiche colturali rispettose dell'ambiente in frutticoltura. *L'Informatore Agrario* 19:51–60.

Marangoni, B. and D. Scudellari. 1989. Razionalizzazione degli interventi agronomici per migliorare la qualità delle pesche. *Rivista di Frutticoltura* 6:21–30.

McCarthy, M.G., P.R. Dry, P.F. Hayes, and D.M. Davidson. 1992. Soil management and frost control, p. 148–177. In: B.G. Coombe and P.R. Dry (eds.). *Viticulture, Volume 2: Practices*. Winetitles, Adelaide, Australia.

Moncomble, D. and A. Descotes. 1992. Couverture du sol et lutte contre les mauvaises herbes. *Phytoma - la Défense des Végétaux* 445:33–34.

Nogales, R., M.D. Mingorance, C. Elvira, and E. Benitez. 1996. Agricultural use of municipal solid waste composts and vermicomposts: effects on soil properties. *Orcazette* 4(3):5–8.

Paino, V., J.P. Peillex, O. Montlahuc, A. Cambon, and J.P. Bianchini. 1996. Municipal tropical compost:effects on crops and soil properties. *Compost Science & Utilization* 4 (2):62–69.

Paoletti, M.G., D. Pimentel, B.R. Stinner, and D. Stinner. 1992. Agroecosystem biodiversity:matching production and conservation biology, p. 3–23. In: M.G. Paoletti and D. Pimentel (eds.). *Biotic Diversity in Agroecosystems*. Symposium on Agroecology and Conservation Issues in Tropical and Temperate Regions, Padova, Italy 26–29 Sept. 1990. Elsevier, Amsterdam.

Perelli, M. 1994. *Norme per la disciplina dei fertilizzanti*. ASSOFERTILIZZANTI, Roma: p. 81.

Pinamonti, F. 1998a. Compost mulch effects on soil fertility, nutritional status and performance of grapevine. *Nutrient Cycling in Agroecosystems* 51:239–248.

Pinamonti, F. 1998b. Esperienze di utilizzo del compost, p. 249–281. In: Consorzio Italiano Compostatori (CIC). *Produzione ed Impiego del Compost di Qualità*. CIC Publication, Bologna, Italy.

Pinamonti, F. and G. Zorzi. 1996. Experiences of compost use in agriculture and in land reclamation projects, p. 517– 527. In: M. De Bertoldi, P. Sequi, B. Lemmes, and T. Papi (eds.). *The Science of Composting*. Proceedings of the International Symposium on The Science of Composting, Bologna, Italy 30 May–2 Jun. 1995. Blackie Academic & Professional, London.

Pinamonti, F., I. Artuso, G. Stringari, and G. Zorzi. 1991. Investigation of compost-mulch in the vineyards of Trentino: effects on the soil and culture, p. 29–36. In: Association Nationale pour la Protection des Plantes (ANPP). *Proceedings of the 3rd International Symposium on No-Tillage and Other Soil Management Techniques in Vines*, Montpellier, France 18–20 Nov. 1991. ANPP Annales 3, Paris.

Pinamonti, F., M. Stefanini, and A. Dalpiaz. 1996. Soil management effects on nutritional status and grapevine performance. *Viticultural and Enological Sciences* 51(2):76–82.

Pool, R.M., R.M. Dunst, and A.N. Lakso. 1990. Comparison of sod, mulch, cultivation and herbicide floor management practices for grape production in non irrigated vineyards. *Journal of the American Society for Horticultural Science* 115:872–877.

Prasad, R. and J.F. Power. 1997. *Soil Fertility Management for Sustainable Agriculture*. CRC Lewis Publishers, Boca Raton, Florida, p. 59.

Press, C.M., W.F. Mahafee, J.H. Edwards, and J.W. Klöpper. 1996. Organic by-product effects on soil chemical properties and microbial communities. *Compost Science & Utilization* 4(2):70–88.

Reg. CEE (Regolamento del Consiglio delle comunità Europee), 2078/92. 1992. *Gazzetta ufficiale delle comunità Europee* N. L 215, 30/7/1992:85–90.

Richards, D. 1983. The grape root system. *Horticultural Reviews* 5:127–168.

Robinson, J.B. 1992. Grapevine Nutrition, p. 178–208. In: B.G. Coombe and P.R. Dry (eds.). *Viticulture, Volume 2: Practices*. Winetitles, Adelaide, Australia.

Sauvage, D. 1995. La lutte contre l'érosion grâce aux mulchs. *Phytoma – la Défense de Végétaux* 478:43–45.

Scienza, A., D. Forti, G. Zorzi, and G. Cargnello. 1987. Use of composted urban wastes in viticulture, p. 533–542. In: M. De Bertoldi, M.P. Ferranti, P. L'Hermite, and F. Zucconi (eds.). *Proceedings of the International Symposium on Compost:Production, Quality and Use*, Udine, Italy, 17–19 Apr. 1986. Elsevier Applied Science, London.

Seguin, G. 1981. Recherches récentes sur le terroirs viticoles, p. 37–40. In: P. Ribéreau-Gayon and P. Sudraud (eds.). *Actualités Œnologiques et Viticoles*. Dunod, Paris.

Sequi, P. 1976. Funzioni nutrizionali della sostanza organica del terreno. *L'Italia Agricola* 113, 3:91–112.

Sequi, P., A. Benedetti, S. Canali, and F. Tittarelli. 1996. Il ruolo del compostaggio nell'agricoltura sostenibile. *Acqua Aria* 3:305–309.

Sikora, J.L. 1996. Effect of compost-fertilizer blends on crop growth, p. 423–430. In: M. De Bertoldi, P. Sequi, B. Lemmes, and T. Papi (eds.). *The Science of Composting*. Proceedings of the International Symposium on The Science of Composting, Bologna, Italy 30 May–2 June 1995. Blackie Academic & Professional, London.

Strabbioli, G. and A. Angeloni. 1987. Effects of composted agricultural residues on apple trees and of urban waste on peach trees, p. 584–597. In: M. De Bertoldi, M.P. Ferranti, P. L'Hermite, and F. Zucconi (eds.). *Proceedings of the International Symposium on Compost: Production, Quality and Use*, Udine, Italy, 17–19 Apr. 1986. Elsevier Applied Science, London.

Tellarini, S. 1994. Fabbisogno di sostanze organiche dei suoli forlivesi e valutazione del mercato potenziale di ammendanti organici, p. 49–74. In: Osservatorio Agroambientale Cesena (OAC). *Proceedings of Giornate di studio Compost: dai rifiuti una risorsa per l'agricoltura*. Indagine sulla disponibilità di rifiuti organici in provincia di Forlì, Italy 14–15 Apr. 1994. OAC Publication.

United States Environmental Protection Agency (U.S. EPA). 1994. Product quality and marketing, p. 98–110. In: USEPA. *Composting Yard and Municipal Solid Waste*. Technomic Publishing Company, Lancaster, Pennsylvania.

Van Der Westhuizen, J.H. 1980. The effect of black plastic mulch on growth, production and root development of Chenin blanc vines under dryland conditions. *South African Journal of Enology and Viticulture* 1:1–6.

Van Huyssteen, L. 1988. Grapevine root growth in response to soil tillage and root pruning practices, p. 44–56. In: J.L. Van Zyl (ed.). *The Grapevine Root and its Environment*. Department of Agriculture and Water Supply Republic of South Africa, Pretoria, Technical Communication 215.

Walsh, B.D., A.F. MacKenzie, and D.J. Buszard. 1996. Soil nitrate levels as influenced by apple orchard floor management systems. *Canadian Journal of Soil Science* 76:343–349.

Walter, B. 1980. Possibilités et limites d'utilisation des déchetes urbains en viticulture. *Bulletin de l'Office International de la Vigne et du Vin* 53 (590):273–284.

Wang, C.T., H.T. Chen, and F.J. Lay. 1991. Effects of organic manures on the yield and quality grapes. *Bulletin of Taichung Discrict Agricultural Improvement Station* (Tatsuen Hsiang, Changhua, Taiwan) 32:41–48.

Widmer, T.L., J.H. Graham, and D.J. Mitchell. 1998. Composted municipal waste reduces infection of Citrus seedlings by *Phytophtora nicotianae*. *Plant Disease* 82(6):683–688.

Zamborlini, M., F. Franzin, and M. Muni. 1990. Indagine su un campione di terreni di buona produttività:dotazione di azoto e di sostanza organica. *Agricoltura e Ricerca* 115:5–10.

Zucconi, F. 1988. Interazione tra apparato radicale e suolo. *Rivista di Frutticoltura* 1-2:105–109.

Zucconi, F. 1993. Allelopathies and biological degradation of soils. *Acta Horticulturae* 324:11–21.

Zucconi, F. and M. De Bertoldi. 1987. Organic waste stabilization throughout composting and its compatibility with agricultural uses, p. 109–137. In: D.L. Wise (ed.). *Global Bioconversions*, Vol. 3. CRC Press, Boca Raton, Florida.

Zucconi, F. and A. Monaco. 1987. Aspetti biologici della stanchezza del terreno e problemi del reimpianto del pesco. *Rivista di Frutticoltura* 2:61–66.

Zucconi, F., M. Forte, A. Monaco, and M. De Bertoldi. 1981. Biological evaluation of compost maturity. *BioCycle* 22(4):27–29.

CHAPTER 9

Compost Utilization in Sod Production and Turf Management

Allen V. Barker

CONTENTS

I. INTRODUCTION

High costs of disposal of unutilized or unrecycled byproducts and residues generated by municipalities, industries, and agriculture are causing people to consider alternatives to disposal (Rosen et al., 1993). Composting is an alternative that could save money in handling, reduce the volumes and masses of wastes, and convert organic byproducts and residues into products with high potential for use in horticultural and agronomic industries (Hyatt, 1995; Stratton et al., 1995; Stratton and Rechcigl, 1997). Furthermore, some materials, such as yard trimmings, are banned

1-56670-460-X/01/$0.00+$.50

for landfill deposition in many states, and composting may arise as the principal means of handling these restricted materials (Steuteville, 1995a, 1995b).

Research and practice have demonstrated that a wide variety of crops can be grown in media or with land-applied soil amendments developed from composts or from uncomposted byproducts and residues (Bolton, 1975; Bryan and Lance, 1991; Buchanan and Gliessman, 1991; Coosemans and van Asshe, 1983; Gouin, 1993; Hartz et al., 1996; Joiner and Conover, 1965; Ozores-Hampton et al., 1994; Purman and Gouin, 1992). Often modifications of cultural conditions are necessary to allow for changes in nutrients, pH, soluble salts, porosity, water-holding capacity, and cation exchange capacity imparted by additions of composted or raw byproducts or residues to media or to soils, but generally organic amendments improve media and soils with respect to physical and chemical properties. Composting improves suitability of raw byproducts and residues in agricultural applications by narrowing the carbon:nitrogen ratios (C:N ratios), stabilizing the form of N, decomposing potentially toxic organic residues, lowering pH of alkaline feedstock, reducing and suppressing pathogens, lessening odors, reducing metal availability, and producing other ameliorating effects such as giving a granular material for easy handling (Angle et al., 1981; Hoitink et al., 1997; Lemmon and Pylypiw, 1992; Muller and Korte, 1975; Parr and Hornick, 1992; Racke and Frink, 1989; Sikora, 1998; Vandervoort et al., 1997).

Drawbacks and concerns about agricultural uses of composts relate to the nature of the feedstocks (raw byproducts and residues) from which composts are made. They have led to underutilization of composts in horticulture and agronomy (Flanagan et al., 1993; Hauck, 1985). Long-term concerns in utilization of composts, particularly those made from biosolids, exist due to fears that metals from contaminated feedstocks may enter the food chain or be phytotoxic to crops (Bolton, 1975; Chaney, 1973; Chaney and Ryan, 1993). Opinions as to the safety of use of metal-contaminated biosolids have evolved after additional research. The current view is that composted or uncomposted biosolids are safe for land application (Chaney, 1973; Chaney, 1990a, 1990b; Chaney and Ryan, 1993), although the controversy continues (McBride, 1995).

Among other concerns about composts are perceptions that pesticides and other organic contaminants added with residues may persist during composting and be retained in the finished product (Muller and Korte, 1975; Racke and Frink, 1989). Much evidence suggests that pesticide residues and many organic contaminants are destroyed during composting (Chaney et al., 1996; Sikora, 1998; Vogtmann and Fricke, 1992). Salinity imparted by the feedstocks or developed during composting is also a major concern with some composts (Gouin, 1993). In turf and sod production, problems with general and specific ion salinity often are solved with management of compost applications (Cisar, 1994; Landshoot, 1995; O'Brien and Barker, 1995a, 1996b).

Compost has many applications in turfgrass management and production in the field, or in container-like conditions of sod production in layers of compost on plastic or on other impermeable flat surfaces (Cisar, 1994; Cisar and Snyder, 1992; Landshoot, 1995; Logsdon, 1991; Neel et al., 1978). Wildflower sods may be produced in compost-based sods for transplanting to landscapes or grown in compost laid on land for establishment of meadows (Mitchell et al., 1994; O'Brien and Barker, 1995a, 1995b, 1997; Pill et al., 1994). This chapter reviews uses of composts in management of turfgrass production in several arenas and also addresses production of sods and meadows of wildflowers, which may be grown in systems similar to those used for turfgrass management.

II. UTILIZATION OF ORGANIC BYPRODUCTS AND RESIDUES

A. Field-Production Systems

1. Uses of Uncomposted Byproducts and Residues

Processed organic byproducts and residues, such as biosolids, fermentation byproducts, and other organic residues, account for less than 1% of the total N fertilizers used in the U.S., although these materials represent large tonnages of potential, slow-release fertilizers (Hauck, 1985). Much of the early research in land applications of byproducts and residues was conducted with organic materials that were not composted prior to land application.

Successful land remediation and turf production with various land-applied biosolids has been well demonstrated (Amundson and Jarrell, 1983; Clapp et al., 1994; Feagley et al.,1994; Hue, 1995). The successful uses of biosolids amendments to land are based on improved soil physical and chemical properties, which impart favorable growing conditions for crops (Avnimelech et al., 1992; Epstein et al., 1976; Feagley et al., 1994; Giusquiani et al., 1995; Kirkham, 1974; Pagliai et al., 1981). Metals and soluble salts, including ammonium (NH_4) salts, are often-cited problems associated with uses of biosolids and have limited their widespread acceptance as soil amendments (Bolton, 1975; Coosemans and van Asshe, 1983; O'Brien and Barker, 1996a, 1996b; Sanders et al., 1986).

Uncomposted food and yard-waste portions (called heavy fractions, after removal of metals and most of the plastic and paper) of municipal solid waste (MSW) were evaluated for sod production (Flanagan et al., 1993). Soil incorporation of the heavy fractions resulted in greater air, water, and total porosity and lesser bulk density of a loamy sand relative to conditions in the unamended soil. Kentucky bluegrass (*Poa pratensis* L.) sod strength, at about 270 days after amendment, was increased relative to strength of sod produced in unamended soil. No inhibition of root growth by the heavy fraction was noted, and the plant density of bluegrass and bermudagrass (*Cynodon dactylon* Pers.) sods was increased by the amendment. Soil amendment with the heavy fraction increased post-transplanting rooting of sods of bermudagrass. A shorter production time for sods in amended soils relative to that in unamended soils was suggested. The benefits of incorporation of the MSW were increased soil fertility with respect to available N, phosphorus (P), potassium (K), calcium (Ca), and zinc (Zn) and increases in soil organic matter. The benefits lasted for 2 years.

N derived from mineralization of fermentation byproducts from industrial production of antibiotics and organic acids was sufficient to produce high quality stands of mixed turfgrasses for 2 or 3 years following land applications without supplemental N fertilizers (Wright et al., 1982). Depending on the amounts added (112 or 224 $Mg \cdot ha^{-1}$), applications of the materials also increased soil pH by about one unit (pH 5 to pH 6). However, recovery of N from excessive byproduct applications of 224 $Mg \cdot ha^{-1}$ was only 5 to 11% of applied N. This high application also inhibited seed germination and growth if Kentucky bluegrass, perennial ryegrass (*Lolium perenne* L.), or Chewings fescue (*Festuca rubra* ssp. *commutata* Gaud.) was seeded immediately following application. Generally, a waiting period between application of raw or immature composts and seeding of turfgrass is prescribed to avoid phytotoxicity to germinating seeds and to seedlings (O'Brien and Barker, 1995b, 1996b, 1997).

Similar to the effects of the raw fermentation byproducts, additions of deinked paper sludge or primary paper mill sludge caused development of poor ground cover and poor Kentucky bluegrass and perennial ryegrass quality during the year following applications to land (Norrie and Gosselin, 1996). However, in this case, the inhibitory effects were attributed to the low nutrient status of the paper sludges and were overcome by N fertilization. By the end of the year following application, the inhibitory effects had disappeared.

2. Use of Composted Byproducts and Residues

Composting of byproducts and residues has been suggested to improve the quality of wastes for land application (Camberato et ai., 1997; Edwards et al., 1995; Sikora, 1998). Production and characteristics of composts for horticultural and agronomic uses have been reviewed recently (Barker, 1997; Stratton et al., 1995; Stratton and Rechcigl, 1997). Several advantages of use of composted materials relative to uncomposted materials were suggested by these authors. Some of the benefits of composting and uses of composed materials follow.

a. Destruction of Organic Contaminants

Contents of persistent organic chemicals are considered important in determinations of quality of composts for land application (Barker, 1997). These chemicals may be introduced into the composts as xenobiotic compounds (organic contaminants) in the original materials (feedstocks), or they may be formed during composting (Barker, 1997). Nevertheless, Chaney et al. (1996) and Vogtmann and Fricke (1992) have shown that mature or immature composts are generally low in xenobiotic compounds and that setting of maximum tolerance levels for these substances is unneeded in agricultural applications.

Composting also has potential for reducing concentrations of various unwanted organic residuals that enter with feedstocks derived from agricultural operations, including materials generated in turfgrass management. Composting will degrade various organic compounds in feedstock if the process of composting is carried out with proper aeration, moisture, C:N ratios, and duration (Sikora, 1998). Also, the temperatures generated in composting result in proliferation of thermophilic microorganisms that might degrade organic compounds more efficiently than mesophilic organisms (Sikora, 1998).

Lawn and turf care pesticides may be carried with uncomposted grass clippings to vegetable and flower gardens or to other sites of use and impart plant injury. Composting destroyed nearly half of the 2,4-dichlorophenoxyacetic acid residues of lawn clippings (Michel et al., 1995). Composting of contaminated feedstocks of mixed broadleaf tree leaves and grass clippings converted diazinon {O,O-diethyl O-[6-methyl-2-(1-methylethyl)-4-pyrimidinyl] phosphorothionate} to less phytotoxic metabolites or to other residues with low bioavailability, although the fraction of mineralization of total diazinon to carbon dioxide was much smaller (11%) than mineralization of total organic carbon (48%) in the feedstock (Michel et al., 1997). Degradation of insecticide and herbicide residuals to low or nondetectable levels occurred in grass clippings taken from test plots of sprayed turfgrass and stored or composted for 21 days (Lemmon and Pylypiw, 1992). After 1 year of composting of grass clippings, herbicide residuals were reduced to concentrations at or below levels of detection (Vandervoort et al., 1997). However, not all organic contaminants in feedstocks are degraded during composting. In laboratory composting of biosolids and wood chip feedstocks for 18 to 20 days, the insecticide, carbaryl (1-naphthyl methylcarbamate), was degraded to 3 to 4% of its original concentration, but a recalcitrant polyaromatic hydrocarbon, phenanthrene, persisted at 89 to 93% of its original concentration (Racke and Frink, 1989). Nevertheless, the general conclusion from research with pesticide degradation by composting was that pesticide concentrations were lowered to nonhazardous levels for crops in soils receiving the composts.

The effects of compost application on the behavior of pesticides on land or in soil are variable and may be imparted onto land with or without turfgrass covers. Generally, compost additions to land lessen pesticide loading in surface waters by restricting runoff in the absence of established turf cover (Malone et al., 1996). Turfgrass covers on land may facilitate pesticide degradation even if compost is not applied. Rapid pesticide degradation can occur in thatch (Sears and Chapman, 1979). Unlike some other crops, turfgrass leaves and thatch are strong sorbents of organic compounds and have substantial effects on retention and transformations of chemicals applied to turf (Lickfeldt and Branham, 1995; McDowell et al., 1985).

The principal effects of composts on activities of pesticides in soil follow the actions of composts on the chemistry and biology of the soil. Modifications of pesticide activity in soil depend on the nature and reactivity of the organic additions and their effects on microbial activities (Alvey and Crowley, 1995; Hance, 1973). Compost additions improved degradation of herbicides [benthiocarb (S-4-chlorobenzyl diethylthiocarbamate) and MCPA (4-chloro-2-methylphenoxyacetic acid)] in soil (Duah-Yentumi and Kuwatsuka, 1980). However, additions of compost suppressed soil mineralization of atrazine (2-chloro-4-ethylamino-6-isopropylamino-*s*-triazine) relative to rates in unamended soils or in soils amended with starch or rice hulls (Alvey and Crowley, 1995). The suppressive effect of compost on atrazine mineralization was attributed to the high N conditions of the compost because all treatments receiving supplemental inorganic N fertilization also gave lower rates of atrazine mineralization than those without N addition (Alvey and Crowley, 1995).

If organic matter and N are limiting factors in mineralization, additions of compost might stimulate microbial activities and, consequently, accelerate pesticide degradation, as was demonstrated with degradation of atrazine in soils amended with farm manures, straw, and N fertilizers (Hance, 1973). On the other hand, organic matter additions, as with compost incorporation into soil, initially may increase stabilization and decrease mineralization of pesticide residuals by increasing soil organic matter and pesticide sorption (Barriuso et al., 1997). Benefits may be accrued from sorption, as sorption of pesticides to organic matter generally restricts their leaching (Bellin et al., 1990; Guo et al., 1993), but some organic amendments produce soluble organic compounds that enhance desorption and water solubility of pesticides (Barriuso et al., 1992).

b. Enhancement of Soil Fertility and Plant Nutrition

Some composts are considered as dilute fertilizers to carry plant nutrients to land (Barker, 1997). The principal nutritional benefit given to crops from composts usually is the N carried in slow-release form, although composts can carry other nutrients based on the feedstock (Chaney, 1990a, 1990b; Dick and McCoy, 1993; Ozores-Hampton et al., 1994; Wen et al., 1995). Since fertilization of established turfgrass is largely with topical applications of dry materials, interest in use of slow-release fertilizers has been strong for purposes of N conservation. For example, ammonia volatilization is a principal environmental and economic concern from surface application of N fertilizers to turfgrass or to bare soil. Losses of N following surface applications of urea to grass sods are rapid and range from 13 to 60% of the total N application depending on the amount and method of application (Petrovic, 1990; Titko et al., 1987; Volk, 1959). Torello et al. (1983) reported substantial N losses with prilled or sprayed urea applied to Kentucky bluegrass turf but noted much smaller losses with slow-release, sulfur-coated urea. In other research, release of N from slowly dissolving sulfur-coated urea formulations was too slow to maintain good, green color of Kentucky bluegrass during the growing season unless applications were large (245 $kg \cdot ha^{-1}$ of N), whereas residual effects from the more rapidly dissolving forms were too low to maintain good grass quality throughout the following growing season (Hummel and Waddington, 1984).

Dry, activated biosolids have been evaluated relative to slow-release chemical N fertilizers and were shown to be of varying values, with summer applications being better than winter applications during which losses of N occurred (Volk and Horn, 1975). N losses from surface-applied farm manures and biosolids also can be substantial due to the volatility of NH_4 in the materials (Beauchamp, 1983; Christensen, 1986). Research with controlled-release fertilizers and organic byproducts led to work evaluating composts as potential sources of slowly but readily available plant nutrients.

Composting allows for stabilization of the nitrogenous content of organic manures so that losses are restricted after their application to land (Lionello and Francesco, 1993; Stratton et al., 1995; Terman, 1979). The mineralization of N from composted biosolids was estimated to be half the rate of mineralization from uncomposted biosolids (O'Keefe et al., 1986). The mineralization rates of N in soil-incorporated composts ranges, over a growing season, from 10 to 30% of the total N present (Hadas and Portnoy, 1994; Mattingly, 1956). The slow mineralization rate of organic N in composts allows them to be considered as environmentally sound alternatives to controlled-release chemical fertilizers that have been used widely in turfgrass management, particularly with respect to conservation of N that could be lost by nitrate leaching from rapidly mineralizing or soluble fertilizers.

The slow mineralization rates of composts can be a drawback in providing N to crops, thus requiring that large applications be made to provide adequate N. Sufficient quantities of compost may not be available at one time. Also, applications may have to be made in several passes over the field. Consideration must be given to the total amount of compost that is applied in one or multiple applications. Gouin (1993) recommended that compost applications to landscapes not exceed 300 $m^3 \cdot ha^{-1}$ (about 10 $Mg \cdot ha^{-1}$) to avoid salinity and nutrient imbalances. Landshoot (1995) recommended that compost applications of between 250 and 500 $m^3 \cdot ha^{-1}$ (2.5 to 5 cm layers) be incorporated into soil to improve stand development and growth of turfgrass. This range of amounts was sufficient to provide nutrients and to increase organic matter in soil and yet to avoid problems associated with large applications of compost to land.

Composts, as with many organic byproducts with fixed ratios of nutrients, may have an unfavorable ratio of N to other macronutrients so that applications to meet N requirements may give too high additions of other nutrients, such as P (Jacobs, 1998). Blending of low rates of compost application with chemical N fertilizers is an alternative to massive applications of compost to supply all of the N (Sikora, 1996). Blending helps to adjust the ratio of N to other nutrients to closely match crop requirements, aids in adjustment of wide C:N ratios, and facilitates N release from composts. Cisar (1994) and Cisar and Snyder (1992) noted that an initially high C:N ratio in some composts restricted release of N, caused a low-N status of foliage of several turfgrass species, and slowed establishment of sods grown directly in compost on plastic sheeting. Frequent fertilization quickly overcame these restrictions. With time, which allowed for mineralization of N from the composts and narrowing of C:N ratio, quality of sod improved with composts thereby lessening the need of further fertilization. Cisar (1994) reported additional benefits of soil amendment with composts. For example, his research demonstrated that pesticide leaching through sand amended with MSW compost was less than that through sand alone. He noted further that salinity of compost did not seem to affect turfgrass growth in pots and that electrical conductivity dropped from 2.65 $dS \cdot m^{-1}$ to 0.65 $dS \cdot m^{-1}$ with rainfall and irrigation. Landshoot (1995) reported that composts with salinities ranging from electrical conductivities of 4 to 8 $dS \cdot m^{-1}$ did not hinder Kentucky bluegrass growth in soil amended with compost (250 or 500 $m^3 \cdot ha^{-1}$), since dilution of the compost with soil seemed to alleviate any deleterious effects due to salinity.

Landshoot (1995) reported that incorporation of compost at 250 or 500 $m^3 \cdot ha^{-1}$ with a starter fertilizer [40N-20P-20K ($kg \cdot ha^{-1}$)] increased the rate of bluegrass establishment in land (clay loam) relative to the rate in unamended land or in topsoil-amended land. Biosolids compost, brewery byproduct compost, farm manure-yard trimmings compost, and paper mill byproduct compost gave best results due to their effects in improving soil physical conditions, such as water-infiltration rates, and their relatively high amounts of available N and P. Composts of yard trimmings and spent mushroom substrate improved soil physical conditions but were not as effective as the other composts for supplying nutrients. Use of starter fertilizers increased the rate of sod establishment with all composts except those made from biosolids or brewery byproducts, which contained enough readily available plant nutrients to mask any effects from the starter fertilizers.

Landshoot (1995) noted that one poultry manure compost and another yard-trimmings compost in his evaluations were unsatisfactory for use in turfgrass establishment. A 30% seedling death rate with the poultry manure compost was attributed to its high NH_4 concentration. High NH_4 concentrations are toxic and are characteristic of immature composts (O'Brien and Barker, 1995a, 1995b, 1996a, 1996b, 1997). The yard-trimmings compost was deemed unsatisfactory because of its high weed seed population, which imparted a 50% weed cover to plots. Weed seeds can be introduced with immature compost of seed-containing feedstocks or with composts in which weeds were allowed to grow and to produce seeds (Landshoot, 1995).

Landshoot (1995) reported that the improved soil structure with respect to water-infiltration rate and improved plant nutrient supply lasted for at least 3 years. After 3 years, compost-amended soils had water-infiltration rates of 25 cm or more per hour compared to less than 2.5 cm per hour with unamended or topsoil-amended soil. Turfgrass growing in unamended soil or in topsoil-amended soil showed symptoms of P deficiency. Landshoot (1995) estimated that composts supplied N for 3 years, but in decreasing amounts with time, and suggested using monitoring of turf color and soil testing to recommend amounts of N fertilization.

Landshoot (1995) suggested also that composts can be used as surface applications to established turfgrass areas. With surface applications, compost should be applied in about 0.65-cm thick layers and then incorporated into soils with aeration tillage with a hollow-tine soil aerator implement and a drag mat to break up the cores and to mix the compost into the soil. Successive surface applications without incorporation should be avoided since the unincorporated compost leads to a dry surface layer that appears to restrict turfgrass rooting in the surface zones. However, incorporation by aeration can be stressful to turfgrass in hot, dry periods; hence, surface applications should be restricted to cool, moist seasons when grass is growing actively (Landshoot, 1995).

In production of ryegrass (*Lolium multiflorum* L.) in fine sands (Myakka fine sand) of south Florida, supplemental fertilization with ammonium nitrate (50 $kg{\cdot}ha^{-1}$ of N) improved stand establishment and forage yields with compost applications supplying N at amounts ranging from 168 to 672 $kg{\cdot}ha^{-1}$ (Stratton and Rechcigl, 1998). Over a 2 year period, yields of compost-amended areas were twice those of areas receiving only ammonium nitrate applied at equivalent N rates. Sullivan et al. (1998) reported similar results with forage tall fescue (*Festuca arundinacea* Schreb.) with compost applications enhancing yields for 3 years after an initial application. Land applications (155 $Mg{\cdot}ha^{-1}$) of yard trimmings-food residues compost with an average N concentration of 17 $g{\cdot}kg^{-1}$ gave twice the dry matter yields of grass produced with wood chips-sawdust-food residues compost with an average N concentration of 8 $g{\cdot}kg^{-1}$ (Sullivan et al., 1998).

Compost applications in addition to supplying nutrients to land can increase availabilities of soil-borne nutrients or add nutrients other than N to soil. Yard waste compost amendments increased the soil solution P of an acid soil (Ultisol, pH 4.85) with high P-fixing capacity (Hue et al., 1994). However, Wen et al. (1997) noted no increases in P accumulation by several non-turf crops and no increase in sodium bicarbonate extractable soil P following applications of biosolids compost, livestock manure compost, or uncomposted biosolids. Wen et al. (1996/1997) noted, however, that K availability from the composts or uncomposted biosolids was equal to that from KCl if enough organic materials were added. Incremental additions of compost also increased the pH and extractable Ca of media. On the other hand, some composts may lead to nutrient immobilization in soil (Barker, 1997), but simultaneous applications of N with composts can alleviate immobilization of N and later facilitate release of N from composts (Stratton and Rechcigl, 1998).

Barker (1997) suggested that composts with less than 1% N (10 $g \cdot kg^{-1}$) are more suitable as mulches than as sources of nutrients. Hartz et al. (1996) with pepper (*Capsicum annuum* L.) production noted no net release of N from field-incubated composted green plant residue and noted N immobilization with the compost incubated in containers in studies of growth of tomato (*Lycopersicon esculentum* Mill.) or marigold (*Tagetes erecta* L.). Supplemental N fertilization was required to sustain growth and to obtain optimum yields with use of the compost, which was 1.1% N (11 $g \cdot kg^{-1}$).

Peacock and Daniel (1992) noted that initial release of N from organic fertilizers may be slow due to the need for microbial degradation. However, microbial inoculation of an organic fertilizer (mixture of soybean [*Glycine max* Merr.] meal, blood meal, bone meal, and potassium sulfate) did not enhance tall fescue or hybrid bermudagrass (*Cynodon dactylon* [L.] Pers. x *Cynodon transvaalensis* Davy.) growth or N absorption relative to use of the uninoculated fertilizer, which released less than 20% of the N relative to an equivalent application of urea. These results suggest that the release of N from composts with inherently low N may be insufficient for turfgrass fertilization and that supplemental N applied with the compost would increase the benefits achieved from compost.

c. Preparation of Seedbeds for Turf or Sod Production

Composts have been used successfully in soil preparation for turfgrass establishment, sod production, and landscape operations. Angle et al. (1981) reported that the hazards of direct application of biosolids to land were lessened by composting. Composting of biosolids reduced the number of pathogenic organisms, lessened the availability of heavy metals, reduced odors, and gave a granular material that was easier to handle relative to uncomposted biosolids. Furthermore, applications of biosolids compost (Angle et al., 1981) improved establishment rate from seed and general appearance of turfgrass of mixed species even with compost applications of 260 $Mg \cdot ha^{-1}$ or larger. The effects were attributed to improved physical conditions such as higher water-holding capacity and chemical properties such as higher pH of acid soils and higher nutrient availabilities. Also, Giusquiani et al. (1995) noted that, following incorporation of MSW compost in land, total pore volume in soils increased and that enzymatic activities indicative of increased microbial activity were enhanced. Applications of MSW compost to a clay loam stabilized aggregated soil structure, prevented crusting, and aided in the reclamation of saline-sodic soils (Avnimelech et al., 1992).

Bevacqua and Mellano (1994) reported higher fresh clipping-weight yields of tall fescue grown in sandy loam amended with biosolids compost in field or in greenhouse culture relative to those obtained with equal applications of dried, uncomposted biosolids. No statistical differences in accumulations of copper (Cu), Zn, nickel (Ni), or cadmium (Cd) in leaf tissues were noted, but with the exception of Cd, measurable increases in these metals were detected in soil. Relative to unamended soil, soil treated with a cumulative application of compost (74 $Mg \cdot ha^{-1}$) over a 2-year period had slightly more acidic pH (7.4 vs. 7.7), higher organic matter concentrations (1.50 vs. 0.77%), higher primary nutrients (N, P, K), and increased salinity (EC 2.44 vs. 1.52 $dS \cdot m^{-1}$) .

O'Brien and Barker (1995a) evaluated composts, made from MSW-biosolids, biosolids-woodchips, farm manures-crops residues, and autumn leaves, in the preparation of seed beds for perennial ryegrass or wildflower (mixed annual and perennial species) establishment. Composts were applied in thin layers (2.5-cm thick) on the soil surface or the same amount was incorporated shallowly (5-cm deep) into soil. Mature composts with low ammoniacal concentrations (<150 $mg \cdot kg^{-1}$ of N) were essential for establishment of ground cover by turfgrass or by wildflowers. Bare spots and weeds dominated the areas where the sod crops were not established due to high NH_4 (>2000 $mg \cdot kg^{-1}$ of N) in immature composts. The areas of suppressed crop growth persisted into the second and several ensuing growing seasons. To avoid the phytotoxicity of immature composts, a delay of 7 to 21 days between application of composts and seeding of perennial ryegrass was necessary to reduce NH_4 contents (O'Brien and Barker, 1996a, 1996b). Composts that were aged for 1 year by storage in piles after they were first considered suitable for land application were more suitable for production of perennial ryegrass or wildflower sods on plastic sheeting than were composts used without the additional curing (O'Brien and Barker, 1995b).

Another problem noted by O'Brien and Barker (1995a) was the proliferation of weeds following surface-applied or soil-incorporated composts. Nutrients from composts stimulated weed growth during the summer season of the first year following spring application of composts so that the areas even without kill from the immature composts were dominated with annual broadleaf and grassy weeds (O'Brien and Barker, 1995a). Establishment of annual wildflower species was unsuccessful because of weed competition. However, early in the second growing season following a winter, turfgrass or perennial wildflower populations dominated weeds in the stands so that the appearance of the turf and wildflower meadows was excellent.

Weeds that grew during the first year arose from germination of soil-borne seeds (O'Brien and Barker, 1995a). Weed control was accomplished if a barrier mulch (newspaper, plastic fabric, or sawdust) was placed under a thin layer (1-cm to 2.5-cm thick) of compost mulch to impede emergence of weeds above the soil surface (Barker and O'Brien, 1995). A application of pre-emergence herbicide to the soil surface immediately before compost application also essentially eliminated problems with weed establishment in compost-mulched soil (Barker and O'Brien, 1995). Weed control in the first year allowed for establishment of an abundance of annual wildflowers, which were lost without weed control. No N fertilization was needed for at least 2 years to sustain meadows after the initial applications of compost if stands were established with either surface-applied or shallowly incorporated composts.

Similar to the effects noted with immature composts on establishment of wildflower meadows, harmful effects of compost use in sod production have been reported. Increasing applications of MSW-biosolids composts up to 60 $Mg{\cdot}ha^{-1}$ reduced density of Kentucky bluegrass cover and increased weeds and bare soil in the first year following treatments (Breslin, 1995). The deleterious effects of the composts persisted for more than a year before marketable sod was produced, but eventually no differences in turf quality were measured among treatments. The problems were associated with immaturity of the composts, which resulted in high NH_4 and soluble salt concentrations in the soils (Breslin, 1995; O'Brien and Barker, 1995a, 1995b). Any metals transmitted to the soil by the composts remained in the top 0- to-5-cm zone of the soil, and no enrichment of metals occurred in bluegrass foliage (Breslin, 1995). No enrichment of groundwater with plant nutrients was detected, and concentrations of trace elements in groundwater were below levels of detection. Angle et al. (1981) also reported that elemental analysis of turfgrass foliage (mixed Kentucky bluegrass, red fescue [*Festuca rubra* L.], and tall fescue seeding) showed no excessive accumulations of nutrients or nonessential or toxic elements following application of biosolids compost to land.

d. *Other Considerations in Use of Composts on Turfgrass*

Thermophilic conditions and intense microbial competition during composting kill or inactivate pathogenic organisms and some viruses, rendering compost much safer for application to land than uncomposted residues, which can serve as an inoculum for infection of crops (United States Environmental Protection Agency, 1998). Furthermore, mature compost may contain natural organic chemicals and microorganisms that kill or suppress disease-causing organisms. Chapter 12 in this book (see Section III) discusses suppression of soil-borne plant pathogens by compost.

Hoitink et al. (1993) note that the nutritional value of organic matter for soil microorganisms dictates the activities of biological control agents in compost. Nitrogen-rich composts stimulate disease incidence, particularly with immature composts in which NH_4 concentrations are usually high. For example, *Fusarium* diseases may be increased by the high NH_4 that is often present in biosolids composts (Lumsden et al., 1983). The phytotoxicity of NH_4 or the effects of NH_4 on stimulation of diseases hindered production of perennial ryegrass or wildflower sods on plastic (O'Brien and Barker, 1995a, 1995b, 1996a, 1996b). Consequently, to lower incidences of diseases in production of sods on plastic or other impermeable surfaces, it is important that mature composts be used. However, excessively stabilized organic matter, such as dark peat, has limited abilities to sustain a general microbiological mass with biological control agents, and root rot prevails (Hoitink et al., 1993, 1997). Therefore, excessively mature composts would have limited effects on disease control in organic layers for production of sods on impermeable surfaces.

Topdressing of composts onto turfgrass is suggested to have suppressive effects on diseases (Nelson, 1996). Monthly topdressed applications formulated from various composts (poultry litter, poultry-cow manure, or biosolids) or organic fertilizers (animal or plant meals) mixed with sand (30% compost or organic fertilizer by volume) suppressed dollar spot (*Sclerotinia homeocarpa* F.T. Bennett) of creeping bentgrass (*Agrostis palustris* Huds.) or annual bluegrass (*Poa annua* L.) turf as effectively as the fungicides propiconazole [1-((2-(dichlorophenyl)-4-prophy-1,3-dioxolan-2-yl)methyl)-1H-1,2,4-triazone] or iprodione [3-(3,5-dichlorophenyl)-N-(1-methylethyl)-2,4-dioxo-1-imidazolidinecarboximide] (Nelson and Craft, 1991). All effective organic amendments remained suppressive for 30 days, but at 60 days after application dollar spot intensity did not differ among treatments. Cook et al. (1964) reported significant suppression of dollar spot with biosolids compost but noted no significant effect with inorganic N fertilizers. Markland et al. (1969) also observed the greatest suppression of dollar spot incidence following applications of biosolids compost among several N sources tested. In another study, composted yard waste and composted leaf topdressings improved the appearance of established creeping bentgrass plots relative to unamended plots, but the composts had no effects on severity of diseases (Grebus et al., 1995). In this case, the improvements in turfgrass appearance with compost application were attributed to increased soil fertility (plant nutrient supply) with the topdressing. Fortification of the composts with biological control agents *Trichoderma hamatum* 382 and *Flavobacterium balustinum* 299R_2 provided higher control of dollar spot than applications of chlorothalonil (tetrachloroisophthalonitrile) or no control measure (Grebus et al., 1995).

Studies on the effects of composts on other plant pests are uncommon and have shown variable results. For example, applications of yard waste composts had little effect on densities of plant-parasitic nematodes (*Meloidogyne incognita*) in soils in which vegetable crops were grown in north Florida (McSorley and Gallaher, 1995). In another case, applications of composted cow manure or activated biosolids increased grubs of the green June beetle (*Cotinis nitida* L.) in tall fescue or Kentucky bluegrass turf plots (Potter et al., 1996).

B. Production of Sods in Composts on Impermeable Surfaces

Land for quality sod production in fields must be levelled, drained, cultivated, and fumigated prior to seeding or transplanting of sprigs. During production times of 180 to 540 days, the soil must be fertilized regularly and treated with pesticides (Neel et al., 1978). Research and practice have suggested that production of sods in layers of compost or other organic materials placed on impermeable surfaces, such as plastic sheeting, is an option to field production of sods (Cisar and Snyder, 1992; Decker, 1975, 1989; Logsdon, 1991; Mitchell et al., 1994). The technique for sod production in composts on impermeable surfaces has several advantages over field production (Decker, 1975, 1989; Flanagan et al., 1993; Logsdon, 1991; Neel et al., 1978).

The production process in compost is rapid, requiring as few as 40 to 60 days rather than the hundreds of days required in commercial field production (Decker, 1989; Flanagan et al., 1993; Neel et al., 1978). The system on impermeable surfaces offers a clean environmentally constructive use of organic byproducts and residues, which otherwise might be disposed of at high costs or wasted. Weeds are not a problem in sod production in compost provided the compost is not contaminated with weed seeds (Logsdon, 1991). Sods produced in compost also can be largely free of nematodes and other plant pathogens (Neel et al., 1978). Sod production in composts derived from biosolids or other residues avoids point sources of contamination, since only a thin layer of material is used in sod production thus spreading the products over a large area of land after transplanting. Inputs of resources are no higher or are less in sod production on impermeable surfaces than in fields (Logsdon, 1991). No more water is required to produce sods on the surfaces than in fields. Fertilization may not be needed because of the nutrients conveyed by the composts.

Since many impermeable surfaces other than plastic sheeting are suitable for sod production in compost, land can be conserved. Parking lots, abandoned roads and airstrips, and other hard-surfaced areas can be employed (Neel et al., 1978), provided the issue of drainage is addressed. Decker (1989) did not note that drainage was a problem in comments about turf production on impermeable surfaces but stated that irrigation was required. In areas such as south Florida, commercial production of sod is mainly on shallow muck lands, which are 20 to 50 km away from where the sod is utilized in landscaping (Neel et al., 1978). Production of sods on impermeable surfaces can allow for the sites of production to be close to areas of utilization and close to the sites of byproduct generation. Ultimately, in south Florida and other areas of intensive sod production on organic soils, alternative sites to sod production on mucks must be found as the muck sites are depleted by oxidation of organic matter and export of sods to urban areas. Concentration of sod production on mineral soils is not a good alternative, since sod production on mineral soils is expensive as a result of the added costs of transport of the heavier sods relative to muck-grown sods. Also, sods produced on muck soil often have outperformed sods from mineral soils with respect to rooting, establishment, and growth rate (King and Beard, 1972; Peacock and Dudeck, 1985). Development of a sod-production system in compost on plastic or paved surfaces near urban areas and the utilization of urban byproducts and residues could conserve land resources, reduce waste disposal costs, lessen distances and costs of transport of sods, and provide for production of sods with characteristics of those from organic soils.

Burns and Boswell (1976) demonstrated that satisfactory bermudagrass sods could be produced in an uncomposted biosolids substrate. However, O'Brien and Barker (1998) reported difficulties in seedling emergence and establishment of perennial ryegrass in lime-stabilized biosolids without additions of acidifying agents, such as aluminum sulfate or elemental sulfur. Composting usually helps to improve the suitability of materials for container production of plants (Barker, 1997; Chen and Hadar, 1987; Gouin, 1993; Stratton et al., 1995). However, fresh, immature, or unstable compost required amendments with peat, sand, or limestone to improve germination of perennial ryegrass and six other plant species in the compost, which was derived from organic matter mechanically separated from domestic refuse (Keeling et al., 1994). Growth-suppressing phytotoxicity of the compost was confined to the low molecular weight fraction, which appeared to be principally acetic acid, whereas the high molecular weight fraction of humic acid-like substances possessed slight growth-stimulating properties. Phytotoxicity diminished with time, and at 180 days of growth, identical total shoot biomasses were obtained with unamended refuse compost and with the compost in a sand substrate.

In other research, bermudagrass and bahiagrass (*Paspalum notatum* Flugge) sprigs were transplanted into 10-cm thick layers of various combinations of composted biosolids, composted sugarcane processing byproducts, composted wood chips, sandy muck, and water-treatment sludge laid on plastic (Neel et al., 1978). In 65 days with bermudagrass and in 51 days with bahiagrass, commercially acceptable sods were produced in the layers. The sods rooted into underlying soil within 7 days after transplanting. Based on root weights and force to uproot the sods, the compost-grown sods rooted better after transplanting into underlying soil than field-grown sods harvested from mineral soils. The better rooting was attributed to the fact that compost-grown sods retained root apical meristems because sods were lifted from the plastic, whereas in harvest of field-grown sods the root apices were cut from the roots, resulting in restrictions in rooting after transplanting. The compost-based media that resulted in the best and worst evaluations differed with the species, but the results showed that composts from a number of byproducts and residues can be used for rapid production of sods on plastic (Neel et al., 1978).

Decker (1989) produced tall fescue sods in 4 or 5 weeks in composted biosolids or feed lot wastes or in spent mushroom compost placed on plastic sheeting. The fescue was 27 cm tall in 4 weeks in the composts and was at least twice the height of fescue produced in a clay loam laid on the plastic. The growth of the fescue was also at least twice the height of Kentucky bluegrass grown on the composts. Decker (1989) attributed the poor relative performance of bluegrass on plastic to the effects of summer heat. He stated that biosolids compost required leaching with 10 cm of water for 2 weeks for establishment of sods in the compost over plastic. No pesticides or other adjuvants were used in the sod production.

Cisar and Snyder (1992) evaluated mixed MSW compost in the production of bahiagrass, St. Augustinegrass (*Stenotaphrum secundatum* Kuntze), and hybrid bermudagrass on plastic sheeting. After about 42 days from seeding or transplanting of sprigs, the grasses had discoloration, which was attributed to immobilization of N or to high salinity (2.85 $dS \cdot m^{-1}$) in the compost. Fertilization did not alleviate the early problems with sod quality, but after 90 and 150 days of growth, fertilized sods had higher quality and density than unfertilized sods. The fertilized sods grown on plastic were ready for harvest in 150 days, whereas sods grown conventionally on fields needed 270 to 720 days to develop into a harvestable product. Compost-grown sod and commercially field-grown sod were equal in tear resistance, but at 7 and 21 days after transplanting on land with a sand soil, compost-grown sods produced higher root masses and longer roots in the underlying soil than the commercially grown sods.

Roberts et al. (1995) utilized composts of biosolids or yard residues in greenhouse experiments with tall fescue, perennial ryegrass, and Kentucky bluegrass. Although performance of individual cultivars and species differed with composts, shoot biomass production was deemed satisfactory in all media. However, shoot growth was not as good in a biosolids compost of pH 5.18 as with another biosolids compost at pH 6.88 or a yard waste compost at pH 8.05. The low pH of the one biosolids composts indicates a lack of maturity (Barker, 1997; O'Brien and Barker,

1995a). Bioaccumulation of metals was not detected in samples of shoot tissues of grasses grown in any of the composts (Roberts et al., 1995).

In field applications, Pill et al. (1994) used 10-cm-thick seedbeds created from MSW-biosolids composts and an iron-rich byproduct from titanium oxide production to establish wildflower plantings. Surface-applied mixtures of the compost and byproduct gave the best wildflower growth relative to growth in surface-tilled, unamended soil or in the materials used alone and left on the surface. With sod production in 4-cm-thick compost layers on plastic, O'Brien and Barker (1997) reported that the best wildflower sods with respect to seed germination, stand establishment, and intensity and diversity of bloom over two seasons occurred in mature biosolids-woodchip compost and in poultry manure-cranberry presscake compost in comparison to production in soils, leaf compost, or immature biosolids-woodchips compost. No fertilization was required with the best performing composts. Sods produced in soil (silt loam topsoil) on plastic were weed infested and required N fertilization, which promoted weed number and biomass more than wildflower growth. N deficiency also restricted biomass and quality of sods produced in leaf compost on plastic. Immature composts with high NH_4 (2000 $mg \cdot kg^{-1}$ of NH_4-N) or high salinity (4 to 8 $dS \cdot m^{-1}$) inhibited seed germination and stand establishment (O'Brien and Barker, 1995b). Composts that were aged for a year after initial composting allowed for better seedling emergence, stand establishment, appearance, and shoot biomass of mixed wildflower or perennial ryegrass sods than composts that were used when they were first considered mature and suitable for agronomic or horticultural uses (O'Brien and Barker, 1995b).

The preceding research showed that sods produced in layers of compost on impermeable surfaces, such as plastic sheeting, become established faster than those harvested from mineral soils. Sods produced in layers of compost have the favorable characteristics of sods produced on organic soils and more. Sods produced on muck or in compost have less mass (weight) than sods produced on mineral soils and are cheaper to transport (Neel et al., 1978). Sods produced in organic soils or on compost become established faster after transplanting than those grown on mineral soils. Kentucky bluegrass sods produced on organic soils rooted more rapidly than those produced on mineral soils (King and Beard, 1972). However, Peacock and Dudeck (1985) noted that St. Augustinegrass sods produced on sand soil initially rooted better than those produced on muck, but turf quality and growth rate were greater with sods produced on the organic soils. Field-grown sods must have thick, intertwining rhizomes or stolons to hold together during transplanting because most of the roots are removed at harvest. In contrast, sods grown in layers of compost on plastic or on other impervious surfaces are held together by the matted roots, which develop more rapidly than stolons or rhizomes; hence, production of tear-resistant sods is more rapid in layers of compost than in soil in fields. Thick-cut (5 cm) commercial sods removed from soils root more slowly than thin-cut ones because roots that grow into the receiving soil are generated by the crowns and not from regrowth of secondary roots. However, thin-cut sods desiccate rapidly (King and Beard, 1972; Madison, 1970; Neel et al., 1978).

III. SUMMARY

Composts have been evaluated for turfgrass management of several species (Table 9.1). Composts are effective substrates for incorporation into land to improve soil fertility of seedbeds through imparting better soil physical properties, such as enhancing aggregation, limiting crusting and compaction, and increasing water-holding capacity and drainage. These physical improvements are evident in coarse-textured or in fine-textured soils, and the effects may last for 2 or 3 years after the application of composts.

Compost additions to soils also improve soil fertility with respect to plant nutrients, largely N and P, which are components of composts or which are made more available in soils receiving composts. The enhancement in soil fertility due to nutrients may also last for 2 or 3 years, but the effects tend to diminish after 1 year. In the initial year, agronomic rates of application of biosolids and farm manure composts may supply enough nutrients for turf establishment without supplemental fertilization. With yard-trimmings compost, autumn-leaf compost, and MSW composts, starter fertilization is required because of the low N concentrations, wide C:N ratios, and slow mineralization of these materials.

Soil acidity may be raised or lowered favorably by incorporated compost amendments, depending on the nature of the soils and composts. For soils with pH below 5.5 at the time of application of compost, addition of lime may be needed (Landshoot, 1995). Composts of lime-stabilized biosolids have liming equivalencies of 25% or higher of that of agricultural limestones (Logan and Harrison, 1995; O'Brien and Barker, 1998). Consequently, application rates of composts from lime-stabilized feedstocks may need to be governed to avoid overliming.

Topdressed applications of composts can improve the performance of turfgrasses, but with surface applications, a method of working the compost into the soil must be used. Shallow incorporation of a few centimeters with aerators and dragging is usually sufficient (Landshoot, 1995). Sometimes, topdressing must be restricted to times of year during which grasses are rapidly growing to prevent imparting stresses to the turf. The principal benefit of topdressing is from the nutrients carried by the compost; hence, relatively nutrient-rich composts should be used for topdressing. In addition to supplying nutrients, some topdressed applications of composts have disease-suppressing effects that are equivalent to those of commercial fungicides, but frequent applications, perhaps in 30-day cycles, of composts are required over a growing season to accomplish disease control.

Immature composts should be avoided for soil incorporation or for topdressing because of potential disease-stimulating effects and phytotoxic levels of NH_4 in these materials. Aging in piles or delay in seeding or sprigging of grasses after compost application helps to avoid these problems.

Composts have been used as a medium for sod production in layers on impermeable surfaces (Table 9.1). Polyethylene plastic sheeting is a common surface on which to grow sods by this system. Many substrates, such as uncomposted byproducts, residues and wastes, peats, sand, and soil, are suitable materials for production of sods on plastic, but compost is the superior material. For this purpose, compost is light weight, often rich in nutrients, and can be essentially weed, insect, and

Table 9.1 Compost Utilization in Turfgrass Management and Sod Production

Species Name			
Common	**Scientific**	**Area Researched**	**Reference**
Annual bluegrass	*Poa annua*	Turf management	Nelson and Craft, 1991
Bermudagrass (hybrid)	*Cynodon dactylon* x *C. transvaalensis*	Turf management	Peacock and Daniel, 1992
Chewings fescue	*Festuca rubra*	Turf management	Wright et al., 1982
Colonial bentgrass	*Agrostis tenuis*	Turf management	Madison, 1970
Creeping bentgrass	*Agrostis palustris*	Turf management	Nelson and Craft, 1991
Creeping bentgrass	*Agrostis palustris*	Turf management	Grebus et al., 1995
Creeping bentgrass	*Agrostis palustris*	Turf management	Markland et al., 1969
Creeping bentgrass	*Agrostis palustris*	Turf management	Cook et al., 1964
Kentucky bluegrass	*Poa pratensis*	Turf management	Madison, 1970
Kentucky bluegrass	*Poa pratensis*	Turf management	Breslin, 1995
Kentucky bluegrass	*Poa pratensis*	Turf management	Torello et al., 1983
Kentucky bluegrass	*Poa pratensis*	Turf management	Flanagan et al., 1993
Kentucky bluegrass	*Poa pratensis*	Turf management	Landshoot, 1995
Italian ryegrass	*Lolium multiflorum*	Turf management	Stratton and Rechcigl, 1998
Perennial ryegrass	*Lolium perenne*	Turf management	Norrie and Gosselin, 1996
Perennial ryegrass	*Lolium perenne*	Turf management	Wright et al., 1982
Tall fescue	*Festuca arundinacea*	Turf management	Sullivan et al., 1998
Tall fescue	*Festuca arundinacea*	Turf management	Peacock and Daniel, 1992
Tall fescue	*Festuca arundinacea*	Turf management	Bevacqua and Mellano, 1994
Bahiagrass	*Paspalum notatum*	Sod production	Neel et al., 1978
Bahiagrass	*Paspalum notatum*	Sod production	Cisar and Snyder, 1992
Bermudagrass	*Cynodon dactylon*	Sod production	Burns and Boswell, 1976
Bermudagrass	*Cynodon dactylon*	Sod production	Flanagan et al., 1993
Bermudagrass	*Cynodon dactylon*	Sod production	Neel et al., 1978
Bermudagrass (hybrid)	*Cynodon dactylon* X *C. transvaalensis*	Sod production	Cisar and Snyder, 1992
Kentucky bluegrass	*Poa pratensis*	Sod production	King and Beard, 1972
Kentucky bluegrass	*Poa pratensis*	Sod production	Decker, 1975, 1989
Perennial ryegrass	*Lolium perenne*	Sod production	O'Brien and Barker, 1995b, 1996b, 1997
Perennial ryegrass	*Lolium perenne*	Sod production	Keeling et al., 1994
St. Augustinegrass	*Stenotaphrum secundatum*	Sod production	Peacock and Dudeck, 1985
St. Augustinegrass	*Stenotaphrum secundatum*	Sod production	Cisar and Snyder, 1992
Tall fescue	*Festuca arundinacea*	Sod production	Decker, 1975, 1989
Tall fescue	*Festuca arundinacea*	Sod production	Roberts et al., 1995

disease free. Use of fertilizers and pesticides is reduced in the on-plastic system relative to soil-based production. However, regular applications of fertilizer may be needed with slow-to-mineralize composts made from carbonaceous materials, such as autumn leaves, yard-trimmings, and MSW without biosolids. With a C:N ratio above 40 in composts, N deficiency is likely without supplemental fertilization. Frequent irrigation is needed for sod production on plastic, but total water consumption does not exceed that of sod produced on land.

The production time for sod in suitable substrates on plastic is 5 to 20 weeks compared to a year or longer on soil. In layers of composts, roots of grasses knit quickly into stable sods that are easily harvested from the plastic without damage. Also, due to lack of root damage, sods produced in compost on plastic establish quickly after placement on land. Problems with this system include a shortage of plentiful, inexpensive composts. The amounts of materials needed are large. Spreading of large amounts of composts over large areas, such as 300 or more $m^3 \cdot ha^{-1}$, may not be competitive with sod production on soil for which compost application rates may be only 10 to 20% of those needed for production on plastic. Placement of even thicknesses of layers on plastic is a challenge on a large-scale basis. However, modern methods of composting help to ensure a supply of compost, and systems have been developed to use as little as 50 $m^3 \cdot ha^{-1}$ with optimum amounts being 100 to 150 $m^3 \cdot ha^{-1}$ (Decker, 1989). Decker (1989) estimated that if 5% of the approximately 100,000 ha of sod production at that time were grown on biosolids compost, the combined biosolids production of Boston, Chicago, New York, Philadelphia, and Washington, DC, could be used.

To determine compost suitability for use in turfgrass management or in sod production, users of compost must know the human and plant pathogen levels in the compost and the composition of the compost with respect to plant nutrients and nonessential elements and pH, as well as the C:N ratio. Allowing for feedstocks to develop into a mature compost helps to reduce pathogenic levels. Heat or chemical pretreatment (e.g., lime-stabilization) and exposure of feedstocks to sunlight diminish pathogens in compostable materials. These treatments have no effects on toxic elements, but separation of potentially contaminating materials from feedstocks has reduced or eliminated hazards of toxic elements in composts. Composting, in addition to diminishing the pathogen prevalence in organic substrates, provides an end product that has well-demonstrated capacities for suppression of soil-borne diseases of turf, sods, and other crops.

Fragments of plastic, glass, and metal may disqualify composts for topdressing, and such composts should be avoided for soil incorporation into lawns, athletic fields, and golf courses. Composts with these fragments are not without value and have uses in turf establishment in landfill covers, roadside landscapes, disturbed lands, or other infrequently accessed areas.

Composting is viewed as a management process to stabilize organic byproducts and residues generated by industrial, agricultural, municipal, and domestic activities. Stabilization of these materials into composts aids in lessening the impacts of the materials on the environment and opens avenues for their use in a number of agricultural applications. Research and practice have demonstrated that composts have wide uses as soil amendments and media for turf and sod production, as well as use in a variety of crop production systems and land remediation (United States Environmental Protection Agency, 1998). Utilization of compost in turf management reaps the benefits of the product and contributes substantially to environmental protection.

REFERENCES

Alvey, S. and D.E. Crowley. 1995. Influence of organic amendments on biodegradation of atrazine as a nitrogen source. *Journal of Environmental Quality* 24:1156–1162.

Amundson, R.G. and W.M. Jarrell. 1983. A comparative study of bermudagrass grown on soils amended with aerobic and anaerobically digested sludge. *Journal of Environmental Quality* 12:508–513.

Angle, J.S., D.C. Wolf, and J.R. Hall III. 1981. Turfgrass growth aided by sludge compost. *BioCycle* 22(6):40–43.

Avnimelech, Y., M. Kochva, Y. Yotal, and D. Shkedy. 1992. The use of compost as a soil amendment. *Acta Horticulturae* 302:217–236.

Barker, A.V. 1997. Uses and composition of composts, p. 140–162. In: J.E. Rechcigl (ed.). *Uses of Byproducts and Wastes in Agriculture*. American Chemical Society, Washington, D.C.

Barker, A.V. and T.A. O'Brien. 1995. Weed control in establishment of wildflower sods and meadows. *Proceedings of the Annual Meeting, Northeastern Weed Science Society* 40:56–60.

Barriuso, E., U. Baer, and R. Calvet. 1992. Dissolved organic matter and adsorption-desorption of dimefuron, atrazine, and carbetamide in soils. *Journal of Environmental Quality* 21:359–367.

Barriuso, E., S. Houot, and C. Serra-Wittling. 1997. Influence of compost addition to soil on the behaviour of herbicides. *Pesticide Science* 49(1):65–75.

Beauchamp, E.G. 1983. Nitrogen loss from sewage sludges and manures applied to agricultural lands, p. 181–194. In: J.R. Freney and J.R. Simpson (eds.). *Gaseous Loss of Nitrogen from Plant- Soil Systems*. Martinus Nijhoff/Dr. W. Junk Publishers, The Netherlands.

Bellin, C.A., G.A. O'Connor, and Y. Lin. 1990. Sorption and degradation of pentachlorophenol in sludge-amended soils. *Journal of Environmental Quality* 19:603–608.

Bevacqua, R.F. and V.J. Mellano. 1994. Cumulative effects of sludge compost on crop yields and soil properties. *Communications in Soil Science and Plant Analysis* 25:395–406.

Bolton, J. 1975. Liming effects on the toxicity to perennial ryegrass of a sewage sludge contaminated with zinc, nickel, copper and chromium. *Environmental Pollution* 9:295–304.

Breslin, V.T. 1995. Use of MSW compost in commercial sod production. *BioCycle* 36(5):68–72.

Bryan, H.H. and C.J. Lance. 1991. Compost trial on vegetables and tropical crops. *BioCycle* 32(3):36–37.

Buchanan, M. and S.R. Gliessman. 1991. How compost fertilization affects soil nitrogen and crop yield. *BioCycle* 32(12):72–77.

Burns, R.E. and F.C. Boswell. 1976. Effect of municipal sewage sludge on rooting of grass cuttings. *Agronomy Journal* 68:382–384.

Camberato, J.J., E.D. Vance, and A.V. Someshwar. 1997. Composition and land application of paper manufacturing residuals, p. 185–202. In: J.E. Rechcigl and H.C. MacKinnon (eds.). *Agricultural Uses of Byproducts and Wastes*. ACS Symposium Series 668, American Chemical Society, Washington, D.C.

Chaney, R.L. 1973. Crop and food chain effects of toxic elements in sludge and effluents, p. 129–146. In: M. DeBertoldi, M.P. Ferranti, P. L'Hermite, and F. Zucconi (eds.). *Composts: Production, Quality and Use*. Elsevier Applied Science, New York.

Chaney, R.L. 1990a. Twenty years of land application research. Part I. *BioCycle* 31(9):54–59.

Chaney, R.L. 1990b. Twenty years of land application research. Part II. *BioCycle* 31(10):68–73.

Chaney, R.L. and J.A. Ryan. 1993. Heavy metals and toxic organic pollutants in MSW-composts: research results on phytoavailability, bioavailability, fate, etc., p. 451–506. In: H.A.J. Hoitink and H.M. Keener (eds.). *Science and Engineering of Composting*. Renaissance Publishers, Worthington, Ohio.

Chaney, R.L., J.A. Ryan, and G.A. O'Connor. 1996. Organic contaminants in municipal biosolids: risk assessment, quantitative pathways, anaylsis, and current research topics. *The Science of the Total Environment* 185:187–216.

Chen, Y., and Y. Hadar. 1987. Composting and use of agricultural wastes in container media, p. 71–77. In: M. DeBertoldi, M.P. Ferranti, P. L'Hermite, and F. Zucconi (eds.). *Composts: Production, Quality and Use*. Elsevier Applied Science, New York.

Christensen, B.T. 1986. Ammonia volatilization loss from surface applied animal manure, p. 193–203. In: A. DamKofoed, J.H. Williams, and P L'Hermite (eds.). *Efficient Land Use of Sludge and Manure*. Elsevier Applied Science Publishers, New York.

Cisar, J.L. 1994. MSW compost for turf. *Grounds Maintenance* 29(3):52, 54, 58.

Cisar, J. and G.H. Snyder. 1992. Sod production on a solid-waste compost over plastic. *HortScience* 27:219–222.

Clapp, C.E., W.E. Larson, and R.H. Dowdy (eds.). 1994. *Sewage Sludge: Land Utilization and the Environment*. Soil Science Society of America miscellaneous publication, ASA, CSSA, and SSSA, Madison, Wisconsin.

Cook, R.N., R.E. Engel, and S. Bachelder. 1964. A study on the effect of nitrogen carriers on turfgrass disease. *Plant Disease Reporter* 48:254–255.

Coosemans, J. and C. van Asshe. 1983. Possibilities of sewage sludge as a fertilizer in agriculture. *Acta Horticulturae* 150:491–502.

Decker, H.F. 1975. Sewage sod system saves time. *Weeds Trees and Turf* 14(6):40–41.

Decker, H.F. 1989. Growing sod over plastic: Turf in five weeks. *Landscape Management* 28(7):68, 70.

Dick, W.A., and E.L. McCoy. 1993. Enhancing soil fertility by addition of compost, p. 622–644. In: H.A.J. Hoitink and H.M. Keener (eds.). *Science and Engineering of Composting*. Renaissance Publishers, Worthington, Ohio.

Duah-Yentumi, S. and S. Kuwatsuka. 1980. Effect of organic matter and chemical fertilizers on the degradation of benthiocarb and MCPA herbicides in soil. *Soil Science and Plant Nutrition* 26:541–549.

Edwards, J.H., E.C. Burt, R.L. Raper, and R.H. Walker. 1995. Issues affecting applications of noncomposted organic wastes to agricultural land, p. 225–249. In: D.L. Karlen, R.J. Wright, and W.D. Kemper (eds.). *Agricultural Utilization of Urban and Industrial Byproducts*. American Society of Agronomy Special Publication 58. ASA, CSSA, and SSSA, Madison, Wisconsin.

Epstein, E., J.M. Taylor, and R.L. Chaney. 1976. Effects of sewage sludge and sewage sludge compost applied to soil — physical and chemical properties. *Journal of Environmental Quality* 5:423–426.

Feagley, S., M.S. Valdez, and W. Hudnall. 1994. Papermill sludge, phosphorus, and lime effect on clover growth on a mine soil. *Journal of Environmental Quality* 23:759–765.

Flanagan, M.S., R.E. Schmidt, and R.B. Reneau Jr. 1993. Municipal solid waste heavy fraction for production of turfgrass sod. *HortScience* 28:914–916.

Giusquiani, P.L., M. Pagliai, G. Gigliotti, D. Businelli, and A. Benetti. 1995. Urban waste compost: effects on physical, chemical, and biochemical soil properties. *Journal of Environmental Quality* 24:175–182.

Gouin, F.R. 1993. Utilization of sewage sludge compost in horticulture. *HortTechnology* 3:161–163.

Grebus, M., C. Musselman, J. Rimelspach, and H. Hoitink. 1995. *Suppression of Turf Diseases with Biocontrol Agent-Fortified Compost Amended Topdressings*. Ohio Agricultural Research and Development Center Special Circular 148.

Guo, L., T.J. Bicki, A.S. Felsot, and T.D. Hinesly. 1993. Sorption and movement of alachlor in soil modified by carbon-rich wastes. *Journal of Environmental Quality* 22:186–194.

Guzman, R. 1996. Composts replaces soil amendments at country club. *BioCycle* 37(5):75–76.

Hadas, A. and R. Portnoy. 1994. Nitrogen and carbon mineralization rates of composted manures incubated in soil. *Journal of Environmental Quality* 23:1184–1189.

Hance, R.J. 1973. The effects of nutrients on the composition of the herbicides atrazine and linuron incubated with soil. *Pesticide Science* 4:817–822.

Hartz, T.K., F.J. Costa, and W.L. Schrader. 1996. Suitability of composted green waste for horticultural uses. *HortScience* 31:961–964.

Hauck, R.D. 1985. Slow release bioinhibitor-amended nitrogen fertilizers, p. 293–322. In: O.P. Englestad (ed.). *Fertilizer Technology and Use*. Soil Science Society of America, Madison, Wisconsin.

Hoitink, H.A.J., M.J. Boehm, and Y. Hadar. 1993. Mechanisms of suppression of soilborne plant pathogens in compost-amended substrates, p. 601–621. In: H.A.J. Hoitink and H.M. Keener (eds.). *Science and Engineering of Composting*. Renaissance Publishers, Worthington, Ohio.

Hoitink, H.A.J., A.G. Stone, and D.Y. Han. 1997. Suppression of plant diseases by composts. *HortScience* 32:184–187.

Hue, N.V. 1995. Sewage sludge, p. 199–247. In: J.E. Rechcigl (ed.). *Soil Amendments and Environmental Quality*. Lewis Publishers, Boca Raton, Florida..

Hue, N.V., H. Ikawa, and J.A. Silva. 1994. Increasing plant-available phosphorus in an ultisol with a yard-waste compost. *Communications in Soil Science and Plant Analysis* 25:3291–3303.

Hummel, N.W., Jr. and D.V. Waddington. 1984. Sulfur-coated urea for turfgrass fertilization. *Soil Science Society of America Journal* 48:191–195.

Hyatt, G.W. 1995. Economic, scientific, and infrastructure basis for using municipal composts in agriculture, p. 19–72. In: D.L. Karlen, R.J. Wright, and W.D. Kemper (eds.). *Agricultural Utilization of Urban and Industrial Byproducts*. American Society of Agronomy Special Publication 58. ASA, CSSA, and SSSA, Madison, Wisconsin.

Jacobs, L.W. 1998. Nutrient management planning for co-utilization of organic byproducts, p. 283–287. In: S. Brown, J.S. Angle, and L. Jacobs (eds.) *Beneficial Co-Utilization of Agricultural, Municipal and Industrial Byproducts*. Kluwer Academic Publishers, Dordrecht, The Netherlands.

Joiner, J.N. and C.A. Conover. 1965. Characteristics affecting desirability of various media components for production of container-grown plants. *Proceedings of the Soil and Crop Science Society of Florida* 25:320–328.

Keeling, A.A., I.K. Paton, and J.A.J. Mullett. 1994. Germination and growth of plants in media containing unstable refuse-derived compost. *Soil Biology & Biochemistry* 26:767–772.

King, J.W. and J.B. Beard. 1972. Postharvest cultural practices affecting the rooting of Kentucky bluegrass sods grown on organic and mineral soils. *Agronomy Journal* 64:259–262.

Kirkham, M.B. 1974. Disposal of sewage sludge on land: effects on soil, plants, and groundwater. *Compost Science* 15:6–10.

Landshoot, P. 1995. Improving turf soils with composts. *Grounds Maintenance* 30(6):33, 35, 37, 39.

Lemmon, C.R. and H.M. Pylypiw. 1992. Degradation of diazinon, chloropyrifos, isofenphos, and pendamethalin in grass and compost. *Bulletin of Environmental Contamination and Toxicology* 48:409–415.

Lickfeldt, D.W. and B.E. Branham. 1995. Sorption of nonionic organic compounds by Kentucky bluegrass leaves and thatch. *Journal of Environmental Quality* 24:980–985.

Lionello, B. and D.Z. Francesco. 1993. Soil fertility improvement and pollution risks from the uses of composts referred to N, P, K and C balance. *Acta Horticulturae* 302:51–62.

Logan, T.J. and B.J. Harrison. 1995. Physical characteristics of alkaline stabilized sewage sludge (N-Viro soil) and their effects on soil physical properties. *Journal of Environmental Quality* 24:153–164.

Logsdon, G. 1991. Compost use in sod production. *BioCycle* 32(3):64–65.

Lumsden, R.D., J.A. Lewis, and P.D. Millner. 1983. Effect of composted sewage sludge on several soil borne pathogens and diseases. *Phytopathology* 75:1543–1548.

Madison, J.H. 1970. Rooting of sod by *Poa pratensis* L. and *Agrostis tenuis* Sibth. *Crop Science* 10:718–719.

Malone, R.W., R.C. Wagner, and M.E. Byers. 1996. Runoff losses of surface-applied metribuzin as influenced by yard waste compost amendments, no-tillage, and conventional tillage. *Bulletin of Environmental Contamination and Toxicology* 57:536–543.

Markland, F.E., E.C. Roberts, and L.R. Frederick. 1969. Influence of nitrogen fertilizers on Washington creeping bentgrass, *Agrostis palustris* Huds. II. Incidence of dollar spot, *Sclerotinia homeocarpa. Agronomy Journal* 61:701–705.

Mattingly, G.E.G. 1956. Studies on composts prepared from waste materials. III. Nitrification in soil. *Journal of the Science of Food and Agriculture* 7:601–605.

McBride, M.B. 1995. Toxic metal accumulation from agricultural use of sludge: are USEPA regulations protective. *Journal of Environmental Quality* 24:5–18.

McDowell, L.L., G.H. Willis, S. Smith, and L.M. Southwick. 1985. Insecticide washoff from cotton plants as a function between application and rainfall. *Transactions of the American Society of Agricultural Engineers* 28:1896–1900.

McSorley, R. and R.N. Gallaher. 1995. Effect of yard waste compost on plant-parasitic nematode densities in vegetable crops. *Journal of Nematology* 27:545–549.

Michel, F.C., Jr., C.A. Reddy, and L.J. Forney. 1995. Microbial degradation and humification of the lawn care pesticide 2, 4-dichlrophenoxyacetic acid during the composting of yard trimmings. *Applied and Environmental Microbiology* 61:2566–2571.

Michel, F.C., Jr., C.A. Reddy, and L.J. Forney. 1997. Fate of carbon-14 diazinon during the composting of yard trimmings. *Journal of Environmental Quality* 26:200–205.

Mitchell, W.H., C.J. Molnar, and S.S. Barton. 1994. Using composts to grow wildflower sod. *BioCycle* 35(2):62–63.

Muller, W.P. and F. Korte. 1975. Microbial degradation of benzo(a)pyrene, monolinuron, and dieldrin in waste composting. *Chemosphere* 3:195–198.

Neel, P.L., E.O. Burt, P. Busey, and G.H. Snyder. 1978. Sod production in shallow beds of waste materials. *Journal of the American Society for Horticultural Science* 103:549–553.

Nelson, E.B. 1996. Enhancing turfgrass disease control with organic amendments. *Turf Grass Trends* 5(6):1–15.

Nelson, E.B. and C.M. Craft. 1991. Suppression of dollar spot on creeping bentgrass and annual bluegrass turf with compost-amended topdressings. *Plant Disease* 76:954–958.

Norrie, J. and A. Gosselin. 1996. Paper sludge amendments for turfgrass. *HortScience* 31:957–960.

O'Brien, T.A. and A.V. Barker. 1995a. Evaluation of field-applied fresh composts for production of sod crops. *Compost Science & Utilization* 3(3):53–65.

O'Brien, T.A. and A.V. Barker. 1995b. Evaluation of fresh and year old composts for production of wildflower and grass sods on plastic. *Compost Science & Utilization* 3(4):69–77.

O'Brien, T.A. and A.V. Barker. 1996a. Evaluation of ammonium and soluble salts on grass sod production in compost. I. Addition of ammonium and nitrate salts. *Communications in Soil Science and Plant Analysis* 27:57–76.

O'Brien, T.A. and A.V. Barker. 1996b. Evaluation of ammonium and soluble salts on grass sod production in compost. II. Delaying seeding after compost application. *Communications in Soil Science and Plant Analysis* 27:77–85.

O'Brien, T.A. and A.V. Barker. 1997. Evaluating composts to produce wildflower sods on plastic. *Journal of the American Society for Horticultural Science* 122:445–451.

O'Brien, T.A. and A.V. Barker. 1998. Acidification of lime-stabilized biosolids in production of synthetic topsoils. *Communications in Soil Science and Plant Analysis* 29:1107–1114.

O'Keefe, B.E., J. Axley, and J.J. Meisinger. 1986. Evaluation of nitrogen availability indexes for a sludge compost amended soil. *Journal of Environmental Quality* 15:121–128.

Ozores-Hampton, M., B. Schaffer, H.H. Bryan, and E.A. Hanlon. 1994. Nutrient concentrations, growth, and yield of tomato and squash in municipal solid-waste-amended soil. *HortScience* 29:785–788.

Pagliai, M., G. Guidi, M. LaMarca, M. Giachetti, and G. Lucamante. 1981. Effect of sewage sludges and composts on soil porosity and aggregation. *Journal of Environmental Quality* 10:556–561.

Parr, J.F. and S.B. Hornick. 1992. Agricultural use of organic amendments: a historic perspective. *American Journal of Alternative Agriculture* 7(4):181–189.

Peacock, C.H. and P.F. Daniel. 1992. A comparison of turfgrass response to biologically amended fertilizers. *HortScience* 27:883–884.

Peacock, C.H. and A.E. Dudeck. 1985. A comparison of sod type and fertilization during turf establishment. *HortScience* 20:108–109.

Petrovic, A.M. 1990. The fate of nitrogenous fertilizers applied to turfgrass. *Journal of Environmental Quality* 19:1–14.

Pill, W.G., W.S. Smith, J.J. Frett, and D. Devenney. 1994. Wildflower establishment in seedbeds created from an industrial co-product and co-composted municipal wastes. *Journal of Environmental Horticulture* 12(4):193–197.

Potter, D.A., A.J. Powell, P.G. Spicer, and D.W. Williams. 1996. Cultural practices affect root-feeding white grubs (Coleoptera: Scarabaeidae) in turfgrass. *Journal of Economic Entomology* 89 (1):156–164.

Purman., J.R. and F.R. Gouin. 1992. Influence of compost aging and fertilizer regimes on the growth of bedding plants, transplants, and poinsettia. *Journal of Environmental Horticulture* 10:52–54.

Racke, K.D. and C.R. Frink. 1989. Fate of organic contaminants during sewage sludge composting. *Bulletin of Environmental Contamination and Toxicology* 42:526–533.

Roberts, B.R., S.D. Kohorst, H.F. Decker, and D. Yaussy. 1995. Shoot biomass of turfgrass cultivars grown on composted waste. *Environmental Management* 19:735–739.

Rosen, C.J., T.R. Halbach, and B.T. Swanson. 1993. Horticultural uses of municipal solid waste composts. *HortTechnology* 3:167–173.

Sanders, J.R., S.P. McGrath, and T.M. Adams. 1986. Zinc, copper and nickel concentrations in ryegrass grown on sewage-sludge contaminated soils of different pH. *Journal of the Science of Food and Agriculture* 37:961–968.

Sears, M.K. and R.A. Chapman. 1979. Persistence and movement of four insecticides applied to turfgrass. *Journal of Economic Entomology* 72:272–274.

Sikora, L.J. 1996. Effect of compost-fertilizer blends on crop growth, p. 423–429. In: M. deBertoldi, P. Sequui, B. Lemmes, and T. Papi (eds.). *The Science of Composting*. Blackie Academic and Professional, London.

Sikora, L.J. 1998. Benefits and drawbacks to composting organic byproducts, p. 69–77. In: S. Brown, J.S. Angle, and L. Jacobs (eds.). *Beneficial Co-Utilization of Agricultural, Municipal and Industrial Byproducts*. Kluwer Academic Publishers, Dordrecht, The Netherlands.

Steuteville, R. 1995a. The state of garbage in America. Part I. *BioCycle* 36(4):54–63.

Steuteville, R. 1995b. The state of garbage in America. Part II. *BioCycle* 36(5):30–36.

Stratton, M.L., A.V. Barker, and J.E. Rechcigl. 1995. Composts, p. 249–309. In: J.E. Richcigl (ed.). *Soil Amendments and Environmental Quality*. Lewis Publishers, Boca Raton, Florida.

Stratton, M.L. and J.E. Rechcigl. 1997. Organic mulches, wood products, and composts as soil amendments, p. 43–95. In: A. Wallace (ed.). *Handbook of Soil Conditioners: Substances which Enhance the Physical Properties of Soils*. Marcel Dekker, New York.

Stratton, M.L. and J.E. Rechcigl. 1998. Compost applications to ryegrass, p. 210–217. In: K.C. Das and E.S. Graves (ed.). *Proceedings 1998 Conference, Composting in the Southeast*. University of Georgia, Athens.

Sullivan, D.M., S.C. Fransen, A.I. Bary, and C.G. Cogger. 1998. Slow-release nitrogen from composts: the bulking agent is more than fluff, p. 319–325. In: S. Brown, J.S. Angle, and L. Jacobs (eds.). *Beneficial Co-Utilization of Agricultural, Municipal and Industrial Byproducts*. Kluwer Academic Publishers, Dordrecht, The Netherlands.

Terman, G.L. 1979. Volatilization losses of nitrogen as ammonia from surface-applied fertilizers, organic amendments, and crop residues. *Advances in Agronomy* 31:189–223.

Titko, S., III, J.R. Street, and T.J. Logan. 1987. Volatilization of ammonia from granular and dissolved urea applied to turfgrass. *Agronomy Journal* 79:535–540.

Torello, W.A., D.H. Wehner, and A.J. Turgeon. 1983. Ammonia volatilization from fertilized turfgrass stands. *Agronomy Journal* 75:454–456.

United States Environmental Protection Agency. 1998. An analysis of composting as an environmental remediation technology. *Solid Waste and Emergency Response* (5305W), EPA 530-R-98-008, USEPA, Washington, D.C.

Vandervoort, C., M.J. Zabik, B. Braanham, and D.W. Lickfeldt. 1997. Fate of selected pesticides applied to turfgrass: effect of composting on residues. *Bulletin of Environmental Contamination and Toxicology* 58:38–45.

Vogtmann, H., and K. Fricke. 1992. Organic chemicals in compost: how relevant are they for the use of it, p. 227–236. In: D.V. Jackson, J.M. Merillot, and P. L'Hermite (eds.). *Composting and Compost Quality Assurance Criteria*. Commission of the European Communities, Luxembourg.

Volk, G.M. 1959. Volatile losses of ammonia following applications of urea to turf or bare soils. *Agronomy Journal* 51:746–749.

Volk, G.M. and G.C. Horn. 1975. Response curves of various turfgrasses to applications of several controlled-release nitrogen sources. *Agronomy Journal* 67:201–204.

Wen, G., T.E. Bates, and R.P. Voroney. 1995. Evaluation of nitrogen availability in irradiated sewage sludge, sludge compost and mature compost. *Journal of Environmental Quality* 24:527–534.

Wen, G., T.E. Bates, R.P. Voroney, J.P. Winter, and M.P. Schellenbert. 1997. Comparison of phosphorus availability with application of sewage sludge, sludge compost, and manure compost. *Communications in Soil Science and Plant Analysis* 28:1481–1497.

Wen, G., J.P. Winter, R.P. Voroney, and T.E. Bates. 1996/1997. Potassium availability with application of sewage sludge, and sludge and manure composts in field experiments. *Nutrient Cycling in Agroecosystems* 47(3):233–241.

Wright, W.R., P.S. Schauer, and R.E. Huling. 1982. Utilization of industrial fermentation residues for turfgrass production. *Journal of Environmental Quality* 11:233–236.

CHAPTER 10

Composts as Horticultural Substrates for Vegetable Transplant Production

Susan B. Sterrett

CONTENTS

I. INTRODUCTION: INDUSTRY NEEDS

The use of transplants is the most reliable method to ensure adequate crop establishment of commercial plantings of various high-value vegetable crops. Other advantages include reduced cost over direct seeding when using expensive hybrid seed, improved land use efficiency, extension of a short growing season, and improved early weed control (Swiader et al., 1992). For some crops, e.g., cauliflower (*Brassica oleracea* L. Botrytis group), concentration of crop maturity is improved with transplants. In comparison with direct seeding, transplants can also result in greater early yield when prices tend to be higher (Swiader et al., 1992). Thus, thc added cost of transplants is often justifiable.

Transplant production for commercial vegetable growers is a highly competitive industry in which the ability to deliver the specified quantity and quality of transplants at a specified time is critical to customer satisfaction. This also is a highly mechanized industry (Figure 10.1). A Florida survey indicated that all of the major

1-56670-460-X/01/$0.00+$.50

Figure 10.1 Vegetable transplant production in containerized flats in commercial greenhouse. (Photo courtesy of C. S. Vavrina, University of Florida.)

commercial transplant growers seed directly into containerized flats (Vavrina and Summerhill, 1992). Therefore, quick, uniform seedling emergence and rapid, consistent plant growth is essential for efficient commercial transplant production.

Vegetable transplants are grown at high densities in polystyrene containers with molded cells in the shape of an inverted pyramid ranging in size upward from 25 mm. Recently, growers have started using plugs with even more restricted root volume to improve production efficiency by increasing number of plants per tray and reducing the need for additional growing space (Vavrina, 1995). Since containers used for transplants grown for retail sales are similar to those used for bedding plants, chapter 6 in this book is more relevant to vegetable transplant production for retail sales.

Mechanization of container filling requires a consistent medium of small particle size. Moisture retention by the medium is critical to maintain a rapid growth rate but aeration is also essential (Bunt, 1976). Water-holding capacity of soil media on a weight or volume basis increases as the pot size decreases, slowing free drainage (Waters et al., 1970). Therefore, a very well drained medium is needed for the production of transplants.

Soilless, peat moss-based growing media were developed to improve reliability of plant growth, reduce incidence of disease, and avoid injury from salinity (Boodley and Sheldrake, 1972; Matkin and Chandler, 1957). High concentrations of soluble salts were historically of concern when soil used for transplant production had been composted with large quantities of manure (Thompson and Kelly, 1957). The disadvantages cited for soil-based composts included land area required for mixing, stockpiles, and storage; variability of composts; shrinkage; odor and flies; salinity, and other compositional problems (Matkin et al., 1957). However, peat moss is a nonrenewable resource with availability of quality product declining and price increasing; hence the interest in finding low-cost, readily available organic medium amendments to replace peat moss. With increasing emphasis on recycling as a means of reducing the loading pressures in landfills, alternate sources of organic matter are becoming more readily available in some areas (especially near municipalities and wastewater treatment facilities). The same disadvantages of soil-based composts are also a concern with composts made from these alternative organic materials and must be considered in the evaluation of these components as growing medium amendments.

II. ALTERNATIVE ORGANIC MEDIA AMENDMENTS

Composting has been used to stabilize a wide variety of organic wastes or byproducts prior to their use as growing medium amendments. Waste disposal has become a national concern in the U.S., with an average of more than 2 kg (≈5 lb.) of solid waste produced per person per day (Goldstein, 1997). Composting would convert high volume wastes, including municipal solid wastes (MSW), yard trimmings, and biosolids into usable organic amendments, thereby reducing the pressure on local landfills (Ozores-Hampton et al., 1998a). Future availability of composts will likely reflect the increasing economic and, as in Florida, legislative pressures to reduce loading in landfills (Tonnessen, 1998).

A. Feedstocks for Composting

Several organic wastes or organic byproducts have been used as compost feedstocks and evaluated as growing medium components for the production of vegetable transplants. Lunt and Clark (1969) noted that noncomposted pine (*Pinus* sp.) bark is an excellent growing medium for ornamentals, but it is not well suited to vegetable transplant production. However, Barragry and Morgan (1978) reported that composting pine bark to overcome nitrogen (N) immobilization improved growth of tomato (*Lycopersicon esculentum* Mill.) compared to growth in noncomposted bark. Composts produced from various feedstocks, including biosolids (Falahi-Ardakani et al., 1987a, 1987b; Sterrett and Chaney, 1982; Sterrett et al., 1982, 1983), MSW (Purman and Gouin, 1992; Roe and Kostewicz, 1992; Vavrina, 1994), yard trimmings (Ozores-Hampton et al., 1998a, 1998b; Roe and Kostewicz, 1992; Roe et al., 1997), and spent mushroom compost (Lohr et al., 1984; Rathier, 1982; Vavrina et al., 1996) have also been evaluated as potential vegetable transplant growing media amendments (Table 10.1).

Wastes that have a high carbon to nitrogen (C:N) ratio are often composted with either animal manure or biosolids (municipal sewage sludge treated to kill pathogens and reduce vector attraction) to speed stabilization during the composting process. Composts containing biosolids are regulated by U.S. EPA Clean Water Act Section 503. The federal regulations focus on safety rather than compost quality and are based upon heavy metal concentrations. Sterrett et al. (1983) reported elevated

Table 10.1 Evaluation of Feedstocks for Composts Used in Growing Media for Vegetable Transplants

Compost	Rate (% Volume)	Growth Response[z]	Factor(s) Affecting Response	Reference
Hardwood bark	50, 80	Cucumber, tomato dry wt. < peat-lite	Immaturity	Bearce and Postlethwaite, 1982
Biosolids (BS)-LM[y]	33, 50	Tomato & cabbage dry wt. and stem diameter = peat-lite	Compost stable, leached at transplanting	Sterrett et al., 1983
BS-HM[y]	33, 50	Tomato & cabbage dry wt. < LM compost & peat-lite	Heavy metal toxicity	Sterrett et al., 1983
BS	33, 50, 67	Tomato, pepper, cabbage dry wt. = peat-lite (33, 50%); pepper dry wt. < peat-lite (67%)	Physical characteristics, media leached at planting	Sterrett and Chaney, 1982
BS	33	Broccoli, cabbage, eggplant, lettuce, pepper, and tomato dry wt. increased linearly over 8 week period	Compost stable, leached at planting	Falahi-Ardakani et al., 1987a
BS	33	Dry wt. of tomato and lettuce grown in compost: peat:vermiculite + N fertilizer = peat-lite; dry wt. of plants in compost: peat:perlite < peat-lite	K requirements satisfied, in part, by vermiculite; other nutrients supplied by compost	Falahi-Ardakani et al., 1988
BS	25, 50	Tomato dry wt. >100% white peat; shoot radius of tomato > with 50% compost than with peat	C:N ratio $\simeq$ 20:1, P & K supplied by compost	Pinamonti et al., 1997
BS	25, 50, 75, 100	Germination of tomato delayed with increased % compost (one trial) and % germination < peat-lite. Shoot and root dry weight < peat-lite and decreased with increased % compost	High soluble salts (unleached)	Vavrina, 1994
BS	10	% emergence slightly < peat-lite, plant height > peat-lite	Dilution of soluble salts	Vavrina, 1995
Municipal solid waste (MSW)	25, 50, 75	% germination $\simeq$ peat-lite, germination delayed slightly in one trial	Some concern about consistency between batches of compost	Vavrina, 1994
MSW	100	% germination and height of tomato < peat-lite	Fertility program needed adjustment relative to peat-lite	Vavrina, 1995
MSW	100	Pepper germination slower than in peat-lite but final % germination of pepper, rape, radish, and dill = peat-lite	Stabilized compost but soluble salts high	Roe and Kostewicz, 1992
BS/MSW (2:1 ratio)	25, 50, 75, 100	% germination ≤ peat-lite and slightly decreased with increasing % compost	Unknown	Vavrina, 1994

Table 10.1 Evaluation of Feedstocks for Composts Used in Growing Media for Vegetable Transplants (Continued)

Compost	Rate (% Volume)	Growth Response[z]	Factor(s) Affecting Response	Reference
BS/MSW (2:1 ratio)	100	% germination and height of tomato varied between batches of compost but was < peat-lite	Inconsistencies in process or feedstock	Vavrina, 1995
BS/MSW;[x] co-compost aged 30 or 90 days	33, 50	Lettuce and cabbage dry wt. = peat-lite	Compost stable, stored in windrows	Purman and Gouin, 1992
BS/yard trimmings (YT)	100	Radish and dill % germination < peat-lite	Stabilized compost but soluble salts high	Roe and Kostewicz, 1992
MSW + YT	100	Pepper germination slower than peat-lite but faster than 100% MSW; final % germination of pepper, rape, radish, and dill = peat-lite	Stabilized compost but soluble salts high	Roe and Kostewicz, 1992
YT + BS	18, 35, 52, 70	Tomato dry wt., leaf area, and stem diameter > peat-lite	C:N ratio = 16.9:1; supplemental nutrients	Ozores-Hampton et al., 1998a, 1998b
YT	100	Tomato and watermelon % germination and dry weight > peat-lite	Stabilized compost	Roe and Kostewicz, 1992
YT + poultry manure (PM)	100	% germination of tomato, watermelon, and lettuce delayed with YT and 25% PM. Plant dry wt. in YT + 25% PM > than in peat-lite after 4 weeks	Total N and soluble salts in YT+PM > in YT alone	Roe and Kostewicz, 1992
YT + grass clippings	100	% germination of tomato, watermelon = peat-lite	Increased N promoted vegetative growth	Roe and Kostewicz, 1992
BS + YT	100	Tomato and cucumber % emergence = peat-lite	Increased aeration in peat-lite	Roe et al., 1997
BS + YT + mixed waste paper	50, 100	Tomato, pepper, cucumber % emergence = peat-lite (50%); tomato, pepper, cucumber % emergence < peat-lite (100%)	Increased aeration in peat-lite	Roe et al., 1997
BS + YT + refuse-derived fuel	50, 100	Tomato, pepper, cucumber % emergence = peat-lite (50%); tomato and pepper % emergence < peat-lite (100%)	NH_4-N considered high	Roe et al., 1997
BS + YT + refuse-derived fuel residuals	100	Cucumber % emergence < peat-lite	High soluble salts; NH_4-N, Zn, Cu elevated	Roe et al., 1997
Spent mushroom compost (SMC)	33	Tomato height ≥ peat-lite	Fertilizer application to peat-lite at 4 weeks	Rathier, 1982

Table 10.1 Evaluation of Feedstocks for Composts Used in Growing Media for Vegetable Transplants (Continued)

Compost	Rate (% Volume)	Growth Response[z]	Factor(s) Affecting Response	Reference
SMC	25–50	Tomato, lettuce, and cucumber height and dry wt. in 25% aged compost = peat-lite; in 50% compost and fresh compost < peat-lite	High soluble salts prior to leaching; NH_4 toxicity in fresh compost	Lohr et al., 1984
SMC	100	Tomato germination < peat-lite; plants in SMC more succulent	Lower H_2O holding capacity; organic N source interfered with hardening off of transplants	Vavrina et al., 1996
Papermill sludge	33, 50, 67	Tomato, cucmber, eggplant, and cabbage stunted, off color	Incomplete stabilization, high pH, high soluble salts	Hornick et al., 1984
Grape branches fragmented to 3–4 mm (prunings/husks/seeds)	100	Tomato dry wt., stem diameter, and height > peat:perlite:sand	Improved physical properties, aeration, and available P	Kostov et al., 1996
Poultry manure	100	Inhibition of tomato and watermelon germination	Excess NH_4-N	Roe and Kostewicz, 1992

[z] Compared to peat-lite medium (peat moss and usually vermiculite; may contain perlite and/or polystyrene). A commercial peat-lite medium was used in most studies, but several commercial media are represented in these trials.

[y] LM (low metals) = biosolids with concentrations of Zn, Pb, Cd, and Ni meeting EPA Section 503 guidelines; HM (high metals) = biosolids contaminated with Zn, Pb, Cd, and Ni from industrial sources, exceeding the guidelines.

[x] Garbage ground and separated magnetically and by density to remove metals and glass.

concentrations of zinc (Zn), copper (Cu), manganese (Mn), nickel (Ni), and cadmium (Cd) in tomato and cabbage (*Brassica oleracea* L. Capitata group) transplants grown in media containing compost made with industrially contaminated biosolids compared to those grown in compost from a residential (low metal) source or in peat–vermiculite (peat-lite). Biosolids that meet the alternative pollution limits (<41 $mg \cdot kg^{-1}$ arsenic [As], 39 $mg \cdot kg^{-1}$ Cd, 1500 $mg \cdot kg^{-1}$ Cu, 300 $mg \cdot kg^{-1}$ lead [Pb], 17 $mg \cdot kg^{-1}$ mercury [Hg)], 420 $mg \cdot kg^{-1}$ Ni, 100 $mg \cdot kg^{-1}$ selenium [Se], and 2800 $mg \cdot kg^{-1}$ Zn) can be marketed for general use without cumulative site loadings for the regulated metals provided pathogen levels in the product are reliably reduced to nondetectable levels by heat and time (Chaney et al., 1999).

B. Compost Characteristics Affecting Plant Response

The most frequently cited problems with using compost in the growing medium for vegetable transplants include unstable or immature compost, high soluble salt concentrations, and poor water-holding capacity (Table 10.1). Because the potential usefulness of composts as organic amendments to vegetable transplant media is directly related to consistent, uniform growth, it is important to understand the factors that may adversely affect plant response.

As defined by Ozores-Hampton et al. (1998a), stability of compost refers to the degree to which the compost consumes N and oxygen in significant quantities to support biological activity and generates heat, carbon dioxide, and water vapor, a process that can cause plant stunting and yellowing of leaves. A desirable C:N ratio is between 15:1 and 20:1 (Hoitink and Fahy, 1986; Hornick et al., 1984; Inbar et al., 1990; Ozores-Hampton et al., 1998a; Roe et al., 1997; Rosen et al., 1993).

Compost maturity refers to the absence of phytotoxic substances that can cause delayed seed germination or seedling and plant death (Ozores-Hampton et al., 1998a). Numerous studies have described delayed or reduced germination due to the presence of phytotoxic substances in compost (Chanyasak et al., 1983; Keeling et al., 1994; Lunt, 1959; Roe and Kostewicz, 1992; Vavrina, 1994; Wong and Chu, 1985; Zucconi et al., 1981). Lohr et al. (1984) noted that symptoms of ammonium toxicity were observed in tomato, cucumber (*Cucumis sativus* L.), and lettuce (*Lactuca sativa* L.) grown in fresh spent mushroom compost. Delayed or inconsistent germination greatly increases transplant production costs and interrupts production schedules.

Variation between sources or between batches of compost from the same source can result in unpredictable plant response. Vavrina (1995) reported significant differences between compost sources and between batches from the same source in emergence and height of tomato after 6 weeks of growth. Immaturity of the compost or differences in composting conditions may have contributed to this response. Gouin (1993) noted that compost piles that are allowed to become anaerobic produce compost containing methanol or methane and acetic acid (foul odors) with a corresponding depression in pH (3.0 to 3.5). High levels of acetic acid (6000 to 28,000 $mg \cdot kg^{-1}$) in immature MSW compost have inhibited germination of several vegetable crops (Keeling et al., 1994). Wong and Chu (1985) reported increased retardation of root elongation by aqueous extracts of refuse compost with increasing ammonia (NH_3) and ethylene oxide concentrations. Lunt (1959) reported delayed germination with addition of municipal wastes to cropland, with the lag period being dose dependant. Zucconi et al. (1981) noted that toxicity from decomposing organic matter is temporary; both germination and root elongation are improved with mature compost over fresh or immature compost.

Storage of the composted, cured product may also affect plant growth. Roe and Kostewicz (1992) reported a strong NH_3 odor from packaged poultry litter compost in a study comparing media composed of various combinations of yard trimmings, poultry litter, and grass clippings to a commercial peat-lite meduim in compartmentalized flats (5 cm^2 cells). Total germination inhibition of tomato and watermelon (*Citrullus lanatus* [Thunb.] Matsum. & Nakai) was reported for all media containing poultry litter compost. Available N is in the ammonium (NH_4) form after initial composting, with conversion to nitrate occurring over time while stored in static piles (Vega-Sanchez et al., 1987). Hence, the premature use of immature compost or packaging of compost (i.e., anaerobic conditions) could disrupt the microbial activity needed for the conversion of organic N to nitrate or promote reversion to

NH_4. In a study designed to compare maturity of stored compost, cured compost (stored 6 weeks, 3 months, or 7 months in stockpiles 3.5 to 4.5 m [~12 to 15 ft] in height) was obtained from three locations (top, middle, and bottom) within each stockpile (Sterrett and Chaney, unpublished data). Growth of tomato transplants in media containing compost from the top of a recently constructed stockpile was similar to growth in the peat-lite control (Figure 10.2.) The dry weight of tomato transplants was consistently lower when grown in media containing compost stored in the bottom third of the pile (Figure 10.3; Sterrett and Chaney, unpublished data). Foul odors were particularly evident from the bottom of the stockpiles when broken to obtain the compost samples. This study indicates the necessity of storing mature compost under aerobic conditions, either in short stacks or in stacks that are frequently turned (Hoitink et al., 1991).

Figure 10.2 Tomato transplants grown in peat-lite (extreme left) or in biosolids compost:vermiculite:peat (2:1:1) using cured compost stockpiled for 6 weeks (A), 3 months (B), or 7 months (C). Compost was taken from top (1), middle (2), or bottom (3) of the stockpile.

High soluble salt concentrations have frequently been cited as a problem with composts containing limed-raw sewage sludge (Chaney et al., 1980; Gouin, 1993; Hornick et al., 1984; Marcotrigiano et al., 1985) and MSW co-composted with biosolids (Roe and Kostewicz, 1992; Rosen et al., 1993; Sanderson, 1980; Shiralipour et al., 1996; Vavrina, 1994). Reusing wood chips removed during the compost screening process or substituting finished compost as the bulking material in subsequent production of biosolids compost can result in substantially elevated soluble salt concentrations (Gouin, 1993). Elevated salt concentrations have also been reported for spent mushroon composts (Lohr et al., 1984; Rathier, 1982), municipal

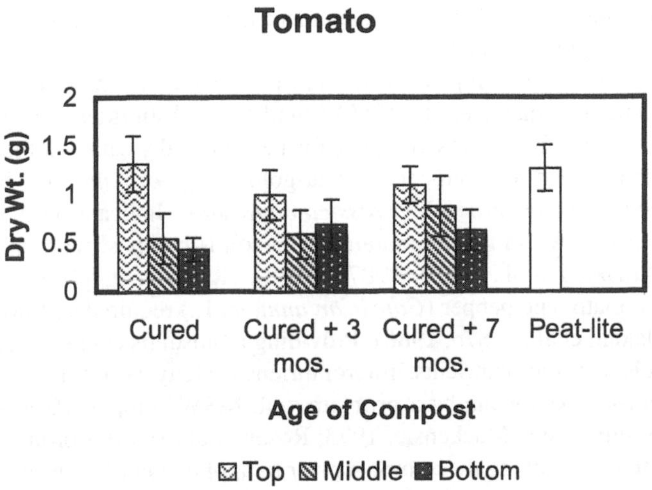

Figure 10.3 Growth of tomato transplants grown in media containing cured compost stockpiled for 6 weeks, 3 months, or 7 months located in the top, middle, or bottom third of the stockpile. Tomatos were transplanted 1 Nov. and harvested 23 Nov. 1982. Vertical bars represent ±1 standard error of the mean, n = 12 (except peat-lite, where n = 6).

leaf composts (Sawhney, 1976), and gelatin waste compost (Gouin and Shanks, 1981). Preplant leaching has been used to ameliorate this problem in small-scale research projects (Chaney et al., 1980; Hornick et al., 1984; Sawhney, 1976; Sterrett and Chaney, 1982; Sterrett et al., 1983). However, this practice is difficult to incorporate into commercial production where flats are mechanically filled with medium and then seeded as part of the filling process. Purman and Gouin (1992) reported lower soluble salt concentrations with MSW co-composted with polymer-dewatered, low metal biosolids than with lime-dewatered, ferric chloride treated biosolids. This suggests that biosolid dewatering technique may have a significant impact on the effectiveness of the resulting compost as a transplant medium component.

Both water-holding capacity and the ratio of water to air held in the root medium after drainage are important for the production of quality transplants. Generally, a very fine particle size is used in transplant production to facilitate uniform filling of the production trays. Since water can be applied as needed, the primary concern is adequate aeration (Argo, 1998). Compost made from biosolids and woodchips or sawdust has a low water-holding capacity relative to peat moss (Falahi-Ardakani et al., 1988; Gouin, 1993) or biosolid/MSW compost (Gouin, 1993). Siminis and Manios (1990) reported that the bulk density of a sphagnum-peat-based medium increased with increasing MSW compost addition, while available water-holding capacity decreased. As a result, Rosen et al. (1993) recommended that MSW compost should be limited to 20% (by volume) in a peat-based potting medium. Vavrina et al. (1996) noted that spent mushroom compost dried out more quickly than the peat-lite standard, suggesting that the addition of vermiculite may be needed to increase water-holding capacity and provide a source of K. It is readily apparent from these studies that adjustments in water management may be needed when using organic composts in the growing medium.

The contribution of compost to the nutrient balance of the transplant medium can be substantial. Chaney et al. (1980) found that substitution of as little as 33% (by volume) digested biosolids compost for peat-lite fully satisfied the phosphorus (P) and micronutrient requirements of marigolds (*Tagetes patula* L.) but only part of the N needs. Lettuce, broccoli (*Brassica oleracea* L. Italica group), and cabbage have also been grown in media containing biosolids compost with only additional N (Falahi-Ardakani et al., 1987a, 1987b, 1988). However, eggplant (*Solanum melongena* L), tomato, and pepper (*Capsicum annuum* L.) required additional N and K (Falahi-Ardakani et al., 1987b, 1988). Providing a biosolids compost meets the U.S. EPA 503 rule for micronutrients, micronutrient toxicity is not a serious concern. However, boron toxicity can be a problem with MSW compost (Gogue and Sanderson, 1975; Purves and Mackensie, 1973; Rosen et al., 1993). Boron toxicity could not be avoided by leaching the compost prior to application (Purves and Mackensie, 1973).

Another aspect that warrants consideration is the ease/safety of handling compost media. Rosen et al. (1993) describe several possible handling procedures for MSW in which glass, metals, and plastic may or may not be removed prior to grinding. Because of safety issues in the handling of the medium during mixing and tray filling, as well as during the field transplanting operation, presence of these inert materials would make the growing medium undesirable for vegetable transplant production.

III. DEVELOPMENT OF STANDARDS

The studies describing benefits of using composts with one or more organic feedstock are numerous. However, as previously discussed, there are also numerous concerns regarding product consistency, handling, and storage. To realize an economic benefit from compost as a transplant medium component, growers need assurance that a given compost source can reliably provide a uniform, stable, readily available product in sufficient quantity. Standardization of the compost product is needed. Specifications should be established based upon reliable and readily interpretable test procedures. These guidelines need to extend beyond state and federal safety regulations that limit heavy metal content and insure pathogen destruction in order to address the various issues that can adversely affect seed germination and plant growth.

The replacement of peat moss, which is a stable but increasingly expensive organic amendment, with composts, which are becoming increasingly available but must be managed closely, will take a concerted effort on the part of the compost industry. Craul and Switzenbaum (1996) described a set of criteria developed to insure consistency of product for a large urban construction project in which a quantity of "constructed topsoil" was needed. Their parameters included specific targets for C:N ratio, stability of product, odor, particle size, pH, cation exchange capacity, and nutrient content. Others have used specific tests or bioassays to monitor compost quality or phytotoxicity (Roe et al., 1997; Rosen et al., 1993; Zucconi et al., 1981). Specific criteria need to be developed for transplant media components. Specification of the most appropriate tests to be included in quality guidelines is beyond the scope of this chapter. However, the need for standardization of quality of commercially available compost products should be obvious. The future of the commercial transplant industry depends on the ability of growers to efficiently produce high quality, uniform transplants on a tight timetable. Compost that is inexpensive, is readily available in adequate quantities, meets state and federal safety guidelines, and is consistently uniform in stability and maturity could be an asset in this highly competitive industry. The challenge to the compost industry is to develop the guidelines needed to insure the consistent production, management, and storage of high-quality compost products.

REFERENCES

Argo, W.R. 1998. Root medium physical properties. *HortTechnology* 8:481–485

Barragry, A.R. and J.V. Morgan. 1978. Effect of mineral and slow-release nitrogen combinations on the growth of tomato in a coniferous bark media. *Acta Horticulturae* 82:43–53.

Bearce, B.C. and D.K. Postlethwaite. 1982. Growth of tomato and cucumber in hardwood bark mixes of three fertilizer frequencies. *HortScience* 17:479.

Boodley, J.W. and R. Sheldrake, Jr. 1972. *Cornell Peat-Lite Mixes for Commercial Plant Growing*. Cornell Information Bulletin 43, Cornell University, Ithaca, New York.

Bunt, A.C. 1976. *Modern Potting Composts*. The Pennsylvania State University Press, University Park, Pennsylvania.

Chaney, R.L., J.B. Munns, and H.M. Cathey. 1980. Effectiveness of digested sewage sludge compost in supplying nutrients for soiless potting media. *Journal of the American Society for Horticultural Science* 105:485–492.

Chaney, R.L., J.A. Ryan, and S.L. Brown. 1999. Environmentally acceptable endpoints for soil metals, p. 111–154. In: W.C. Anderson, R.C. Loehl, and B.P. Smith (eds.). *Environmental Availability in Soils: Chlorinated Organics, Explosives, Metals*. American Academy of Environmental Engineering, Annapolis, Maryland.

Chanyasak,V., A. Katayama, M.F. Hiria, S. Mori, and H. Kubota. 1983. Effects of compost maturity on growth of komatsuna (*Brassica rapa* var. *perviridis*) in Neubaur's pot. *Soil Science and Plant Nutrition* 29:239–250.

Craul, P.J. and M.S. Switzenbaum. 1996. Developing biosolids compost specifications. *BioCycle* 37(12):44–47.

Falahi-Ardakani, A., J.C. Bouwkamp, F.R. Gouin, and R.L. Chaney. 1987a. Growth response and mineral uptake of vegetable transplants growing in composted sewage sludge amended medium. I. Nutrient supplying power of the medium. *Journal of Environmental Horticulture* 5:107–112.

Falahi-Ardakani, A., F.R. Gouin, J.C. Bouwkamp, and R.L. Chaney. 1987b. Growth response and mineral uptake of vegetable transplants growing in composted sewage sludge amended medium. II. Influenced by time of application of N and K. *Journal of Environmental Horticulture* 5:112–116.

Falahi-Ardakani, A., J.C. Bouwkamp, F.R. Gouin, and R.L. Chaney. 1988. Growth response and mineral uptake of lettuce and tomato transplants grown in media amended with composted sewage sludge. *Journal of Environmental Horticulture* 6:130–132.

Gogue, G.J. and K.C. Sanderson. 1975. Municipal compost as a medium amendment for chrysanthemum culture. *Journal of the American Society for Horticultural Science* 100:213–216.

Goldstein, N. 1997. The state of garbage in America. *BioCycle* 38(4):60–67.

Gouin, F.R. 1993. Utilization of sewage sludge compost in horticulture. *HortTechnology* 3:161–163.

Gouin, F.R. and J.B. Shanks. 1981. Composted gelatin waste aids crops. *BioCycle* 22(4):41–45.

Hoitink, H.A.J. and P.C. Fahy. 1986. Basis for control of soil borne plant pathogens with composts. *Annual Review of Phytopathology* 24:93–114.

Hoitink, H.A.J., Y. Inbar, and M.J. Boehm. 1991. Status of compost-amended potting mixes naturally suppressive to soil borne diseases of floriculture crops. *Plant Disease* 75:869–873.

Hornick, S.B., L.J. Sikora, S.B. Sterrett, J.J. Murray, P.D. Millner, W.D. Burge, D. Colacicco, J.F. Parr, R.L. Chaney, and G.B. Willson. 1984. *Utilization of Sewage Sludge Compost as a Soil Conditioner and Fertilizer for Plant Growth*. Agricultural Information Bulletin 464. Agricultural Research Service, United States Department of Agriculture (USDA).

Inbar, Y., Y. Chen, Y. Hadar, and H.A.J. Hoitink. 1990. New approaches to compost maturity. *BioCycle* 31(12):64–69.

Keeling, A.A., I.K. Paton, and J.A. Mullet. 1994. Germination and growth of plants in media containing unstable refuse-derived compost. *Soil Biology and Biochemistry* 26:767–772.

Kostov, O., Y.Tzvetkov, N. Kaloianova, and O. van Cleemput. 1996. Production of tomato seedlings on composts of vine branches and grape prunings, husks, and seeds. *Compost Science and Utilization* 4(2):55–61.

Lohr, V.I., R.G. O'Brien, and D.L. Coffey. 1984. Spent mushroom compost in soiless media and its effects on the yields and quality of transplants. *Journal of the American Society for Horticultural Science* 109:693–697.

Lunt, H.A. 1959. *Digested Sewage Sludge for Soil Improvement*. Connecticut Agricultural Experiment Station Bulletin 622.

Lunt, O.R. and B. Clark. 1969. Horticultural applications for bark and wood fragments. *Forest Production Journal* 9(4):39A–42A.

Matkin, O.A., and P.A. Chandler. 1957. The U.S. type soil mixes, p. 68–85. In: K.F. Baker (ed.). *The U.C. System for Producing Healthy Container-Grown Plants*. California Agricultural Experiment Station, Extension Service Manual 23.

Matkin, O.A., P.A. Chandler, and K.F. Baker. 1957. Components and development of mixes, p. 86–107. In: K.F. Baker (ed.). *The U.C. System for Producing Healthy Container-Grown Plants*. California Agricultural Experiment Station, Extension Service Manual 23.

Marcotrigiano, M., F.R. Gouin, and C.B. Link. 1985. Growth of foliage plants in composted raw sewage sludge and perlite media. *Journal of Environmental Horticulture* 3:98–101.

Ozores-Hampton, M., T.A. Obreza and G. Hochmuth. 1998a. Using composted waste on Florida vegetable crops. *HortTechnology* 8:130–137.

Ozores-Hampton, M.P., C. Vavrina, and T.A. Obreza. 1998b. Yard trimmings-biosolids co-compost can substitute for sphagnum peatmoss in tomato (*Lycopersicon esculentum Mill.)* transplant media. *HortScience* 33:488 (Abstract).

Pinamonti, F., G. Stringari, and G. Zorzi. 1997. Use of compost in soiless cultivation. *Compost Science and Utilization* 5(2):38–46.

Purman, J.R. and F.R. Gouin. 1992. Influence of compost aging and fertilizer regimes on the growth of bedding plants, transplants, and poinsettia. *Journal of Environmental Horticulture* 10:52–54.

Purves, D. and E.J. Mackensie. 1973. Phytotoxicity due to boron in municipal compost. *Plant and Soil* 40:231–235.

Rathier, T.M. 1982. Spent mushroom compost for greenhouse crops. *Connecticut Greenhouse Newsletter* 109:1–6.

Roe, N.E. and S.R. Kostewicz. 1992. Germination and early growth of vegetable seed in composts, p.91–207. In: C.S. Vavrina (ed.). *Proceedings of the National Symposium on Stand Establishment of Horticultural Crops*, 16–20 Nov. 1992, Fort Myers, Florida.

Roe, N.E., P.J. Stoffella, and D. Graetz. 1997. Composts from various municipal solid waste feedstocks affect vegetable crops. I. Emergence and seedling growth. *Journal of the American Society for Horticultural Science* 122:427–432.

Rosen, C.J., T.R. Halbach, and B.T. Swanson. 1993. Horticultural uses of municipal solid waste composts. *HortTechnology* 3:167–173.

Sanderson, K.C. 1980. Use of sewage-refuse compost in the production of ornamental plants. *HortScience* 15:173–178.

Sawhney, B.L. 1976. Leaf compost for container-grown plants. *HortScience* 11:34–35.

Shiralipour, A., B. Faber, and M. Chrowstowski. 1996. Greenhouse broccoli and lettuce growth using cocomposted biosolids. *Compost Science and Utilization* 4(3):38–43.

Siminis, H.I. and V.I. Manios. 1990. Mixing peat with MSW composts. *BioCycle* 31(11):60–61.

Sterrett, S.B. and R.L. Chaney. 1982. Growth of vegetable transplants and bedding plants in sewage sludge compost media. *HortScience* 17:479 (Abstract).

Sterrett, S.B., R.L. Chaney, C.W. Reynolds, F.D. Schales, and L.W. Douglass. 1982. Transplant quality and metal concentrations in vegetable transplants grown in media containing sewage sludge compost. *HortScience* 17:920–922.

Sterrett, S.B., C.W. Reynolds, F.D. Schales, R.L. Chaney, and L.W. Douglass. 1983. Transplant quality, yield, and heavy-metal accumulation of tomato, muskmelon, and cabbage grown in media containing sewage sludge compost. *Journal of the American Society for Horticultural Science* 108:36–41.

Swiader, J.M., G.W. Ware, and J.P. McCollum. 1992. *Producing Vegetable Crops*. 4th edition. Interstate Publishers, Inc., Danville, Illinois.

Thompson, H.C. and W.C. Kelly. 1957. *Vegetable Crops*. 5th edition. McGraw-Hill Book Company, New York.

Tonnessen, D. (ed.). 1998. *Compost Use in Florida*. Florida Center for Solid and Hazardous Waste Management, Florida Department of Environmental Protection, Tallahassee.

Vavrina, C.S. 1994. Municipal solid waste materials as soiless media for tomato transplant production. *Proceedings of the Florida State Horticulture Society* 107:118–120.

Vavrina, C.V. 1995. Municipal solid waste materials as soiless media for tomato transplants: field production results. *Proceedings of the Florida State Horticulture Society* 108:232–234.

Vavrina, C.S., M. Ozores-Hampton, K. Armbrester, and M. Pena. 1996. *Spent Mushroom Compost and Biological Amendments as an Alternative to Soilless Media*. Southwestern Florida Research and Extension Center Station Report–Veg.96.3.

Vavrina, C.S. and W. Summerhill. 1992. Florida vegetable transplant producers survey, 1989–1990. *HortTechnology* 2:480–483.

Vega-Sanchez, F., F.R. Gouin, and G.B. Willson. 1987. The effects of curing time on physical and chemical properties of sewage sludge and on the growth of selected bedding plants. *Journal of Environmental Horticulture* 5:66–70.

Waters, W.E., W. Llewellyn, and J. Nesmith. 1970. The chemical, physical, and salinity characteristics of twenty-seven soil media. *Proceedings of the Florida State Horticulture Society* 83:483–488.

Wong, M.H. and L.M. Chu. 1985. The response of edible crops treated with extracts of refuse compost of different ages. *Agricultural Wastes* 14:63–74.

Zucconi, F., A. Pera, M. Forte, and M. DeBertoldi. 1981. Evaluating toxicity of immature compost. *BioCycle* 22(2):54–57.

CHAPTER 11

Compost Economics: Production and Utilization in Agriculture

George K. Criner, Thomas G. Allen, and Raymond Joe Schatzer

CONTENTS

1-56670-460-X/01/$0.00+$.50

I. INTRODUCTION

Composting and agriculture generally are considered to be a natural fit. Increasing environmental constraints on the disposal of animal manures and a growing understanding of the agronomic benefits of compost utilization suggest that on-farm composting is an obvious win-win solution for farmers. Moreover, rising disposal fees improve the potential profitability of managing municipal wastes with composting and increase the potential for opportunities to compost organic wastes in a farm setting. As suggested by other chapters in the book, a good deal of attention has focused on improving the science and technology of composting. However, for composting to continue its growth as a waste management alternative it will be necessary to develop a clear understanding of the economic implications of managing solid waste with composting and of using compost in agricultural and horticultural applications. From the planning and construction of a compost facility, through the transportation and processing of wastes, to the marketing and use of the end product, understanding the costs and the benefits associated with composting is crucial to effective decision making.

Given the broad range of issues that must be considered, such as feedstock material, composting technologies, environmental regulations, and product markets, the decision to establish a composting operation is complicated. In addition, the determining factors can vary depending upon the point of view of the decision-maker. Commercial composters determine project feasibility based on anticipated profitability. Municipalities and other decision-makers in the public sector may base their decision on finding the least cost treatment approach within the context of an integrated waste management system. Farmers might consider composting as a potentially profitable use of otherwise nonproductive land, idle labor and machine time, or as an own-farm source of nutrients to enhance crop production. Regardless, composting projects must be evaluated within a framework that supports sound decision making. No matter how committed one is to the environmental aspects of composting, treating organic residuals via composting can be technically complicated and relatively expensive.

This chapter is intended to assist decision-makers in their analysis of composting projects by providing a framework for understanding and comparing the anticipated costs and revenues, as they pertain both to compost production and compost utilization. Beyond the financial analysis, potential composters must be prepared to consider broader issues that affect the economic feasibility of the project, such as competition for feedstock from other waste disposal methods, competition in the market for compost products, and regulatory issues. This chapter examines these issues by reviewing current issues and trends in composting, identifying the relevant economic considerations, providing a generic overview of the steps involved in conducting a financial analysis of composting, and presenting a specific case example of an economic analysis of the decision to utilize composting in a horticultural application.

II. SOLID WASTE AND COMPOSTING IN THE UNITED STATES

In 1996, Americans generated nearly 190 million Mg (210 million tons) of municipal solid waste (MSW) (U.S. EPA, 1998a). MSW is the nonliquid, nonhazardous waste from households, institutions (e.g., high schools), and commercial establishments. Although municipal waste disposal issues generally attract the greatest public attention, the volume of mining and other resource exploration generated waste, agricultural waste, and industrial waste are all much larger than MSW generation (Figure 11.1).

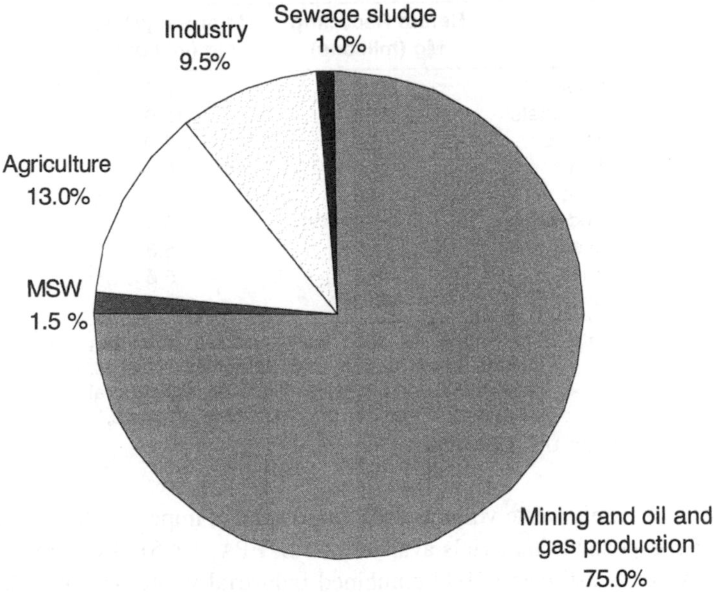

Figure 11.1 Waste generation in the U.S. 1996. (From Miller, G.T., 1999. *Living in the Environment: Principles, Connections, and Solutions*. 11th edition. Brooks/Cole Publishing Company, Pacific Grove, California. With permission.)

In addition to the types of solid waste shown in Figure 11.1, there are also various other classifications. There are medical wastes, radioactive wastes, special wastes, and hazardous wastes. MSW itself is commonly broken down by materials (e.g., newspaper, food, white goods such as stoves), or in categories (e.g., food packaging, apparel), or by size (bulky items such as a stove or couch, vs. nonbulky items such as normal paper and food items). Unfortunately, MSW measurement is sometimes a moving target. Many communities and some states have banned certain wastes from their waste stream. For example, if one is comparing per capita waste generation between two states, it is important to know if one of the states has banned leaf and yard waste and would therefore have a smaller MSW-collected waste stream, all other things being equal. Removing leaf and yard waste from the waste stream keeps the material from landfills and incinerators. This helps reduce leachate and methane generation at landfills and nitrogen oxide formation at incinerators (U.S. OTA, 1989).

Today, there are fewer, but larger, landfills. Nationwide, capacity does not appear to be a problem, but regional disposal issues still arise. As a result, municipalities continue to search for cost effective alternatives for waste disposal, including composting technologies. As a form of recycling, the U.S. Environmental Protection Agency (U.S. EPA) has estimated that composting can handle from 30 to 60% of a community's waste stream (U.S. EPA, 1994). Presently, a large quantity of organic material (paper, yard waste, wood, food waste) still remains in the waste stream after recovery and recycling efforts (Table 11.1).

Table 11.1 Total Waste Generation and Net Waste Disposal in the U.S., 1996

	Before Recycling Mg (millions)	After Recycling Mg (millions)
Paper	72.5	42.9
Yard waste	25.4	15.6
Plastics	18.0	17.0
Metals	14.6	8.8
Wood	9.8	9.4
Food waste	19.9	19.4
Glass	11.2	8.3
Other	6.3	5.5
Total	190.2	138.2

Note: The municipal solid waste stream does not include construction and demolition wastes, industrial wastes, sewage biosolids, agricultural wastes, or a number of other waste products.

From U.S. EPA, 1997.

Knowing a state's waste volume and composition is important for estimating the volume of compostable materials available (U.S. EPA, 1995). As an example, consider that Maine's estimated 1997 combined industrial waste was over 2.7 million Mg (3 million tons), while MSW was estimated at 1.48 million Mg (1.64 million tons). Maine's MSW is thought to be roughly 50% residentially generated and 50% commercially generated. State departments of environmental protection or agriculture may maintain inventories of available organic matter. For instance, the Maine Department of Agriculture, Food, and Rural Resources periodically publishes an inventory of waste products that may be usable on farms (Seekins and Mattei, 1990). Most, if not all, of these materials are well suited for composting. When planning a composting facility it is important to conduct an inventory of available organic material in order to mix the most cost effective compost recipe.

Table 11.2 presents an example of the composition of the nonbulky portion of MSW generated from domestic sources (NBDSW — "non-bulky domestic solid waste"). A large portion of NBDSW (27.8%) is comprised of food. Aside from paper-based products, food is the single largest item in the municipal waste stream.

This is of interest with respect to composting MSW. Note that if all paper is combined with food waste, the portion readily available for composting is 60.8%, nearly two thirds of all residential waste set by the curb.

Table 11.2 may be used to illustrate some important issues that potential composters must consider when analyzing local MSW as feedstock. First, Maine's current curbside waste probably would be somewhat different (e.g., lower levels of newspaper) because the data were collected before many Maine communities initiated curbside recycling. Also, since leaf and yard waste in a given autumn week can constitute up to 20% of the waste stream (U.S. OTA, 1989), this category was excluded from the analysis. The large volumes of leaf and yard waste on some garbage collection routes, by random chance alone, could skew the data collection and analysis. This exclusion also made sense given Maine's trend toward banning leaf and yard waste from the MSW waste stream.

Table 11.2 Annual Average Percent Weight of Waste Categories for Maine Municipalities

Waste Category	Percent NBDSW[z]
Paper	33.0
Plastic	6.7
Glass	4.1
Metal	3.3
Other	
Food waste	27.8
Cat litter-pet bedding	3.9
Diapers	3.8
Other	4.2
Total other waste	52.9
Grand total	100.0

Note: Data were collected in 1991 and 1992 from 14 Maine communities, with samples being taken in each of the four seasons (fall, winter, spring, summer). Leaf and yard waste was not included in the analysis.

[z] NBDSW = nonbulky domestic solid waste.

From Criner, G.K. et al., 1994. *Maine's Household Garbage.* Maine Agricultural and Forest Experiment Station Bulletin 841. Orono, Maine. With permission.

Another point from Table 11.2 concerns the levels of metal, glass, and plastics in the waste stream. Since Maine has an expanded bottle deposit law, considerable glass, aluminum cans and glass bottles are redeemed for deposit. Thus Maine's percentages of beverage container packaging will likely appear lower when compared to other states.

Finally, note that cat litter and pet bedding (primarily consisting of cat litter) constitute just slightly more of the waste stream than disposable diapers, by weight (Table 11.2). Combined, these items account for more than 7% of the curbside nonbulky disposable waste stream, and both of these items would be eligible for composting. However, both have potential pathogen issues that need to be addressed and would likely impact handling, monitoring, and perhaps permitting, and any non-biodegradable plastic from the diapers would need to be removed for some compost applications.

A. Market Trends for Compost

Compost has a variety of uses, ranging from its traditional use by gardeners as a soil enhancement to more innovative uses for wetland restoration, landfill cover, and industrial pollution control. A report by the U.S. EPA lists eight separate market segments for compost and the potential demand for each segment (Table 11.3). In all, the potential demand for compost far outweighs the supply of compost that could be available if the entire applicable waste stream were composted. By far, the largest potential demand is agricultural. It is estimated that the agricultural market for compost could reach 684 million m^3 (895 million yd^3) per year. If the entire organic portion of the U.S. waste stream were composted it would generate approximately 20 million m^3 (26 million yd^3) per year. By contrast, the potential market demand by nurseries is less than 0.8 million m^3 (1 million yd^3) per year.

Table 11.3 Potential Market Size for Compost Products and Current Market Penetration

Market Segment	Potential Market Size m^3 (millions)	Estimated Market Penetration (%)
Agriculture	684.3	<2
Silviculture	79.5	<1
Sod production	15.3	<1
Residential retail	6.1	80
Nurseries	0.7	1 to 50[z]
Delivered topsoil	2.8	<5
Landscapers	1.5	<20
Landfill cover	0.5	<5

[z] Market penetration for field nurseries, <1%; container nurseries, <50%.
From U.S. EPA, 1998a, 1998b.

In recent years, composters have been able to increase their market penetration by producing compost with the particular characteristics required by end users and by blending composts to improve product consistency. For example, BFI Organics, Inc. (Falmouth, ME) markets special blends of manufactured soils for use in turf building, erosion control, and others. Value-added compost products are produced for the bagged retail market by combining compost with fertilizers and other products to stimulate plant growth. It is estimated that 80 to 85% of compost is successfully marketed. According to a survey by The Composting Council, landscapers are the largest users of compost, accounting for 31% of compost used. Other significant market segments include food crop production (25%), landfills (14%), and nurseries (11%) (U.S. EPA, 1998a).

A key issue for composters is the price that their product can command in various markets. As with most products, the prevailing price depends on consistency, promotion, packaging, and distribution channels. For most composters, marketing strategy includes two choices. One option is to produce high-quality composts, blended and packaged for the retail market. This strategy aims at producing a product that will command a sufficiently high price to make its production profitable. The alternative strategy is to produce lower quality composts for the bulk markets that generate less revenue and rely on tipping fees to cover the composter's operating costs.

The higher value retail and nursery markets are already well established and have less room for additional market penetration against competing products. The market demand for compost, and the price that consumers are willing to pay for it also varies depending upon the raw materials used to produce it. Average prices reported from various sources include $35/Mg ($32/ton) for compost from yard trimmings, $43/Mg ($39/ton) for source-separated organics compost, and $3.30/Mg ($3/ton) for compost derived from MSW (U.S. EPA, 1998b). Based on information from the 1997 *BioCycle* survey of biosolids composters, the estimated average price of biosolids compost is $7.70/Mg ($7/ton) (Goldstein and Block, 1997). These prices reflect average revenue from finished compost. Depending upon the feedstock material, the degree of weight reduction that occurs during composting must be factored in to calculate revenue per unit weight of raw material input. For example, at $43/Mg ($39/ton) for compost from source separated organics, a 50% reduction in weight during composting means that the composter is receiving $21.50 per Mg ($19.50 per ton) of waste that gets processed.

In New England, approximately 70 to 80% of compost goes to large public works projects which utilize low value bulk compost. Most farm composters are able to produce only moderate volumes of bulk product for on-farm use or for a local market (Fabian et al., 1993). In that case, the farm composter may rely on revenue from tipping fees to maintain the project's viability. Depending upon the region of the country, the prevailing tipping fees that can be charged may or may not be sufficient to sustain the project. In 1997, the national average tipping fee for landfills was $42/Mg ($38/ton), while in the Northeast the average tipping fee was $59.50/Mg ($54/ton). For individual states, the fees ranged from $33/Mg ($30/ton) or less in many states to $85/Mg ($77/ton) in New Jersey (U.S. EPA, 1998b).

III. ECONOMIC ISSUES RELATED TO COMPOST PRODUCTION SYSTEMS

Composting is based on the biological decomposition of organic materials by microorganisms. It is a process that can be controlled by manipulating the factors critical to decomposition including nutrient balance, surface area, moisture content, oxygen level, and temperature. These can be manipulated indirectly by adjusting the mix of feedstocks before composting or through direct mechanical means such as shredding, dewatering and forced aeration.

Three general types of composting technologies are widely used in waste management: windrows, aerated static piles, and in-vessel systems. The approaches vary primarily in the methods and level of technology employed to control the composting process. This, in turn, determines such factors as capital cost, labor intensity, and the amount of facility space required to process a given volume of waste.

A. Windrows

The windrow approach is the simplest, most common, and least expensive method of composting. Windrow composting is generally considered to have the lowest capital cost where labor is required for occasional turning or mixing to vent excess heat and moisture and to build and reform the windrows. It is a very robust process that is used to successfully process yard and agricultural wastes, and wastewater treatment biosolids.

The principal drawback to the windrow approach is the extensive space requirements for processing large volumes of wastes. The extensive land area necessary for windrow composting is a result of the relatively long time that it takes for raw waste to be transformed into a stabilized product. A large area for windrow composting is also desirable from an odor nuisance standpoint. The larger the composting area, the greater the opportunity for offensive odors to dissipate. Depending upon the frequency of turning the material and other factors, the process can take anywhere from 4 to 18 months.

The relatively simplistic design is both an advantage and disadvantage of the windrow system. Because of the lower volume of some of these facilities, some states allow a simple permitting process. Some windrow operations are covered to control precipitation, have leachate collection and treatment systems, and have sophisticated monitoring programs. In general, however, windrow composting is the least technologically sophisticated of the composting approaches.

B. Aerated Static Piles

This approach is slightly more sophisticated than the windrow method of composting and is amenable to greater automation. Instead of manually turning the piles, mechanical blowers and a network of pipes are used to move air throughout the pile to control both the internal temperature and oxygen level. With remote sensors, the timing and duration of aeration can be automatically controlled. Although a windrow composting site does not require electricity service at the site, aerated static piles usually require electrical service to operate the pumps and blowers.

The forced aeration method requires more energy than windrows, but it can be done with less space because the process takes less time. A stabilized product can be produced in 2 to 20 weeks. In some cases, the aerated static pile approach also produces fewer objectionable odors than do windrows. As with windrow composting, the static piles usually need to be reformed periodically due to volume reduction.

C. In-Vessel Composting

In this approach, the composting process is usually monitored and controlled within an enclosed environment. This method makes possible a high level of computerized automation. It is by far the most costly and capital intensive option, but also requires the least amount of space. Composting time can be reduced to as little as 1 or 2 weeks and facility throughput is usually larger than in other systems.

D. Operational Issues

In addition to the costs that are inherent to each composting technology, other operational concerns have economic implications. Additional costs to address or alleviate problems arise from additional project engineering, capital equipment, or operating expenses.

1. Odors

The process of composting has the potential to create objectionable odors. Their impact and the measures employed to mitigate the problem depends on several factors, including the mix of feedstock materials, the compost technology employed, and the proximity of the compost facility to residential areas. Odors can also be reduced by managing the mix of feedstocks and taking steps to ensure that adequate oxygen is available to prevent the decomposition process from turning anaerobic. Feedstocks containing high levels of nitrogen, such as grass clippings and food wastes, should be balanced with high-carbon wood and paper wastes.

Windrows produce minimal odors except on those periodic occasions when the windrows are turned. Moreover, because windrow facilities typically require large land areas, they are often located in less densely populated areas and present less of a nuisance. Some windrow facility operators make it a practice to alert neighbors when the windrows are scheduled to be turned. In-vessel composting technologies are normally used to process larger volumes of waste and therefore may produce objectionable odors. Since composting takes place in an enclosed space, it presents a better opportunity to manage odors through filtration methods.

Often, the odor problem is primarily due to the raw materials or recipe ingredients used by the facility. Wastewater treatment biosolids, for example, when arriving by the trailer-truckload can generate significant odors. Immediate mixing of these odorous materials with bulking agents (e.g., wood chips) can help to reduce the odor problem.

2. Leachate

Leachate is the liquid residue that seeps from compost piles with high-moisture feedstocks. Because leachate might contain heavy metals and organic chemicals, some facilities are required to implement steps to collect leachate to prevent it from flowing into the local watershed or ground water. This can be done by constructing a paved surface to capture the flow of leachate or by collecting surface runoff in sedimentation ponds.

3. Pathogens and Heavy Metals

Various pathogenic organisms can be present in the compost if the process is not allowed to generate sufficiently high temperatures to kill the organisms. The potential for pathogens also varies with the types of waste in the feedstock. Co-composting with biosolids increases the potential presence of pathogens. The presence of heavy metals also is influenced by the feedstock material. Heavy metals can limit the marketability of MSW compost in general and specifically for use in the production of food crops.

IV. ECONOMIC CONSIDERATIONS FOR COMPOSTERS

Compost is a manufactured product and the efficiency by which the product is produced depends on how well the process is designed and controlled. The cost of producing the product is influenced by the level of technology employed and by the characteristics of the raw material. The marketability of the product is determined by its quality, its price, and the price and availability of competing products.

Project costs may normally be divided into two categories, capital costs and operating and maintenance costs. Capital costs, as defined here, include costs for items that have an expected useful life of over 1year, or are up-front costs such as engineering and attorney's fees. Operation and maintenance costs are those costs associated with running and managing a composting facility. These include wages and salaries of the plant manager and employees, as well as utilities, supplies, equipment repair, insurance, taxes, and all other nondurable inputs. Revenues that should be considered in the analysis include any government grants received, tipping fees charged to external users, revenues from sales of compost, and any avoided costs associated with the project. This section discusses the various costs and how they should be included in the economic assessment.

A. Capital Costs

Several factors determine the extent of capital costs. The types and volumes of materials to be composted, the availability of land, and the desired quality of the compost product affect the decision of which compost technology to use. Facility location relative to residential neighbors may entail additional costs to ameliorate offensive noise and odors. Intangibles such as design, engineering, and legal fees also are considered capital costs. Capital costs can be included in the analysis on a yearly basis by annualizing the costs in one of two ways. Depreciation is the annual amount charged to write off the cost of a capital asset over its expected useful life. It does not include any interest costs that may be associated with acquisition of the asset. Amortization defines the periodic payments needed to spread the cost of the asset over the course of its expected useful life, including both the principal and interest costs associated with the asset. All computer spreadsheet applications include simple functions for calculating a periodic (e.g., monthly, yearly, etc.) amortization charge based on the capital cost of the asset, the periodic interest rate, and the number of periods over which the capital cost is to amortized. In the Microsoft Excel© spreadsheet, the periodic payment is determined by the function "=pmt(rate,nper,pv)", where "rate" is the periodic interest rate, "nper" is the number of periods over which the principal is amortized, and "pv" is the present value or total principal value to be amortized.

For example, suppose a firm purchased a $30,000 tractor for use in its composting operation. If this tractor was expected to last 7 years and the capital was borrowed at 8% interest, the annual amortized value is $5,762 [=pmt(.08,7,–30000)]. Trade-ins, salvage value, and other factors will complicate the analysis although this method will provide a good guideline to the annual capital cost of the item.

1. Land

Site related costs depend upon local real estate values and how central the location is relative to sources of feedstock. More remote locations tend to have lower real estate values, but this is at least partially offset by the higher transportation costs that will ensue. Since land does not depreciate like equipment and buildings, the cost of ownership is usually considered to be the interest cost associated with the land purchase. If the necessary land is already under ownership, the cost should be based on the foregone income that could be obtained by renting the land. Alternatively, a municipality may attach zero cost to land it already owns.

2. Site Preparation

These costs will vary widely depending on the size and type of facility and the characteristics of the land. Creating the necessary infrastructure for the facility will include grading, access roads, fencing, and gates. Depending upon soil condition, paving may be required to manage surface water. In addition, facilities that will process MSW typically require additional site engineering to manage leachate. These up-front costs should be amortized over the life of the facility.

3. Buildings and Equipment

This category of costs is the one most impacted by the decision of which level of composting technology to use. Lower technology approaches that focus on composting only a moderate amount of yard trimmings can operate with only a front-end loader for windrow turning and a small equipment shed for tool storage. Higher volume windrow operations can accelerate the compost process by employing grinders and shredders to reduce particle size. In addition, specialized windrow turning equipment is available to process material more quickly than front-end loaders. Also available are several models of shredder-turners that are powered by a tractor power-take-off. Screening equipment is often a necessary piece of equipment for ensuring a high-quality finished product. Aerated static piles, although requiring less turning, involve a greater expenditure for piping, blowers, and other aerating equipment.

Facilities for composting MSW require the greatest investment in plant and equipment. Material that is not source-separated will need to be processed on-site to remove inorganic contaminants, requiring costly separating equipment. Additional processing of feedstock for composting requires shredders, grinders, and trommels. Many MSW facilities also employ conveyors to move material between different processing equipment. Depending upon plant location and government regulations, additional investment may be required in odor control technologies. Capital costs for each piece of equipment needs to be amortized over the proper life expectancy.

4. Training

Regardless of the facility size and technology employed, managing the compost process effectively is crucial to producing a quality compost product. Overall training costs are determined by the amount of facility staffing and the level of technology involved. While staff training is generally regarded as an operating cost, and it is likely to be an ongoing expense, providing proper training prior to start-up will help to avoid a long and costly trial-and-error period at the outset. These initial training costs should be amortized over the expected life of the facility.

5. Permits and Legal, Engineering, and Consulting Fees

In addition to the engineering and legal services necessary for actual land acquisition and facility development, professional services are frequently needed for the preparation of permit applications. Permitting agencies generally require site layout, engineering, and financing details as part of the approval requirements. Information on specific activities for process monitoring, effluent and odor control, and product marketing are also components of permit applications. Fees paid for these professional services should be amortized over the life of the facility. Any permitting process which allows for public comment should be approached with adequate preparation. Since many compost recipes include manures or sludges (and thus pathogens and/or heavy metals), opponents may attack the proposed composting operation on the grounds of "stop the toxins" or similarly inflammatory statements.

B. Operating and Maintenance Costs

In addition to raw materials, operating and maintenance costs are those costs associated with the day-to-day operation of the facility. These include wages and salaries, utility expenses, taxes, insurance, office and other supplies, and equipment repair. These are expenses that must be estimated as part of the economic analysis of the project. Sample costs for a commercial composter in the northeastern U.S. are provided in Table 11.4. With average tipping fees of $59.50/Mg ($54/ton) in the northeastern U.S., and operating costs of $55/Mg ($50/ton), this facility can generate a small profit from tipping fee revenues with additional profits coming from sales of compost product.

1. Labor Costs

The cost for labor varies with the scale of operation and the technology utilized. Small-scale leaf composting using the windrow approach may operate with only one person to receive the material, turn it four times per year, and water the piles as needed. More extensive composting by facilities that employ more technology

Table 11.4 Estimated Production Costs for a Commercial Biosolids Composter in the Northeastern U.S., 1999

Expense Category	$ per Mg	% of Total
Purchased bulking agent	7.17	13
Labor, mgt., maintenance, & fuel	19.84	36
Depreciation (bldgs. & machines)	14.33	26
Testing, monitoring, & compliance	2.76	5
General & administrative	5.51	10
Purchasing activities	2.20	4
Sales activities	3.31	6
Total	55.12	100

may require more labor. Even large-scale yard trimmings composting may need a supervisor to oversee the operation, equipment operators to manage the incoming materials and turn the piles, maintenance workers, and staff to monitor and manage the compost process.

2. Repairs, Fuel, Parts, and Supplies

The significance of the costs in this category varies depending upon the type of composting facility. A rule of thumb suggests that these expenses are equal to approximately 15% of initial capital equipment costs. In-vessel facilities have higher equipment operating and maintenance costs due to their higher reliance on processing equipment. Supplies for odor management systems also can be significant depending on the type of system employed. Fuel and maintenance costs for windrow systems depend primarily on the equipment used to mix the feedstock and turn the windrows.

3. Utilities

Utility payments, especially for electricity, vary by type of composting facility. Windrow systems require virtually no electricity other than that required in auxiliary buildings. Aerated static pile facilities require electricity to operate the blowers. In-vessel systems, with their high reliance on automation and need for odor management, have the greatest demand for energy.

4. Other

These costs include payments for insurancc, taxes, laboratory testing, and other miscellaneous fees.

Table 11.5 presents information compiled from various sources covering 13 different composting operations and published by the U.S. Environmental Protection Agency (U.S. EPA, 1998b). The average total costs, including capital and operating expenses, range from $23.87/Mg ($21.65/ton) for yard waste windrow composting operations to approximately $55.12/Mg ($50.00/ton) for municipal mixed waste composting. The mixed waste costs shown here are comparable to a recent study of

Table 11.5 Total Estimated Costs for Various Composting Strategies in the U.S.

Composting Strategy	Compost Technology & Materials	Avg. Vol: Mg/Year	Capital: Dollars/Mg	Operating: Dollars/Mg	Total Costs: Dollars/Mg
		Yard Waste			
Atlantic County, NJ	Windrow	19,958	$11.24	$13.01	$24.25
Lehigh County, PA	Windrow	15,422	$11.46	$8.93	$20.39
Bluestone SWA, IA	Windrow	63,503	$4.63	$7.72	$12.35
Bozeman, MT	Windrow	1,814	$1.65	$7.17	$8.82
Average[z]					$23.87
Institutional Onsite	**Low-Tech**				
Kelley Air Force Base	Windrow	726	$64.96	$27.56	$92.52
Georgia Diagnostic & Classification Center	Windrow	943	$12.11	$29.67	$41.79
	High-Tech				
Natural Resources Canada	In-vessel	85	$68.64	$132.21	$200.84
Rikers Island Correctional Center	In-vessel	3,629	$41.91	$63.38	$105.29
Average[z]					$54.01
		Mixed Waste			
Sumpter County, FL	In-vessel	10,024	$26.72	$43.20	$69.93
Wright County, MN	Aerated/ Windrow	44,815	$29.01	$36.82	$65.83
Truman, MN	N/a	16,511	$46.09	$14.47	$60.57
Columbia County, WI	In-vessel	16,983	$16.49	$31.21	$47.70
Sevierville, TN	In-vessel/ tunnel	51,891	$26.01	$17.34	$43.36
Average[z]					$54.99

[z] The average cost is weighted based on Mg of material composted per year.

From: U.S. EPA, 1998b.

MSW composting, which found an average total cost of $59.52/Mg ($54.00/ton) (Renkow and Rubin, 1998). The category labeled "institutional onsite" includes institutional facilities that are composting primarily food wastes and select paper wastes. The facilities listed in Table 11.5 are a military base, a correctional center, and large government agencies that are composting their on-site food wastes, but the category can also include universities, schools, hospitals, and other institutions that generate large quantities of organic materials.

Data for facilities that compost wastewater treatment biosolids are lacking in Table 11.5. A recent survey of biosolids composters by *BioCycle* magazine included information on processing costs (Goldstein and Gray, 1999). The figures provided are not directly comparable to those in Table 11.5 due to differences in the measurement of waste volumes and the lack of capital cost data in the report. Whereas most compost costs are measured in terms of dollars per unit weight of feedstock input without accounting for moisture content, the biosolids composting operations report volumes as dry weight of output. The reported operating costs for windrow projects averaged $103.62/Mg ($94/ton) dry weight (range from $22.05/Mg [$20/ton] to $220.46/Mg [$200/ton]). Aerated static pile systems reported costs that averaged $136/Mg ($124/ton) (from $16.53/Mg [$15/ton] to $330.69/Mg [$300/ton]). The in-vessel systems produced compost at an average cost of $239.20/Mg ($217/ton) (ranging from $55.12/Mg [$50/ton] to $661.39/Mg [$600/ton]).

V. ECONOMIC ANALYSIS OF THE HORTICULTURAL USE OF COMPOST

The economics of compost use on horticultural cropping systems depends upon many factors related to yields, inputs used, and prices paid and received. Each individual grower will need to analyze carefully each of these factors to make the correct decision concerning the use of compost in his or her operation. Before using results from any other situation, the grower must examine the facts associated with the situation and compare them to his or her own operation. Chances are the grower's situation will differ from the results reported for another situation. The grower should only use the reported results as a guide to develop an estimate of the economics associated with his or her individual operation.

A grower typically is most interested in the profit that can be obtained using a particular cropping system. We can examine the economics of a cropping system by comparing the expected costs with the expected returns. The costs and returns from a cropping system can vary significantly from one operation to the next and even by field on a particular farm. Profits probably will vary from season to season on the same field. Costs and return variability occurs because of differences in the weather, initial soil characteristics, cultural practices followed, machinery size and type, prices paid and received, size of the field, yields, labor availability, management capabilities, and capital resources. Conclusions concerning the profitability of compost use for a particular cropping system will depend upon all these factors.

The grower can approach the economics or profitability of these factors by developing a budget of the expected costs and returns associated with the use of compost within a particular cropping system. A budget of expected costs and returns is termed an enterprise budget. An enterprise budget typically contains three parts: operating or variable costs, ownership or fixed costs, and revenues (Kay and Edwards, 1999). Details on constructing enterprise budgets or cost and return estimates can be found in most farm management books. If the grower is interested in the total operation's profit, then the enterprise budgets can be combined to create a whole business budget. A whole business budget shows the total costs and returns for the operation and provides an estimate of the expected total business profit. The whole business budget provides an estimate of the income statement for the operation. If the grower is interested in the cash flow associated with the cropping system, then the enterprise budget information can be used to construct a summary cash flow statement of cash inflows and outflows.

If the grower is interested in just the impact on the profits of adding compost to the current cropping system, a partial budget could be developed, which provides less information than the enterprise or the whole business budget. The partial budget only examines the costs and returns that change when moving from one practice or set of practices to another. Since the partial budget only looks at changes, less information is needed than is required to develop either an enterprise budget or whole business budget. The partial budget provides an estimate of the expected change in profit from what the grower is currently obtaining.

Before the grower can use any of these budget techniques, he or she must collect the relevant price and quantity information. Whichever budgeting format the grower decides to use to analyze the situation, he or she must remember that developing a budget is only "pencil farming." The budget is only an estimate of the future and in hindsight may not closely resemble what actually happened.

Other chapters of this book provide information on the physical advantages and disadvantages of using compost on horticultural crops. The grower will find that information useful in developing the specific information required for developing a budget for the operation. Since every grower will have a different situation, it is difficult to provide a definitive answer on the economics of compost use. To provide the reader with an idea of the type of information a grower must collect or make estimates for, we will examine the information needed to develop an estimate of the expected profit from using compost on a field of watermelons (*Citrullus lanatus* [Thunb.] Matsum. & Nakai).

The cultural practices and inputs used impact the costs of producing a watermelon crop. We will assume the farmer is producing a commercial crop of watermelons with irrigation on a sandy loam soil. Inputs the farmer must consider include compost, seed, herbicides, insecticides, fertilizer, water, and machinery. The amount and type of compost used may influence the needed quantities of each of these inputs.

The farmer would need to answer many questions. How much compost will be used? When will the compost be applied? Where will he or she obtain the compost? Will using compost reduce the need for herbicide and by how much? Alternatively, can he or she change to a less costly herbicide or do away with it entirely? The same questions must be answered with respect to insecticides. What about the amount of seed sown? Will the compost influence the germination rate? If so, by how much? Will the seeding rate need to be increased or decreased? How much fertilizer can the compost replace, if any? What nutrients will be available from the compost? Does he or she need any additional nutrients with compost?

The farmer would need to look at what effect the use of compost has on the use of machinery and labor. How is the compost applied to the soil? Can the application be done by machine or will additional hand labor be needed? Will the compost be left on top of the soil as mulch or will it be incorporated into the soil? How will tillage and spraying operations be impacted? Finally, the harvesting cost will be impacted if there is a change in yield because of the use of the compost.

The addition of the compost may influence the yield or the quality of the yield obtained. For example, anecdotal evidence from central Maine corn (*Zea mays* L.) producers suggests that their use of compost during the 1997 and 1998 growing seasons resulted in good yields in the summer of 1999 despite near drought conditions.

Sufficient evidence to quantify this type of benefit rarely exists. The compost may decrease the length of the growing season such that the farmer can take advantage of an earlier marketing period with a higher price. As suggested in others parts of the book, yield impacts vary by cropping system and location. However, the yield impact from the use of compost and the cost of obtaining and applying it probably will have the largest influence on the profitability of its use. The farmer must be comfortable with the yield estimate made relative to the amount of compost used.

The last pieces of information the farmer needs to estimate costs and returns are the prices of the output and the inputs. Estimates for the prices of inputs will probably be more exact than for the output (watermelons) unless the farmer will contract the output for a set price before production begins. The price of the compost and its application cost will have a major influence upon the profitability of its use. If the timing of harvest or the quality of the product is influenced by the use of the compost, then the farmer will need to adjust the expected price of the product.

Once the farmer has obtained this information, he or she can develop the enterprise budget for watermelons with and without the use of compost. The profit obtained from both cropping systems can be compared to see which provides the highest return. The farmer may want to do some sensitivity analysis to see how the results change as yields and prices vary. Table 11.6 shows a sample watermelon budget with and without the use of compost. We have assumed that the main differences between the two situations are the addition of the compost, which costs about $75 per hectare, and a 5% increase in yield on the composted acreage. With these changes, the return above the specified costs decreases by about $15 per hectare. If we assumed the compost could replace some of the fertilizer, then the use of the compost may result in about the same level of return above specified costs. We may expect some carryover impact of the compost on the soil such that the next crop may also receive a benefit that might help offset some of the costs. As previously mentioned, the individual producer must examine his or her situation and develop an appropriate cost and return estimate.

VI. CONCLUSIONS

Economics has been and will continue to be a key factor in waste management decisions. As a disposal alternative for most of the organic portion of the U.S. waste stream, composting merits attention for many noneconomic reasons. It reduces the volume of raw wastes, stabilizes organic materials, and produces an end product that has proven beneficial characteristics in horticultural applications. In addition, research is uncovering a range of other innovative uses of compost including erosion control, wetland mitigation, bioremediation, land reclamation, biofilters, storm water filtrates, and fungicides. Economics, however, will largely dictate the extent to which composting replaces existing products and approaches currently available for these purposes.

In the competition for market share, compost will have to perform as effectively as alternative products at a similar price. Unfortunately, some of the benefits of compost use cannot be measured directly. The horticultural example of using

Table 11.6 Costs and Returns for Watermelon Production on One Hectare, With and Without Compost

			With Compost		Without Compost	
Operating Inputs	**Unit**	**Price $/Unit**	**Quantity Units**	**Value $/ha**	**Quantity Units**	**Value $/ha**
Watermelon seed	kg	26.46	1.68	44.45	1.68	44.45
Nitrogen (N)	kg	0.55	89.60	49.39	89.60	49.39
Phosphorous (P_2O_5)	kg	0.42	33.60	14.08	33.60	14.08
Potassium (K_2O)	kg	0.29	56.00	16.05	56.00	16.05
Compost	Mg	5.51	8.96	49.39	0.00	0.00
Compost application	ha	24.70	1.00	24.70	0.00	0.00
Hoeing labor	hour	6.50	19.76	128.44	19.76	128.44
Harvest labor	hour	6.50	62.34	404.56	59.28	385.32
Rent fertilizer spreader	ha	6.18	2.00	12.35	2.00	12.35
Insecticide	ha	25.32	2.00	50.64	2.00	50.64
Custom aerial application	ha	9.88	2.00	19.76	2.00	19.76
Annual operating capital	dollar	0.09	112.99	9.89	75.28	6.59
Machinery labor	hour	6.50	3.59	23.34	3.59	23.34
Machinery fuel, lube, repairs	dollar		34.67	34.67	34.67	34.67
Total operating costs				881.70		785.07
Fixed Costs						
Machinery interest, depreciation, taxes, and insurance				55.31		55.31
Total fixed costs				55.31		55.31
Production						
Watermelons	Mg	121.28	14.11	1711.43	13.44	1629.94
Total receipts				1711.43		1629.94
Returns above total operating cost				829.73		844.87
Returns above all specified costs				774.42		789.56

Note: Watermelons are loaded on a semi-truck in the field. N, P_2O_5, and K_2O amounts are actual kilograms of material.

compost as an input in watermelon production shows slightly smaller economic returns when compared to a similar production situation without the use of compost. It should be noted, however, that the analysis is based on 1 year of production. Some observers have suggested that the benefits of compost use only accrue after several years of application or appear under certain conditions (e.g., low rainfall) (McConnell et al., 1994). Although those benefits may be difficult to measure directly, they include less fertilizer leaching, and therefore decreased fertilizer application, due to the improved moisture retention of compost treated soils. Soil tilth has been shown to improve after several years of compost application leading to greater root growth and better plant resistance to drought (U.S. OTA, 1989).

Much of the interest in large-scale composting derives from its potential as an environmentally positive means for treating organic wastes. The extent to which it is adopted for that purpose, however, will depend on competition from alternative waste disposal options. Although composting is more environmentally "friendly" than landfilling, the relatively flat trends in landfill tip fees has slowed the growth in the number of composting operations. Where landfilling tip fees are higher, and large volumes of compostable waste are centrally located (e.g., a commercial food processing facility) the economics of composting are more favorable.

At present, there is not a lot of pressure on waste disposal facilities to change their current practices. Several emergent trends, however, may have an impact on future waste management decisions. The continuing development of agricultural land and adjacent properties tends to increase public opposition to land spreading sewage treatment biosolids. In addition, it forces municipalities to transport the biosolids greater distances to suitable application sites, thereby raising the cost of land spreading and increasing the economic attractiveness of composting. In recent years, there has been considerable consolidation of waste management companies. The result may be an increase in landfill tip fees that could spur greater interest in composting. Also, the recent emphasis on nutrient management to preserve water quality could lead to greater use of composting of organic wastes.

Finally, increasing market demand and the resulting higher prices for compost products could provide the economic incentive for development and expansion of composting projects. Although landscapers and nurseries continue to be the largest end users of compost, growing numbers of landscape architects are specifying compost as a component of manufactured soils (Goldstein and Gray, 1999). Efforts to increase penetration in traditional compost markets, coupled with the growing number of innovative uses for compost, could also lead to increased demand.

REFERENCES

Criner, G.K., J.D. Kaplan, S. Juric, and N.R. Houtman. 1994. *Maine's Household Garbage*. Maine Agricultural and Forest Experiment Station Bulletin 841. Orono, Maine.

Fabian, E.E., T.L. Richard, D. Kay, D. Allee, and J. Regestein. 1993. *Agricultural Composting: A Feasibility Study for New York Farms*. Staff report, Cornell University, Ithaca, New York.

Goldstein, N. and D.Block. 1997. Biosolids composting holds it own. *BioCycle* 38(12):64–74.

Goldstein, N. and K. Gray. 1999. Biosolids composting in the United States. *BioCycle* 40(1):63–75.

Kay, R.D. and W.M. Edwards. 1999. *Farm Management*. 4th edition. McGraw-Hill, New York.

McConnell, D., A. Shiralipour, and W. Smith. 1994. Compost impact on soil/plant properties, p. 89–91. In: *Composting Source Separated Organics*. The JG Press, Inc., Emmaus, Pennsylvania.

Miller, G.T. 1999. *Living in the Environment: Principles, Connections, and Solutions*. 11th edition. Brooks/Cole Publishing Company, Pacific Grove, California.

Renkow, M. and A.R. Rubin. 1998. Does municipal solid waste composting make economic sense? *Journal of Environmental Management* 53:339–347.

Seekins, B. and L. Mattei. 1990. *Update: Usable Waste Products for the Farm*. Maine Department of Agriculture, Food, and Rural Resources, Augusta, Maine.

U.S. Environmental Protection Agency (U.S. EPA). 1994. *Composting Yard Trimmings and Municipal Solid Waste*. Solid Waste and Emergency Response, Report No. EPA530-R-94-003. Washington, D.C.

U.S. Environmental Protection Agency (U.S. EPA). 1995. *Decision-Makers' Guide To Solid Waste Management, Volume II*. Office of Solid Waste, Report No. EPA530-R-95-023. Washington, D.C.

U.S. Environmental Protection Agency (U.S. EPA). 1997. *Municipal Solid Waste Factbook – Internet Version*. Office of Solid Waste. http://www.epa.gov/reg5oopa/students/municipal_solid_waste_factbook.htm verified 9 Oct. 2000.

U.S. Environmental Protection Agency (U.S. EPA). 1998a. *Characterization of Municipal Solid Waste in the United States: 1997 Update*. Office of Solid Waste, Report No. EPA530-R-98-007. Washington, D.C.

U.S. Environmental Protection Agency (U.S. EPA). 1998b. *Organic Materials Management Strategies*. Solid Waste and Emergency Response, Report No. EPA530-R-97-003. Washington, D.C.

U.S. Office of Technology Assessment (U.S. OTA). 1989. *Facing America's Trash: What's Next for Municipal Solid Waste?* U.S. Government Printing Office, Washington, D.C.

Section III

Benefits of Compost Utilization in Horticultural Cropping Systems

CHAPTER 12

Spectrum and Mechanisms of Plant Disease Control with Composts

Harry A. J. Hoitink, Matthew S. Krause, and David Y. Han

CONTENTS

I. INTRODUCTION

During the 1960s, nurserymen across the U.S. explored the possibility of using composted tree bark as a peat substitute to reduce potting mix costs. Improved plant growth and decreased losses caused by Phytophthora root rots were observed as secondary benefits in the nursery industry. Today composts are recognized to be as effective as fungicides for the control of such root rots (Hardy and Sivasithamparam, 1991; Hoitink et al., 1991; Ownley and Benson, 1991). Therefore, the ornamental plant industry relies heavily on compost products for control of diseases caused by these soil-borne plant pathogens. Composts have replaced methyl bromide in this industry (Quarles and Grossman, 1995). In field applications of composts similar results have been obtained (Hoitink and Fahy, 1986; Lumsden et al., 1983; Schüler et al., 1993). Examples of diseases controlled by composts were reviewed by Hoitink

1-56670-460-X/01/$0.00+$.50

and Fahy (1986). A summary of the types of diseases suppressed by various types of composts is presented in Table 12.1.

Composts must be of consistent quality to be used successfully in biological control of diseases of horticultural crops, particularly if used in container media (Inbar et al., 1993). The rate of respiration is one of several procedures that can be used to monitor stability of composts (Iannotti et al., 1994). Variability in compost stability is one of the principal factors limiting its widespread utilization. Maturity is less important in ground bed or field agriculture as long as the compost is applied sufficiently ahead of planting to allow for additional stabilization; however, lack of maturity frequently causes problems here as well.

Table 12.1 Summary of Literature on Suppression of Plant Diseases by Various Types of Peats and Composts

	Disease Suppressed			
Peat or Compost Type[z]	Pythium + Phytophthora Root Rots	Rhizoctonia[y] Diseases	Fusarium[y] Wilts	References
Spagnum peat H_4	–	–	–	Boehm and Hoitink, 1992; Chen et al., 1988b; Mandelbaum and Hadar, 1990.
Sphagnum peat H_2, H_3	+	–	–	Boehm and Hoitink, 1992; Tahvonen, 1982; Wolffhechel, 1988.
Pine bark	+	+	+	Boehm and Hoitink, 1992; Chen et al., 1988b; Ownley and Benson, 1991; Trillas-Gay et al., 1986.
Hardwood bark	+	+	+	Chen et al., 1988a, 1988b; Kuter et al., 1983; Nelson et al., 1983; Trillas-Gay et al., 1986.
Yard/green wastes	+	+		Grebus et al., 1993; Rÿckeboer et al., 1998; Schüler et al., 1993; Tuitert et al., 1998.
Grape pomace	+	+		Gorodecki and Hadar, 1990; Mandelbaum and Hadar, 1990.
Cow manure	+	+		Gorodecki and Hadar, 1990; Hoitink and Fahy, 1986.
Biosolids	+	+		Chen et al., 1988a, 1988b; Kuter et al., 1988; Lumsden et al., 1983.

[z] Indicates peat decomposition level on the von Post scale (Puustjärvi and Robertson, 1975) or raw materials from which compost was prepared.

[y] Requires inoculation with biocontrol agents or long-term curing of composts for consistent induction of suppression.

Effects of chemical properties of composts on soil-borne disease severity often are overlooked (reviewed by Hoitink et al., 1991). Highly saline composts enhance Pythium and Phytophthora diseases unless they are applied months ahead of planting to allow for leaching. Composts prepared from municipal biosolids have a low carbon to nitrogen (C/N) ratio. They release considerable amounts of nitrogen (N) and enhance Fusarium wilt (Hoitink et al., 1987). On the other hand, composts from high C/N materials such as tree barks immobilize N and suppress Fusarium diseases if colonized by an appropriate microflora (Trillas-Gay et al., 1986). High ammonium and low nitrate nutrition increases Fusarium wilts (Schneider, 1985). Perhaps biosolids composts enhance Fusarium diseases because they predominantly release ammonium (NH_4).

II. FATE OF BIOCONTROL AGENTS DURING COMPOSTING

The composting process is often divided into three phases. The initial phase occurs during the first 24 to 48 hr as temperatures gradually rise to 40 to 50°C, and sugars and other easily biodegradable substances are destroyed. During the second phase, when high temperatures of 55 to 70°C prevail, less biodegradable cellulosic substances are destroyed. Thermophilic microorganisms predominate during this part of the process. Plant pathogens and seeds are killed by the heat generated during this phase (Bollen, 1993; Farrell, 1993). Compost piles must be turned frequently to expose all parts to high temperature to produce a homogeneous product free of pathogens and weed seeds. Unfortunately, most beneficial microorganisms also are killed during the high temperature phase of composting.

Curing begins as the concentration of readily biodegradable components in wastes declines. As a result, rates of decomposition, heat output and temperatures decrease. At this time, mesophilic microorganisms that grow at temperatures <40°C recolonize the compost from the outer low-temperature layer into the compost windrow or pile. Therefore, suppression of pathogens and/or disease is largely induced during curing, because most biocontrol agents also recolonize composts after peak heating.

Bacillus spp., *Enterobacter* spp., *Flavobacterium balustinum, Pseudomonas* spp., other bacterial genera and *Streptomyces* spp., as well as *Penicillium* spp., several *Trichoderma* spp., *Gliocladium virens*, and other fungi have been identified as biocontrol agents in compost-amended substrates (Chung and Hoitink, 1990; Hadar and Gorodecki, 1991; Hardy and Sivasithamparam, 1991; Hoitink and Fahy, 1986; Nelson et al., 1983; Phae et al., 1990). The moisture content of compost critically affects the potential for bacterial mesophiles to colonize the substrate after peak heating. Dry composts (< 34% moisture, w/w) become colonized by fungi and are conducive to Pythium diseases. In order to induce suppression, the moisture content must be high enough (at least 40 to 50%, w/w) so that bacteria as well as fungi colonize the substrate after peak heating. Water must often be added to composts during composting and curing to avoid the dry condition. Compost pH also affects the potential for beneficial bacteria to colonize composts. A pH < 5.0 inhibits bacterial biocontrol agents (Hoitink et al., 1991).

Variability in suppression of Rhizoctonia damping-off and Fusarium wilt encountered in substrates amended with mature composts is due in part to random recolonization of compost by effective biocontrol agents after peak heating. Field compost more consistently suppresses Rhizoctonia diseases than the same compost produced in a partially enclosed facility where few microbial species survive heat treatment (Kuter et al., 1983). Compost produced in the open near a forest (field compost), an environment that is high in microbial species diversity, is colonized by a greater variety of biocontrol agents than the same compost produced in an in-vessel system (Kuter et al., 1983). Frequently, however, Rhizoctonia and other diseases are observed for some time after composts are first applied (Kuter et al., 1988; Lumsden et al., 1983). Three approaches can be used to solve this problem. First, curing of composts for 4 months or more renders composts more consistently suppressive (Kuter er al., 1988). The second approach is to incorporate composts into field soils for several months before planting (Lumsden et al., 1983). The third approach is to inoculate composts with specific biocontrol agents (Kwok et al., 1987).

A specific strain of *Flavobacterium balustinum* and an isolate of *Trichoderma hamatum* have been identified that induce consistent levels of suppression to diseases caused by a broad spectrum of plant pathogens, if inoculated into compost after peak heating, but before significant levels of recolonization have occurred. Patents have been issued to The Ohio State University for this process (Hoitink, 1990). In Japan, Phae et al. (1990) isolated a *Bacillus* strain that induces predictable biological control in composts. It has been recognized for decades that single strains are not as effective in biological control in field applications as are mixtures of microorganisms (Garrett, 1955). The same applies to container media (Kwok et al., 1987).

III. MECHANISMS OF SUPPRESSION IN COMPOSTS

Two classes of biological control mechanisms known as "general" and "specific" suppression have been described for compost-amended substrates. The mechanisms involved are based on competition, antibiosis, hyperparasitism, and the induction of systemic acquired resistance in the host plant. Propagules of plant pathogens such as *Pythium* and *Phytophthora* spp. are suppressed through the "general suppression" phenomenon (Boehm et al., 1993; Chen et al., 1988a, 1988b; Cook and Baker, 1983; Hardy and Sivasithamparam, 1991; Mandelbaum and Hadar, 1990). Many types of microorganisms present in compost-amended container media function as biocontrol agents against diseases caused by *Phytophthora* and *Pythium* spp. (Boehm et al., 1993; Hardy and Sivasithamparam, 1991). Propagules of these pathogens, if inadvertently introduced into compost-amended substrates, do not germinate in response to nutrients released in the form of seed or root exudates. The high microbial activity and biomass caused by the general soil microflora in such substrates prevents germination of spores of these pathogens and infection of the host (Chen et al., 1988a; Mandelbaum and Hadar, 1990). Propagules of these pathogens remain dormant and are typically not killed if introduced into compost-amended soil (Chen et al., 1988a; Mandelbaum and Hadar, 1990).

An enzyme assay, that determines microbial activity based on the rate of hydrolysis of fluorescein diacetate (FDA), predicts suppressiveness of potting mixes to Pythium diseases (Boehm and Hoitink, 1992; Chen et al., 1988a; Mandelbaum and Hadar, 1990; You and Sivasithamparam, 1994). Similar information has been developed for soils on "organic farms" where soil-borne diseases tend to be less prevalent (Workneh et al., 1993). The length of time that the suppressive effect lasts also may be determined with FDA activity (Boehm and Hoitink, 1992). This is known as the "carrying capacity" of the substrate relative to biological control.

The mechanism of biological control for *Rhizoctonia solani* in compost-amended substrates is different from that of *Pythium* and *Phytophthora* spp. because only a narrow group of microorganisms is capable of eradicating *R. solani*. This type of suppression is referred to as "specific suppression" (Hoitink et al., 1991). *Trichoderma* spp, including *T. hamatum* and *T. harzianum*, are the predominant parasites recovered from composts prepared of lignocellulosic wastes (Kuter et al., 1983; Nelson et al., 1983). Parasites are microorganisms capable of colonizing plant pathogens resulting in lysis or death. These fungi interact with various bacterial strains in the biological control of Rhizoctonia damping-off (Kwok et al., 1987). Notably, *Penicillium* spp. are the predominant parasites recovered from sclerotia of *Sclerotium rolfsii* in composted grape (*Vitis* spp.) pomace, a high sugar and low cellulose content waste (Hadar and Gorodecki, 1991). *Trichoderma* spp. were not recovered from this compost and were not effective when introduced. The composition of the feedstock, as expected, appears to have an impact on the microflora in composts active in biological control.

IV. BIOLOGICAL ENERGY AVAILABILITY VS. SUPPRESSIVENESS

The decomposition level of organic matter in compost-amended substrates has a major impact on disease suppression. For example, *R. solani* is highly competitive as a saprophyte (Garrett, 1962). It can utilize cellulose and colonize fresh wastes but not low cellulose mature compost (Chung et al., 1988). *Trichoderma*, an effective biocontrol agent of *R.solani,* is capable of colonizing immature as well as mature compost, but it grows to higher populations in fresh compost (Chung et al., 1988; Nelson et al., 1983). In fresh, undecomposed organic matter, biological control does not occur because both the pathogen and the biocontrol agent grow as saprophytes. Therefore, *R. solani* (the pathogen) remains capable of causing disease here. Presumably, synthesis of lytic enzymes involved in parasitism of pathogens by *Trichoderma* is repressed in fresh organic matter due to high glucose concentrations in such waste (de la Cruz et al., 1993). The same processes may occur in antibiotic production, which also plays an important role in biocontrol.

In mature compost, where concentrations of free nutrients such as glucose are low (Chen et al., 1988a), sclerotia of *R. solani* are killed by the parasite, and biological control prevails (Chung et al., 1988; Nelson et al., 1983). The foregoing reveals that composts must be adequately stabilized to reach that decomposition level where biological control is feasible. In practice, this occurs in composts (tree barks, yard wastes, etc.) that have been (1) stabilized far enough to avoid phytotoxicity and (2) colonized by the appropriate specific microflora. Practical guidelines that define this critical stage of decomposition in terms of biological control are not yet available. Industry presently controls decomposition level by maintaining constant conditions during the entire process and adhering to a given time schedule.

Composted pine (*Pinus* spp.) bark produced by such a process has been utilized with great success in floriculture, indicating that this approach to quality control is quite acceptable (Hoitink et al., 1991).

Excessively stabilized organic matter, the opposite end of the decomposition scale, does not support adequate activity of biocontrol agents. As a result, suppression is lacking and soil-borne diseases are severe, as in highly mineralized soils where humic substances are the predominant forms of organic matter (Workneh et al., 1993). The length of time that soil-incorporated composts support adequate levels of biocontrol activity has not yet been determined. Presumably, it varies with soil temperature, soil characteristics and the type of organic matter from which the compost was prepared. Loading rates and farming practices of course also play a role.

We have studied the carrying capacity of soil organic matter in potting mixes prepared with sphagnum peat to bring a partial solution to this problem (Boehm and Hoitink, 1992; Boehm et al., 1993). Sphagnum peat typically competes with compost as a source of organic matter in horticulture. Both the microflora and the organic matter in peat itself can affect suppression of soil-borne diseases. The literature on that effect is reviewed briefly here.

Dark, more decomposed sphagnum peat, harvested from a 1.2 m or greater depth in most peat bogs, is low in microbial activity and consistently conducive to *Pythium* and *Phytophthora* root rots (Boehm and Hoitink, 1992; Hoitink et al., 1991). On the other hand, light, less decomposed sources of sphagnum peat, harvested from near the surface of peat bogs, have a higher microbial activity (FDA activity) and suppress root rot. Unfortunately, the suppressive effect of light peat on Pythium root rots is of short duration (Boehm and Hoitink, 1992; Tahvonen, 1982; Wolffhechel, 1988). Light peats are used most effectively for short production cycles (6 to 10 week crops), such as in plug and flat mixes used in the ornamentals industry. Composts have longer lasting effects (Boehm and Hoitink, 1992; Boehm et al., 1993; You and Sivasithamparam, 1994).

As previously mentioned, the rate of hydrolysis of FDA predicts suppressiveness of peat mixes and of compost-amended substrates to Pythium root rot (Boehm and Hoitink, 1992). As FDA activity in suppressive substrates declines to < 3.2 μg FDA hydrolyzed min^{-1} g^{-1} dry weight mix, the population of *Pythium ultimum* increases, infection takes place and root rot develops. During this collapse in suppressiveness, the composition of bacterial species also changes (Boehm et al., 1993; Boehm et al., 1997). A microflora typical of suppressive soils, which includes *Pseudomonas* spp. and other rod-shaped Gram negative bacteria as the predominant rhizosphere colonizers, is replaced by pleomorphic Gram-positive bacteria (e.g., *Arthrobacter*) and putative oligotrophs (Boehm et al., 1997). The microflora of the conducive substrate resembles that of highly mineralized niches in soil (Kanazawa and Filip, 1986).

Non-destructive analysis of soil organic matter, utilizing Fourier transform infra red spectroscopy (FT-IR) and cross polarization magic angle spinning–^{13}carbon nuclear magnetic resonance spectroscopy (CPMAS–^{13}C NMR), allows characterization of biodegradable components of soil organic fractions (Chen and Inbar, 1993; Inbar et al., 1989). CPMAS–^{13}CNMR allows quantitative analysis of concentrations of readily biodegradable substances such as carbohydrates (hemicellulose, cellulose, etc.) vs. lignins and humic substances in soil organic matter (reviewed by Chen and Inbar, 1993). Boehm et al. (1997) reported that the carbohydrates in Sphagnum peat decline as suppressiveness is lost. During the same time, bacterial genera capable of causing biological control are replaced by those that cannot provide control. Biocontrol agents inoculated into the more decomposed substrate are not able to induce sustained biological control of Pythium root rot. The same phenomenon has been observed for Phytophthora root rot of avocado (*Persea americana* Mill.) on mulched trees in the field (You and Sivasithamparam, 1994). Therefore, biocontrol of these diseases is determined by the carrying capacity of the substrate that regulates species composition and activity and, in turn, the potential for sustenance of biological control.

V. COMPOST FOR CONTROL OF FOLIAR DISEASES

Composts incorporated into soils or potting mixes may also reduce the severity of foliar diseases of plants. Tränkner (1992) reported that powdery mildew on small grains was less severe on compost-amended field soil than on unamended field soil. Zhang et al. (1996) demonstrated that only part of the root system of a cucumber (*Cucumis sativus* L.) plant had to be exposed to compost to protect the entire root system against Pythium root rot. They also showed that anthracnose of cucumber was less severe in some batches of composts than on plants in peat mixes.

Several types of bacteria and fungi have been identified that can induce these systemic effects in plants (Maurhofer et al., 1994; Wei et al., 1991; Zhang et al., 1998). This microflora in compost activates the synthesis of pathogenesis-related (PR) proteins in plants, although much of the activation does not occur until after the plant becomes infected with the pathogen (Zhang et al., 1996). This shows that effective batches of composts prime the plant to better protect itself against pathogens. Unfortunately, this effect of composts is highly variable in nature. Suppression of soil-borne plant pathogens, on the other hand, has become a predictable phenomenon, as previously described.

During the past decade, a series of projects have been published on the control of plant diseases of above ground plant parts with water extracts, also known as steepages, prepared from composts (Weltzien, 1992; Yohalem et al., 1994). Steepages often are prepared by soaking mature composts in water (still culture; 1:1, w/w) for 7 to 10 days. The steepage is filtered and then sprayed on plants. Efficacy unfortunately also varies with compost type, batch of steepage produced, crops, and the disease under question. Sackenheim (1993), utilizing plate counting procedures, reported that aerobic microorganisms predominate in steepages. The microflora included strains of bacteria and isolates of fungi already known as biocontrol agents. He developed a number of enrichment strategies, that included nutrients as well as microorganisms, to improve efficacy of the steepages. Even so, steepages do not provide reproducible results.

Control induced by compost steepages has been attributed to systemic acquired resistance (SAR) induced in plants by elicitors present in the extracts (Zhang et al., 1998). These extracts activate the production of PR proteins in plants to a degree not different from that induced by salicylic acid. The mechanism by which steepages induce resistance may differ, therefore, from that induced by the microflora growing on roots of plants produced in composts.

VI. DISEASE SUPPRESSION — FUTURE OUTLOOK

Success in biological control of diseases with composts is possible only if all factors involved in the production and utilization of composts are defined and kept consistent. Most composts are variable in quality. Therefore, composted pine bark remains the principal compost used for the preparation of potting mixes or soils naturally suppressive to soil-borne plant pathogens. Composted manures, yard wastes, and food wastes are steadily gaining in popularity, and offer the same potential (Gorodecki and Hadar, 1990; Grebus et al., 1994; Inbar et al., 1993; Marugg et al., 1993; Schüler et al., 1993; Tuitert et al., 1998).

Controlled inoculation of composts with biocontrol agents is a procedure that must be developed on a commercial scale to induce consistent levels of suppression to pathogens such as *R. solani* (Grebus et al., 1993; Hoitink et al., 1991; Phae et al., 1990; Rÿckeboer et al., 1998). Recently, tree bark was proposed as a food base for the culture of biocontrol agents and as a carrier of such agents for use in agricultural applications (Steinmetz and Schönbeck, 1994). However, this new field of biotechnology is still in its infancy. Major research and development efforts will need to be directed toward this approach for disease control. Recycling through composting increasingly is chosen as the preferred strategy for waste treatment. This also applies to farm manures. For this reason, composts are becoming available in greater quantities. Peat, on the other hand, is a limited resource that cannot be recycled. In conclusion, future opportunities for both natural and controlled-induced suppression of soil-borne plant pathogens appear bright.

REFERENCES

Boehm, M.J. and H.A.J. Hoitink. 1992. Sustenance of microbial activity and severity of Pythium root rot of poinsettia. *Phytopathology* 82:259–264.

Boehm, M.J., L.V. Madden, and H.A.J. Hoitink. 1993. Effect of organic matter decomposition level on bacterial species diversity and composition in relationship to Pythium damping-off severity. *Applied Environmental Microbiology* 59:4171–4179.

Boehm, M.J., T. Wu, A.G. Stone, B. Kraakman, D.A. Iannotti, G.E. Wilson, L.V. Madden, and H.A.J. Hoitink. 1997. Cross-polarized magic-angle spinning ^{13}C nuclear magnetic resonance spectroscopic characterization of soil organic matter relative to culturable bacterial species composition and sustained biological control of Pythium root rot. *Applied Environmental Microbiology* 63:162–168.

Bollen, G.J. 1993. Factors involved in inactivation of plant pathogens during composting of crop residues, p. 301–318. In: H. A. J. Hoitink and H. M. Keener (eds.). *Science and Engineering of Composting: Design, Environmental, Microbiological and Utilization Aspects.* Renaissance Publications, Worthington, Ohio.

Chen, W., H.A.J. Hoitink, A.F. Schmitthenner, and O.H. Tuovinen. 1988a. The role of microbial activity in suppression of damping-off caused by *Pythium ultimum*. *Phytopathology* 78:314–322.

Chen, W., H.A.J. Hoitink, and L.V. Madden. 1988b. Microbial activity and biomass in container media predicting suppressiveness to damping-off caused by *Pythium ultimum*. *Phytopathology* 78:1447–1450.

Chen, Y. and Y. Inbar. 1993. Chemical and spectroscopical analyses of organic matter transformations during composting in relation to compost maturity, p. 551–600. In: H.A.J. Hoitink and H.M. Keener (eds.) *Science and Engineering of Composting: Design, Environmental, Microbiological and Utilization Aspects*. Renaissance Publications, Worthington, Ohio.

Chung, Y.R., H.A.J. Hoitink, and P.E. Lipps. 1988. Interactions between organic-matter decomposition level and soil-borne disease severity. *Agriculture Ecosystems and Environment* 24:183–193.

Chung, Y.R. and H.A.J. Hoitink. 1990. Interactions between thermophilic fungi and *Trichoderma hamatum* in suppression of Rhizoctonia damping-off in a bark compost-amended container medium. *Phytopathology* 80:73–77.

Cook, R.J. and K.F. Baker. 1983. *The Nature and Practice of Biological Control of Plant Pathogens*. American Phytopathological Society, St. Paul, Minnesota.

de la Cruz, J., M. Rey, J.M. Lora, A. Hidalgo-Gallego, F. Dominguez, J.A. Pintor-Toro, A. Llobell, and T. Benitez. 1993. Carbon source control on β-glucanases, chitobiase and chitinase from *Trichoderma harzianum*. *Archives of Microbiology* 159:316–322.

Farrell, J.B. 1993. Fecal pathogen control during composting, p. 282–300. In: H.A.J. Hoitink and H.M. Keener (eds.). *Science and Engineering of Composting: Design, Environmental, Microbiological and Utilization Aspects*. Renaissance Publications, Worthington, Ohio.

Garrett, S.D. 1955. A century of root-disease investigation. *Annals of Applied Biology* 42:211–219.

Garrett, S.D. 1962. Decomposition of cellulose in soil by *Rhizoctonia solani* Kühn. *Transactions of the British Mycological Society* 45:114–120.

Gorodecki, B. and Y. Hadar. 1990. Suppression of *Rhizoctonia solani* and *Sclerotium rolfsii* in container media containing composted separated cattle manure and composted grape marc. *Crop Protection* 9:271–274.

Grebus, M.E., K.A. Feldman, C.A. Musselman, and H.A.J. Hoitink. 1993. Production of biocontrol agent-fortified compost-amended potting mixes for predictable disease suppression. *Phytopathology* 83:1406 (Abstract).

Grebus, M.E., M.E. Watson, and H.A.J. Hoitink. 1994. Biological, chemical and physical properties of composted yard trimmings as indicators of maturity and plant disease suppression. *Compost Science and Utilization* 1:57–71.

Hadar, Y. and B. Gorodecki. 1991. Suppression of germination of sclerotia of *Sclerotium rolfsii* in compost. *Soil Biology and Biochemistry* 23:303–306.

Hardy, G.E. St. J. and K. Sivasithamparam. 1991. Suppression of *Phytophthora* root rot by a composted *Eucalyptus* bark mix. *Australian Journal of Botany* 39:153–159.

Hoitink, H.A.J. 1990. Production of disease suppressive compost and container media, and microorganism culture for use therein. *US Patent 4960348*. Feb. 13, 1990.

Hoitink, H.A.J., M. Daughtery, and H.K. Tayama. 1987. Control of cyclamen Fusarium wilt — A preliminary report. *Ohio Florist's Association Bulletin* 693:1–3.

Hoitink, H.A.J. and P.C. Fahy. 1986. Basis for the control of soil-borne plant pathogens with composts. *Annual Review of Phytopathology* 24:93–114.

Hoitink H.A.J., Y. Inbar, and M.J. Boehm. 1991. Status of composted-amended potting mixes naturally suppressive to soil-borne diseases of floricultural crops. *Plant Disease* 75:869–873.

Iannotti, D.A., M.E. Grebus, B.L. Toth, L.V. Madden, and H.A.J. Hoitink. 1994. Oxygen respirometry to assess stability and maturity of composted municipal solid waste. *Journal of Environmental Quality* 23:1177–1183.

Inbar, Y., Y. Chen, and Y. Hadar. 1989. Solid-state carbon-13 nuclear magnetic resonance and infrared spectroscopy of composted organic matter. *Soil Science Society of America Journal* 53:1695–1701.

Inbar, Y., Y. Hadar, and Y. Chen. 1993. Recycling of cattle manure: the composting process and characterization of maturity. *Journal of Environmental Quality* 22:857–863.

Kanazawa, S. and Z. Filip. 1986. Distribution of microorganisms, total biomass, and enzyme activities in different particles of brown soils. *Microbial Ecology* 12:205–215.

Kuter, G.A., H.A.J. Hoitink, and W. Chen. 1988. Effects of municipal sludge compost curing time on suppression of Pythium and Rhizoctonia diseases of ornamental plants. *Plant Disease* 72:751–756.

Kuter, G.A., E.B. Nelson, H.A.J. Hoitink, and L.V. Madden. 1983. Fungal populations in container media amended with composted hardwood bark suppressive and conducive to Rhizoctonia damping-off. *Phytopathology* 73:1450–1456.

Kwok, O.C.H., P.C. Fahy, H.A.J. Hoitink, and G.A. Kuter. 1987. Interactions between bacteria and *Trichoderma hamatum* in suppression of Rhizoctonia damping-off in bark compost media. *Phytopathology* 77:1206–1212.

Lumsden, R.D., J.A. Lewis, and P.D. Millner. 1983. Effect of composted sewage sludge on several soil-borne pathogens and diseases. *Phytopathology* 73:1543–1548.

Mandelbaum, R. and Y. Hadar. 1990. Effects of available carbon source on microbial activity and suppression of *Pythium aphanidermatum* in compost and peat container media. *Phytopathology* 80:794–804.

Marugg, C., M.E. Grebus, R.C. Hansen, H.M. Keener, and H.A.J. Hoitink. 1993. A kinetic model of the yard waste composting process. *Compost Science and Utilization* 1:38–51.

Maurhofer, M., C. Hase, P. Meuwly, J.-P. Métraux, and G. Défago. 1994. Induction of systemic resistance of tobacco to tobacco necrosis virus by the root-colonizing *Pseudomonas fluorescens* strain CHAO: influence of the *gac*A gene and of pyoverdine production. *Phytopathology* 84:139–146.

Nelson, E.B., G.A. Kuter, and H.A.J. Hoitink, 1983. Effects of fungal antagonists and compost age on suppression of Rhizoctonia damping-off in container media amended with composted hardwood bark. *Phytopathology* 3:1457–1462.

Ownley, B.H. and D.M. Benson. 1991. Relationship of matric water potential and air-filled porosity of container media to development of Phytophthora root rot of rhododendron. *Phytopathology* 81:936–941.

Phae, C.G., M. Saski, M. Shoda, and H. Kubota. 1990. Characteristics of *Bacillus subtilis* isolated from composts suppressing phytopathogenic microorganisms. *Soil Science and Plant Nutrition* 36:575–586.

Puustjärvi, V. and R.A. Robertson, 1975. Physical and chemical properties. p. 23–28. In: D.W. Robinson and J.G.D. Lamb (eds.). *Peat in Horticulture*. Academic Press Inc., New York.

Quarles, W. and J. Grossman. 1995. Alternatives to methyl bromide in nurseries — Disease suppressive media. *The IPM Practitioner* 17(8):1–13.

Rÿckeboer, J., K. Deprins, and J. Coosemans. 1998. Compost onderdrukt de kiemplanten-schimmels *Pythium ultimum* en *Rhizoctonia solani*: Veredelde compost doet beter! *Vlacovaria* 3:20–26.

Sackenheim, R. 1993. Untersuchungen über Wirkungen von wässerigen, mikrobiologisch aktiven Extracten aus kompostierten Substraten auf den Befall der Weinrebe *(Vitis vinifera)* mit *Plasmopora viticola, Uncinula necator, Botrytis cenerea* und *Pseudopezicula tracheiphila*. Ph.D. Thesis. Rheinische Friedrich-Wilhelms Universität, Bonn, Germany.

Schneider, R.W. 1985. Suppression of Fusarium yellows of celery with potassium chloride and nitrate. *Phytopathology* 75:40–48.

Schüler, C., J. Pikny, M. Nasir, and H. Vogtmann. 1993. Effects of composted organic kitchen and garden waste on *Mycosphaerella pinodes* (Berk. et Blox) Vestergr., causal organism of foot rot on peas (*Pisum sativum* L.). *Biological Agriculture and Horticulture* 9:353–360.

Steinmetz, J. and F. Schönbeck. 1994. Conifer bark as growth medium and carrier for *Trichoderma harzianum* and *Gliocladium roseum* to control *Pythium ultimum* on pea. *Journal of Plant Disease Protection* 101:200–211.

Tahvonen, R. 1982. The suppressiveness of Finnish light colored Sphagnum peat. *Journal of the Scientific Agricultural Society of Finland* 54:345–356.

Tränkner, A. 1992. Use of agricultural and municipal organic wastes to develop suppressiveness to plant pathogens. p. 35–42. In: E.C. Tjamos, G.C. Papavizas, and R.J. Cook (eds.). *Biological Control of Plant Diseases*. Plenum Press, New York.

Trillas-Gay, M.I., H.A.J. Hoitink, and L.V. Madden. 1986. Nature of suppression of Fusarium wilt of radish in a container medium amended with composted hardwood bark. *Plant Disease* 70:1023–1027.

Tuitert, G., M. Szczech, and G.J. Bollen. 1998. Suppression of *Rhizoctonia solani* in potting mixtures amended with compost made from organic household waste. *Phytopathology* 88:764–773.

Wei, G., J.W. Kloepper, and S. Tuzun. 1991. Induction of systemic resistance of cucumber to *Colletotrichum orbiculare* by select strains of plant growth-promoting rhizobacteria. *Phytopathology* 81:1508–1512.

Weltzien, H.C. 1992. Biocontrol of foliar fungal diseases with compost extracts, p. 430–450. In: J.H. Andrews and S. Hirano (eds.) *Microbial Ecology of Leaves*. Brock Springer Series in Contemporary Bioscience. Springer-Verlag, New York.

Wolffhechel, H. 1988. The suppressiveness of sphagnum peat to *Pythium* spp. *Acta Horticulturae* 221:217–222.

Workneh, F., A.H.C. Van Bruggen, L.E. Drinkwater, and C. Sherman. 1993. Variables associated with a reduction in corky root and Phytophthora root rot of tomatoes in organic compared to conventional farms. *Phytopathology* 83:581–589.

Yohalem, D.S., R.F. Harris, and J.H. Andrews. 1994. Aqueous extracts of spent mushrooms substrate for foliar disease control. *Compost Science and Utilization* 2:67–74.

You, M.P. and K. Sivasithamparam. 1994. Hydrolysis of fluorescein diacetate in an *Persea americana* plantation mulch suppressive to *Phytophtora cinnamoni* and its relationship with certain biotic and abiotic factors. *Soil Biology and Biochemistry* 26:1355–1361.

Zhang, W., W.A. Dick, and H.A.J. Hoitink. 1996. Compost-induced systemic acquired resistance in cucumber to Pythium root rot and anthracnose. *Phytopathology* 84:1138.

Zhang, W., D. Han, W.A. Dick, K.R. Davis, and H.A.J. Hoitink. 1998. Compost and compost water extract-induced systemic acquired resistance in cucumber and arabidopsis. *Phytopathology* 88:450–455.

CHAPTER 13

Weed Control in Vegetable Crops with Composted Organic Mulches

Monica Ozores-Hampton, Thomas A. Obreza, and Peter J. Stoffella

CONTENTS

I. INTRODUCTION

Excessive weed growth causes about a 13% average potential crop production loss annually in the U.S. (Dusky et al., 1988). Weeds reduce crop quality and yield by competing for light, water, and nutrients, and increase harvesting costs.

Herbicides are the most utilized pesticides in any crop production system because they reduce farm labor, increase yields, and reduce production costs (Altieri and Liebman, 1988). Since the combined cost of herbicides and their application is lower than the cost of mechanical weed control, growers have become dependent on chemical control. Nevertheless, long-term herbicide usage can have a potential negative impact on the environment. In the last decade, environmental concerns associated with pesticide usage in agriculture have increased. For example, the Environmental Protection Agency (EPA) has restricted usage of several common herbicides in Florida because of groundwater contamination and potential negative effects on wildlife and humans (Crnko et al., 1992). Major pathways for herbicide

1-56670-460-X/01/$0.00+$.50

movement from croplands include leaching of water-soluble chemicals and surface runoff of chemicals adsorbed to sediment (Schneider et al., 1988).

Weed growth suppression is one of the most important effects of mulches (Food and Agriculture Organization, 1987; Grantzau, 1987), and composted and noncomposted organic mulches were an important weed control method prior to the development of chemical herbicides (Altieri and Liebman, 1988). Weed suppression by mulches can be due to the physical presence of the materials on the soil surface, and/or the action of phytotoxic compounds generated by microbes in the composting process (Niggli et al., 1990; Ozores-Hampton, 1997, 1998; Ozores-Hampton et al., 1999). To suppress weeds physically, a 10- to 15-cm-thick mulch layer is needed (Food and Agriculture Organization, 1987; Marshall and Ellis, 1992). In general, germination of weed seed declines as burial depth increases (Table 13.1). Germination inhibition at greater depths has been attributed to several factors including light, temperature, and moisture (Baskin and Baskin, 1989). Additionally, organic mulches (composted and/or noncomposted) may improve soil physical and biological properties as they decompose by reducing soil erosion, minimizing soil compaction, increasing water- holding capacity, slowing the release of nutrients, increasing microbial activity, and controlling soil temperature (Food and Agriculture Organization, 1987; Foshee et al., 1996). This chapter presents information on the effects of composted organic mulches as an alternative biological weed control method in vegetable crop production systems.

Table 13.1 Depth of Burial in Soil Required to Prevent Weed Seed Germination

Common Name	Scientific Name	Soil Depth (cm)	Reference
Chamico; thornapple	*Datura ferox*	10	Reisman-Berman and Kigel (1991)
Jimsonweed	*Datura stramonium*	10	Reisman-Berman and Kigel (1991)
Great Lakes wheatgrass	*Agropyron psammophilum*	8	Zhang and Maun (1990)
Redstem filaree	*Erodium cicutarium*	9	Blackshaw (1992)
Yellow nutsedge	*Cyperus esculentus*	0.5	Lapham and Drennan (1990)
Giant foxtail	*Setaria faberi*	6	Mester and Buhler (1991)
Bugweed	*Solanum mauritianum*	15	Campbell and van Staden (1994)
Round-leaved mallow	*Malva pusilla*	8	Blackshaw (1990)

From Ozores-Hampton, M.P., 1998. Compost as an alternative weed control method. *HortScience* 33:938–940. With permission.

II. PHYTOTOXIC EFFECTS OF COMPOSTED ORGANIC MULCHES

Composting is a biological decomposition process in which microorganisms convert organic materials into a relatively stable humus-like material. During decomposition, microorganisms assimilate complex organic substances and release inorganic nutrients (Metting, 1993). An adequate composting process should kill pathogens and stabilize organic carbon (C) before the material is used as a soil amendment or mulch.

Traditional compostable organics include animal manures, leaves and grass clippings, paper, wood chips, straw, and textiles. Composts made from waste materials like biosolids, household garbage (municipal solid waste, or MSW), yard trimmings, and food waste have recently become available on a commercial scale. The largest potential user of these compost materials is agriculture (McConnell et al., 1993; Parr and Hornick, 1993). Compost incorporated into soils has increased yields of corn (*Zea mays* L.) (Gallaher and McSorley., 1994), black-eyed pea (*Vigna unguiculata* [L.] Walp.), okra (*Abelmoschus esculentus* L.) (Bryan and Lance, 1991), tomato (*Lycopersicon esculentum* Mill.), squash (*Cucurbita pepo* L.), pepper (*Capsicum annuum* L.), snap beans (*Phaseolus vulgaris* L.), eggplant (*Solanum melongena* L.), (Ozores-Hampton and Bryan, 1993a, 1993b; Ozores-Hampton et al., 1994; Roe et al., 1997), and watermelon (*Citrullus lanatus* [Thunb.] Matsum. & Nakai) (Obreza and Reeder, 1994).

Phytotoxic chemical compounds in composted waste materials can injure plants (Table 13.2). The type and degree of injury are directly related to compost maturity or stability. Compost maturity is the degree to which the material is free of phytotoxic substances. Stability is the degree to which it consumes nitrogen (N) and oxygen (O_2) in significant quantities to support biological activity, while generating heat, carbon dioxide (CO_2), and water vapor (Florida Dept. Agr. and Consumer Services, 1994). Toxic substances obtained from water extracts of composted waste materials inhibited germination and growth of tobacco (*Nicotiana tabacum* L.) seedlings and also induced darkening and necrosis of root cells (Patrick and Kock, 1958). Organic acids such as acetic, propionic, and butyric acids can accumulate in compost with a high C:N ratio, and high concentrations of ammonia can accumulate in compost with a low C:N ratio (Hadar et al., 1985; Jimenez and Garcia, 1989). Crop injury has been linked to use of immature compost (Zucconi et al., 1981b). Toxicity of compost also has been related to composting methodology. Phytotoxins disappeared faster in static piles than with the windrow composting method (Zucconi et al., 1981a).

Identification of phytotoxins in compost extracts from fresh and 5-month-old MSW compost indicated that fresh compost contains acetic, propionic, isobutyric, butyric, and isovaleric acids in larger concentration (DeVleeschauwer et al., 1981). The most phytotoxic organic acid is acetic, which can completely inhibit cress (*Lepidium sativum* L.) seed growth at concentrations above 300 $mg \cdot kg^{-1}$ (DeVleeschauwer et al., 1981) and cucumber (*Cucumis sativus* L.) seed germination at concentrations above 30 $mg \cdot kg^{-1}$ (Shiralipour et al., 1997). Inhibitory (no germination) effects of acetic acids on seed germination of 'Poinset' cucumber was a metabolic phenomenon, and not a result of high ionic strength or pH imbalance (Shiralipour et al., 1997). The concentration of acetic acid in several lots of immature MSW compost ranged between 6000 and 28,000 $mg \cdot kg^{-1}$ (Keeling et al., 1994). Combining immature MSW compost with N did not improve the germination percentage of several vegetable crops, suggesting that phytotoxicity rather than C:N ratio was primarily responsible for poor seed germination and growth inhibition (Keeling et al., 1994). Additionally, application of immature compost can cause the root zone to become anaerobic by reducing soil O_2; increasing soil temperature to levels that are incompatible with normal root function;

Table 13.2 Phytotoxicity of Several Compounds Found in Compost

Phytotoxic Compound	Compost Type/Age[z]	Species Affected	Reference
Acetic acid	Wheat straw, 4 weeks	Barley (*Hordeum vulgare* L.)	Lynch (1978)
Acetic acid	MSW, immature	Cabbage (*Brassica oleracea* L. Capitata group)	Keeling et al. (1994)
		Cauliflower (*Brassica oleracea* L. Botrytis group)	Keeling et al. (1994)
		Cress (*Lepidium sativum* L.)	Keeling et al. (1994)
		Lettuce (*Lactuca sativa* L.)	Keeling et al. (1994)
		Onion (*Allium cepa* L.)	Keeling et al. (1994)
		Tomato (*Lycopersicon esculentum* Mill.)	Keeling et al. (1994)
Ammonia	Biosolids	*Brassica campestris* L.	Hirai et al. (1986)
Ammonia and copper	Spent pig litter, < 24 weeks	Lettuce	Tam and Tiquia (1994)
		Snap beans (*Phaseolus vulgaris* L.)	Tam and Tiquia (1994)
		Tomato	Tam and Tiquia (1994)
Ammonia, ethylene oxide	MSW, <16 weeks	*Brassica parachinensis* L.	Wong (1985)
Organic acid	Cow manure, 12 weeks	Tomato	Hadar et al. (1985)
Organic acid	MSW, < 4 weeks	*Brassica campestris* L.	Hirai et al. (1986)
Organic acids and other compounds	Yard trimming waste, < 17 weeks	Australian pine (*Casuarina equisetifolia* J. R & G. Forst.)	Shiralipour et al. (1991)
		Bahiagrass (*Paspalum notatum* Flugge.)	Shiralipour et al. (1991)
		Brazilian pepper (*Schinus terebinthifolius* Raddi.)	Shiralipour et al. (1991)
		Ear tree (*Enterolobium cyclocarpum* Jacq.)	Shiralipour et al. (1991)
		Punk tree (*Melaleuca leucadendron* L.)	Shiralipour et al. (1991)
		Ragweed (*Ambrosia artemissifolia* L.)	Shiralipour et al. (1991)
		Tomato	Shiralipour et al. (1991)
		Yellow nutsedge (*Cyperus esculentus* L.)	Shiralipour et al. (1991)
Phenolic acids	Pig slurries < 24 weeks	Barley	Maureen et al. (1982)
		Wheat (*Triticum aestivum* L.)	Maureen et al. (1982)

[z] MSW = municipal solid waste.

From Ozores-Hampton, M.P., 1998. Compost as an alternative weed control method. *HortScience* 33:938–940. With permission.

and causing N immobilization by the soil microbial population because of a high C:N ratio (Jimenez and Garcia, 1989).

Laboratory, greenhouse, and field experiments on the use of immature MSW-biosolids compost as a weed control agent indicated that it could reduce weed germination and subsequent weed growth (Ozores-Hampton, 1997). The compost utilized for these experiments was produced from MSW (front-end separated) and biosolids (lime-stabilized and dewatered) co-composted through a three-compartment Eweson digester in an aerobic environment for 3 days, cured for 8 weeks using the windrow composting method, and screened.

To distinguish between compost chemical and physical effects on weed germination and growth, water extracts from immature MSW-biosolids compost were evaluated for effects on weed seed germination (Ozores-Hampton et al., 1996; Ozores-Hampton et al., 1999). Ivyleaf morningglory (*Ipomoea hederacea* L.), barnyardgrass (*Echinochloa crus-galli* L.), common purslane (*Portulaca oleracea* L.), and corn were selected as plant indicators to determine the composting stage with maximum chemical inhibition of seed germination and growth. Extracts were prepared from immature (3-day-old, 4-week-old, 8-week-old), and mature (1-year-old) MSW-biosolids composts by mixing 20 g (dry weight) of compost with 50 mL of water. The 8-week-old compost extract was the most phytotoxic because it decreased percentage germination, root growth, and germination index (GI, a combination of germination percentage and root growth); and increased mean days to germination (MDG) of each indicator species the most.

The extract of 8-week-old compost was evaluated for its effect on germination of 14 economically important weed species (Table 13.3). The extract decreased or inhibited germination of most weed species except yellow nutsedge (*Cyperus esculentus* L.), for which tubers were used as propagules. Germination of cress, wild mustard (*Brassica kaber* [DC.] L.C. Wheeler), lovegrass (*Eragrostis curvula* [Schrad.] Nees.), and dichondra (*Dichondra carolinensis* Michx.) seeds was completely inhibited by 8-week-old compost extract. The extract decreased germination

Table 13.3 Seed Germination of 14 Weed Species Affected by Water Extract from 8-Week-Old Co-Composted Municipal Solid Waste and Biosolids Compost

Common Name	Scientific Name	Control (%)	Compost (%)
Cress	*Lepidium sativum*	100*	0
Wild mustard	*Brassica kaber*	95*	0
Crabgrass	*Digitaria sanguinalis*	61*	2
Barnyardgrass	*Echinochloa crus-galli*	69*	51
Pigweed	*Amaranthus retroflexus*	89*	4
Wild radish	*Raphanus raphanistrum*	13*	1
Florida beggarweed	*Desmodium tortuosum*	56*	11
Curly dock	*Rumex crispus*	43*	3
Ground cherry	*Physalis ixocarpa*	37*	7
Lovegrass	*Eragrostis curvula*	36*	0
Ivyleaf morningglory	*Ipomoea hederacea*	96*	77
Common purslane	*Portulaca oleracea*	78*	66
Dichondra	*Dichondra carolinensis*	95*	0
Yellow nutsedge[z]	*Cyperus esculentus*	32	20

* Mean separation within species by *t*-test ($P < 0.05$).
[z] Tubers were used as propagules.

From Ozores-Hampton, M.P. et al., 1999. Age of co-composted municipal solid waste and biosolids on weed seed germination. *Compost Science and Utilization* 7(1):51–57. With permission.

of crabgrass (*Digitaria sanguinalis* [L.] Scop), pigweed (*Amaranthus retroflexus* L.), wild radish (*Raphanus raphanistrum* L.), curly dock (*Rumex crispus* L.), Florida beggarweed (*Desmodium tortuosum* L.), and ground cherry (*Physalis ixocarpa* L.) by more than 80% compared with a water control treatment. Compost extract decreased barnyardgrass, ivyleaf morningglory, and purslane germination by 15 to 30%. Weed seed germination was inhibited to a greater extent by 8-week-old compost extract compared with extract from mature compost, possibly due to a higher acetic acid concentration (1776 $mg{\cdot}kg^{-1}$ vs. 13 $mg{\cdot}kg^{-1}$) (Ozores-Hampton et al., 1999).

In general, for composts with a high C:N ratio, plant phytotoxicity is associated with the presence of volatile fatty acids (Hadar et al., 1985; Wong and Chu, 1985; Zucconi et al., 1981a, 1981b). Reduction of seed germination due to acetic acids in compost has been reported in cress, onion (*Allium cepa* L.), cabbage (*Brassica oleracea* L. Capitata group), cauliflower (*Brassica oleracea* L. Botrytis group), lettuce (*Lactuca sativa* L.), and tomato (Keeling et al., 1994). Reduction of Florida beggarweed, yellow nutsedge, and ragweed (*Ambrosia artemissifolia* L.) germination has been associated with the presence of volatile fatty acids (Shiralipour et al., 1991). Their compost extracts were made from 3-week-old immature yard trimming waste that was exposed to temperatures over 60°C, simulating a compost pile.

Excessive concentrations of trace metals like copper (Cu) have been associated with plant phytotoxicity (Tam and Tiquia, 1994). Although cadmium (Cd), Cu, lead (Pb), nickel (Ni), and zinc (Zn) concentrations were higher in mature than immature compost, their phytoavailability was low because there was no evidence that they detrimentally affected seed germination, root growth, GI, or MDG for each of the weed species evaluated (Ozores-Hampton et al., 1999). These metals tend to be complexed with organic compounds in compost, and are not water soluble (Eichelberger, 1994). Compost salt concentration was ruled out as a factor that reduced seed germination, since similar electrical conductivities (EC) were obtained from 3-day-old and mature composts (6.6 vs. 6.7 $dS{\cdot}m^{-1}$, respectively) (Ozores-Hampton et al., 1999). Ammonia was associated with the phytotoxic response of plants to spent pig litter (Tam and Tiquia, 1994) and biosolids (Hirai et al., 1986). However, phytotoxicity persisted in sterilized, ammonium-free extracts of MSW compost (Zucconi et al., 1981a).

To evaluate the physical effect of MSW-biosolids compost, immature and mature materials were applied as a mulch, and the effect on seedling emergence and shoot and root dry weight was evaluated in a greenhouse (Ozores-Hampton, 1997; Ozores-Hampton et al., 1997b, 1999). Plastic pots were utilized with different combinations of compost maturities. Ivyleaf morningglory seeds were covered with 7.5 cm of either 3-day-old compost, mature compost, or an artificial medium, or were left uncovered (untreated control). Immature (3-day-old) compost resulted in a 43% decrease in ivyleaf morningglory emergence compared with the control (Table 13.4). Percent emergence responses to artificial medium, mature compost, and the control were similar. Immature compost delayed emergence by 3.4 days compared with the control (Figure 13.1). There was no difference in mean days to emergence (MDE) between artificial medium and mature compost, although emergence was delayed

Table 13.4 Effect of Mature and Immature Compost on Emergence and Seedling Growth of Ivyleaf Morningglory

Treatment	Emergence (%)	MDE[z]	Shoot	Root
		(g dry weight per pot)		
Commercial medium[y]	96.7 a[x]	4.6 b	0.24 b	0.05 b
Mature compost	95.0 a	4.2 b	0.30 a	0.06 b
3-day-old compost	51.7 b	6.8 a	0.04 c	0.02 c
Control (sand)	95.0 a	3.4 c	0.25 b	0.12 a

[z] MDE = Mean days to emergence.
[y] Metro-mix 220 (peat-lite medium).
[x] Mean separation within columns by Duncan's multiple range test, $P \leq 0.05$.

From Ozores-Hampton, M., 1997. *Utilization of Municipal Solid Waste Compost as Biological Weed Control in Vegetable Crop Systems.* Ph.D. Dissertation, University of Florida, Gainesville. With permission.

compared with the control. Shoot and root dry weights were lower for plants that germinated beneath 3-day-old compost compared with mature compost, artificial medium, and the control. Shoot dry weight was higher in plants that germinated beneath mature compost compared with the control or artificial medium, perhaps due to nutrients supplied by the compost. However, higher root dry weights occurred in the control than the mature compost. The delayed and decreased weed seedling emergence and seedling growth caused by the 3-day-old compost may be attributed to both the physical effect of the mulch and to phytotoxic compounds (fatty acids) produced during the composting process (Ozores-Hampton, 1997).

Figure 13.1 Ivyleaf morningglory emergence through immature compost (top-left) as compared with the no compost application (top-right), artificial medium (bottom-left), and mature compost (bottom-right).

The use of immature compost to control weeds in the areas between raised beds of vegetable crops (alley-ways) also has been investigated (Ozores-Hampton, 1997; Ozores-Hampton et al., 1997a). Zucchini squash seeds were planted to plots consisting of three parallel raised beds (0.75 m wide and 0.15 m high) 0.9 m apart covered with white-on-black polyethylene mulch. Four-week-old MSW-biosolids compost was applied to the alley-ways as a mulch in thicknesses of 3.8, 7.5, 11.3, and 15 cm (49, 99, 148, and 198 $t \cdot ha^{-1}$, respectively). Subsequent weed control was compared with that provided by three applications of 1,1′-dimethyl-4,4′-bipyridinium salts (paraquat) at 0.6 $kg \cdot ha^{-1}$ and an untreated control. All compost thicknesses provided excellent weed control compared with the control and herbicide treatments. Compost at 7.5 cm or greater thickness completely inhibited weed germination and growth for 8 months (Figure 13.2). Zucchini yield and fruit size did not differ among treatments. There were no visible signs of zucchini plant stunting, chlorosis, or injury associated with application of immature compost in close proximity (Ozores-Hampton, 1997). DeVleeschauwer et al. (1981) reported that immature compost with a high acetic acid concentration was detrimental to plant growth when it was applied directly to the crop root zone. In our study, the compost was not placed immediately above the crop root zone, and the compost was separated from the raised beds by a layer of polyethylene. Acetic, propionic, and butyric acids were present in our compost at 1221, 34, and 33 $mg \cdot kg^{-1}$ concentrations, respectively, but their migration to crop plants in sufficient concentration to cause phytotoxicity was not detected. Thus, immature compost may be a viable alternative weed control method for alley-ways in vegetable fields, whether applied alone or in combination with chemical herbicides (Ozores-Hampton, 1997).

Immature composts and fresh organic materials may have more potential for reducing herbicide use in row crop production than mature composts. Mature MSW compost applied at 224 $t \cdot ha^{-1}$ reduced weed growth in alley-ways of bell pepper, but herbicides were more effective than the compost (Roe et al., 1993). In another study, fresh newsprint that was fall-applied at 24.4 $t \cdot ha^{-1}$ as a surface residue cover with no additional tillage suppressed winter annual grasses and broadleaf weeds in spring-planted soybean (*Glycine max* [L.] Merrill) crops (Edwards et al., 1994).

After an immature compost mulch reaches a mature state in the field, it can be incorporated into the soil for the following growing season to potentially improve soil productivity. When compost is incorporated into soil, observed benefits to crop production have been attributed to improved soil physical properties due to increased organic matter concentration rather than increased nutrient availability (Ozores-Hampton, 1997). Compost is not considered fertilizer, but significant quantities of nutrients (particularly N, phosphorus [P], and micronutrients) become bioavailable with time as compost decomposes in the soil (Ozores-Hampton et al., 1994).

III. OTHER CONSIDERATIONS WITH COMPOSTED ORGANIC MULCHES

MSW-biosolids compost in alley-ways of vegetables provided insufficient weed control at the mulch/polyethylene interface when compost layers were thin (less that 7.5 cm). Weed growth was vigorous due to nonuniform compost application at the

Figure 13.2 Control plot (top) versus 7.5 cm (135 $t \cdot ha^{-1}$) of MSW compost as mulch (bottom) 240 days after planting a squash crop.

mulch/polyethylene interface and the continuous sloping of the bed shoulders. To achieve more effective weed control, bed shoulders should have a 90 degree angle with the soil surface to allow an even compost thickness or a thicker compost layer (Ozores-Hampton, 1997). Merwin et al. (1995) reported that managing weeds at the edges of the mulched strips and weeds around the bases of apple (*Malus domestica* Borkh.) trees was problematic.

The benefits of composted and noncomposted organic mulches must compensate for their greater expense relative to herbicides. The higher establishment and maintenance costs of organic mulches can be offset by their prolonged efficacy, but a cost analysis should be made before they are recommended as a weed control method. Integrated pest management programs that incorporate alternative weed control methods such as mulch should be considered when possible to help reduce herbicide use in vegetable production.

IV. SUMMARY

Recently, composts made from biosolids, MSW, and/or yard trimmings have become available in large quantity. Once a compost has passed regulatory health and safety standards, vegetable growers are interested in the potential benefits of its use. Compost maturity is a major issue that the composting industry is facing as it attempts to provide a high-quality product to the agricultural community. The potential for using immature compost (mixture of MSW-biosolids) for weed control in the alley-ways between raised beds of vegetable crops has been demonstrated. Suppression of weed germination and growth by immature MSW-biosolids compost was due to the physical presence of the materials on the soil surface, and/or the action of phytotoxic compounds generated by microbes in the composting process.

REFERENCES

Altieri, W. and M.A. Liebman. 1988. The impact, uses, and ecological role of weeds in agroecosystems, p. 1–6. In: M.A. Altieri and M. Liebman (eds.). *Weed Management in Agroecosystems: Ecological Approaches.* CRC Press, Boca Raton, Florida.

Baskin, J.M, and C.C. Baskin. 1989. Seasonal changes in the germination responses of buried seeds of *Barbarea vulgaris. Canadian Journal of Botany* 67:2131–2134.

Blackshaw, R.E. 1990. Influence of soil temperature, soil moisture, and seed burial depth on the emergence of round-leaved mallow (*Malva pusilla*). *Weed Science* 38:518–521.

Blackshaw, R.E. 1992. Soil temperature, soil moisture, and seed burial depth effect on redstem filaree (*Erodium cicutarium)* emergence. *Weed Science* 40:204–207.

Bryan, H.H. and C.J. Lance. 1991. Compost trials on vegetables and tropical crops. *BioCycle* 27(3):36–37.

Campbell, P.L. and J. van Staden. 1994. The viability and germination characteristics of exhumed *Solanum mauritianum* seeds buried for different periods of time. *Plant Growth Regulation* 14:97–108.

Crnko, G.S, W. M. Stall, and J.M. White. 1992. Sweet corn weed control evaluations on mineral and organic soils. *Proceedings of the Florida State Horticultural Society* 105:326–330.

DeVleeschauwer, D.O., P. Verdonock, and P. Van Assche. 1981. Phytotoxicity of refuse compost. *BioCycle* 22(1):44–46.

Dusky, J.A., W.M. Stall, and J.M. White. 1988. Evaluation of herbicides for weed control in Florida production. *Proceedings of the Florida State Horticultural Society* 101:367–370.

Edwards, J.H., R.H. Walker, E.A. Guertal, L.D. Norton, and J.T. Eason. 1994. Options for recycling organics on farm land. *BioCycle* 35(11):66–68.

Eichelberger, A.S. 1994. *Characterization and Mineralization of Municipal Solid Waste Compost.* M.S. Thesis, University of Florida, Gainesville.

Florida Department of Agriculture and Consumer Services. 1994. *Composting Guidelines.* Tallahassee, Florida.

Food and Agriculture Organization (FAO) of the United Nations. 1987. *Soil Management: Compost Production and Use in Tropical and Subtropical Environments.* Soils Bulletin 56.

Foshee, W.G., W.D. Goff, K.M. Tilt, J.D. Williams, J.S. Bannon, and J.B. Witt. 1996. Organic mulches increase growth of young pecan trees. *HortScience* 31:811–812.

Gallaher, R.N. and R. McSorley. 1994. Soil water conservation from management of yard waste compost in a farmer's corn field. Agronomy Research Report AY-94-02. University of Florida, Gainesville.

Grantzau, E. 1987. Bark mulch for weed control in cut-flower perennials. *Zierpflanzenbau* 27:805–806.

Hadar, Y., Y. Inbar, and Y. Chen. 1985. Effect of compost maturity on tomato seedling growth. *Scientia Horticulturae* 27:199–208.

Hirai, M.F., A. Katayama, and H. Kubota. 1986. Effect of compost maturity on plant growth. *BioCycle* 24(6)54–56.

Jimenez, E.I. and V.P. Garcia. 1989. Evaluation of city refuse compost maturity: a review. *Biological Wastes* 27:115–142.

Keeling, A.A., I.K. Paton, and J.A.J. Mullett. 1994. Germination and growth of plants in media containing unstable refuse-derived compost. *Soil Biology and Biochemistry* 26:767–772.

Lapham, J. and D.S.H. Drennan. 1990. The fate of yellow nutsedge (*Cyperus esculentus L.*) seed and seedlings in soil. *Weed Science* 38:125–128.

Lynch, J.M. 1978. Production and phytotoxicity of acetic acid in anaerobic soils containing plant residues. *Soil Biology and Biochemistry* 10:131–135.

Marshall, F. and W. Ellis (eds.). 1992. *Rodale's All-New Encyclopedia of Organic Gardening*. Rodale Press, Emmaus, Pennsylvania.

Maureen, A., E. Ramirez, and J.L. Garraway. 1982. Plant growth inhibitory activity of extracts of raw and treated pig slurry. *Journal of the Science of Food and Agriculture* 33:1189–1196.

McConnell, D.B., A. Shiralipour, and W.H. Smith. 1993. Compost application improves soil properties. *BioCycle* 34(4):61–63.

Merwin, I.A., D.A. Rosenberger, C.A. Engle, D.L. Rist, and M. Fargione. 1995. Comparing mulches, herbicides, and cultivation as orchard groundcover management systems. *HortTechnology* 5:151–158.

Mester, T.C. and D.D. Buhler. 1991. Effects of soil temperature, seed depth, and cyanazine on giant foxtail (*Setaria faberi*) and velvetleaf (*Abutilon theophrasti*) seedling development. *Weed Science* 39:204–209.

Metting, F.B., Jr. 1993. *Soil Microbial Ecology: Applications in Agricultural and Environmental Management*. Marcel Dekker, New York.

Niggli, U., F.P. Weibel, and W. Gut. 1990. Weed control with organic mulch materials in orchards. Results from 8 years of field experiments. *Acta Horticulturae* 285:97–102.

Obreza, T.A. and R.K. Reeder. 1994. Municipal solid waste compost use in tomato/watermelon successional cropping. *Soil and Crop Science Society of Florida Proceedings* 53:13–19.

Ozores-Hampton, M. 1997. *Utilization of Municipal Solid Waste Compost as Biological Weed Control in Vegetable Crop Systems*. Ph.D. Dissertation, University of Florida, Gainesville.

Ozores-Hampton, M.P. 1998. Compost as an alternative weed control method. *HortScience* 33:938–940.

Ozores-Hampton, M. and H.H. Bryan. 1993a. Effect of amending soil with municipal solid waste (MSW) compost on yield of bell peppers and eggplant. *HortScience* 28:463 (Abstract).

Ozores-Hampton, M. and H.H. Bryan. 1993b. Municipal solid waste (MSW) soil amendments: influence on growth and yield of snap beans. *Proceedings of the Florida State Horticultural Society* 106:208–210.

Ozores-Hampton, M.P., B. Schaffer, and H.H. Bryan. 1994. Nutrient concentration, growth, and yield of tomato and squash in municipal solid waste amended soil. *HortScience* 29:785–791.

Ozores-Hampton, M.P., T.A. Bewick, P. Stoffella., D.J. Cantliffe, and T. Obreza. 1996. Municipal solid waste (MSW) compost maturity influence on weed seed germination. *HortScience* 31:577 (Abstract).

Ozores-Hampton, M.P., T.A. Obreza, P.J. Stoffella, and D.A. Graetz. 1997a. Utilization of municipal solid waste-biosolids compost as an alternative weed control agent in commercial vegetable production systems. *Weed Science Society of America Abstracts* 37:78 (Abstract).

Ozores-Hampton, M.P., T.A. Obreza, P.J. Stoffella, and G.E. Fitzpatrick. 1997b. Utilization of municipal solid waste compost mulch for weed control, p. 55. In: *Agronomy Abstracts 1997*. American Society of Agronomy, Madison, Wisconsin (Abstract).

Ozores-Hampton, M.P., P.J. Stoffella, T.A. Bewick, D.J. Cantliffe, and T.A. Obreza. 1999. Age of co-composted municipal solid waste and biosolids on weed seed germination. *Compost Science and Utilization* 7(1):51–57.

Parr, J.F. and S.B. Hornick. 1993. Utilization of municipal wastes, p. 545–559. In: F.B. Metting, Jr. (ed.). *Soil Microbial Ecology: Applications in Agricultural and Environmental Management*. Marcel Dekker, New York.

Patrick, Z.A. and L.W. Kock. 1958. Inhibition of respiration, germination and growth by substances arising during the decomposition of certain plant residues in the soil. *Canadian Journal of Botany* 36:621–647.

Reisman-Berman, O. and J. Kigel. 1991. Dormancy patterns in buried seeds of *Datura ferox* and *D. stramonium*. *Canadian Journal of Botany* 69:173–179.

Roe, N.E., P.J. Stoffella, and H.H. Bryan. 1993. Municipal solid waste compost suppresses weeds in vegetable crop alleys. *HortScience* 28:1171–1172.

Roe, N.E., P.J. Stoffella, and D. Graetz. 1997. Composts from various municipal solid waste feedstocks affect vegetable crops. II. Growth, yields, and fruit quality. *Journal of the American Society for Horticultural Science* 122:433–437.

Schneider, A.D., A.F. Wiese, and O.R. Jones. 1988. Movement of three herbicides in a fine sand aquifer. *Weed Science* 36:432–436.

Shiralipour, A., D.B. McConnell, and W.H. Smith. 1991. Effects of compost heat and phytotoxins on germination of certain Florida weed seeds. *Soil and Crop Science Society of Florida Proceedings* 50:154–157.

Shiralipour, A. D.B. McConnell, and W.H. Smith. 1997. Phytotoxic effect of a short- chain fatty acid on seed germination and root length of *Cucumis sativus* cv. 'Poinset'. *Compost Science and Utilization* 5(2):47–52.

Tam, N.F.Y. and S. Tiquia. 1994. Assessing toxicity of spent pig litter using a seed germination technique. *Resources Conservation and Recycling* 11:261–274.

Wong, M.H. 1985. Phytotoxicity of refuse compost during the process of maturation. *Environmental Pollution (Series A)* 37:159–174.

Wong, M.H. and L.M. Chu. 1985. Changes in properties of a fresh refuse compost in relation to root growth of *Brassica chinensis*. *Agricultural Wastes* 14:115–125.

Zhang, J. and M.A. Maun. 1990. Effects of sand burial on seed germination, seedling emergence, survival, and growth of *Agropyron psammophilum*. *Canadian Journal of Botany* 68:304–310.

Zucconi, F., M. Forte, A. Monaco, and M. de Bertoldi. 1981a. Biological evaluation of compost maturity. *BioCycle* 22(4):26–29

Zucconi, F., A. Pera, and M. Forte. 1981b. Evaluating toxicity of immature compost. *BioCycle* 22(2):54–57.

CHAPTER 14

Nitrogen Sources, Mineralization Rates, and Nitrogen Nutrition Benefits to Plants from Composts

Lawrence J. Sikora and Robin A. K. Szmidt

CONTENTS

I. INTRODUCTION

Composts are categorized as slow-release nitrogen (N) fertilizers because they release or mineralize only a fraction of their total N content. Organic byproducts

1-56670-460-X/01/$0.00+$.50

are amended with a carbon (C) source before composting to achieve the desirable C:N ratio of 30:1 to 35:1. The resulting compost has the majority of its N in the organic form and the rate of release of mineral N from the organic form is generally less than the original organic byproduct. O'Keefe et al. (1986) found that composted biosolids had half the mineralization rate of the uncomposted biosolids. Low mineralization rates are desirable in several instances. Research has long pursued methods to slow the nitrification process or the transformation of ammonium (NH_4^+) to nitrate (NO_3^-) in soils for plants to utilize the N fertilizer before it leaches through the soil profile (Hauck, 1980). By slowing the ammonification process or transformation of organic N to NH_4^+, composts are accomplishing the same goal. Yakovchenko et al. (1996) determined that organic sources of N (manures or legumes) were more efficiently taken up by crops than commercial fertilizer. An additional benefit of compost amendments is that the organic N that is not mineralized in the year of application is "stored" in the soil and will mineralize in future cropping seasons (Sullivan et al., 1998).

Reliable or predictable mineralization rates of a variety of composts are not always available to users. As a consequence, composts are used as mulches or as soil conditioners instead of fertilizers. Often composts and fertilizers are added together to soils and the fertilizer equivalent of compost is ignored because of inadequate information on the compost fertility. With increased emphasis on managing and recording nutrient applications to soils, the necessity to understand, measure, record, and account for N available in composts is important.

Although individual uses of composts are discussed under separate headings, there are common issues, particularly of plant-nutrient availability and form. In some countries there is a statutory requirement to know nutrient content prior to use, particularly where material is to be applied to land. This holds true whether the application is part of a waste-disposal stream or in use of compost as a fertilizer or soil amendment. Most attention has been paid to the supply of N in view of the potential risk of polluting run-off. However, other regulations may apply, for instance in the designation of phosphate sensitive zones. Irrespective of whether there is a statutory obligation, it is good environmental practice to know and optimize use of plant nutrients, including application of compost. Some participative schemes, such as U.K. crop protocols in a partnership between retailers and growers, demand compliance with best practice over and above statutory obligations in order to secure markets.

The balance of nutrients that must be calculated may be simplistic in that total fertilizer value applied should be within specified limits. National standards for use tend to be based on either precedent of using inorganic fertilizer or historical data related to applying noncomposted manures (Krauss and Page, 1997). However, a more comprehensive approach may be to align this to availability, that is, form of nutrients and whether they are slow release or labile. Cheneby and Nicolardot (1992) noted that chemical and physical determination provides little information for evaluating the practical use of a compost. Laboratory incubations were used by these workers to give precise data on N mineralization relating compost formulation to performance in specific situations, such as applications to a range of soil types. However, such kinetic studies may take a significantly long time, perhaps 4 months, by which time data are historical rather than of value in calculating seasonal fertilizer rates. In deciding on rates of application, calculations based on predicted mineralization rate and consequent availability should also consider logistical issues such as soil compaction and risk of groundwater contamination (Sikora, 1998).

II. FACTORS AFFECTING MINERALIZATION OF NITROGEN IN COMPOSTS

Nitrogen mineralization from composts is affected by the same factors that affect the N mineralization of organic N in soils. Physical factors include moisture and temperature. Chemical factors include pH, salts, and the presence of toxic quantities of inorganic or organic compounds.

A. Moisture

Generally, soil organic matter decomposition is curvilinearly related to moisture. Decomposition (mineralization) is slow at high moisture or under very dry conditions. Howard and Howard (1993) formulated a quadratic equation that described moisture effects on carbon dioxide (CO_2) flux, an end product of soil organic matter decomposition. Linn and Doran (1984) reported the effects of soil moisture on CO_2 respiration expressed as percent air-filled pore space (AFP). Respiration was greatest when AFP was 0.6 and declined when AFP was greater or less than 0.6. From low moisture to optimum moisture CO_2 evolution rate is nearly linear (Sikora and Rawls, 2000) and the rate declines in a curvilinear fashion from optimum to saturation. Maximum CO_2 evolution occurred between 30 and 40% saturation. Moisture status of soils is changed by addition of composts (Sikora and Rawls, 2000). With cumulative applications of 66 to 200 Mg ha^{-1}, moisture availability in a gravelly silt loam increased. So as the organic matter content is increased with applications of composts, moisture availability increases making conditions for N mineralization more ideal, and that, in turn, will increase the N mineralization from the compost-amended soil.

B. Temperature

Temperature effects on mineralization of organic matter in soils are often described using Q factors. Similar to extreme moisture effects, temperatures above 35°C or below 10°C reduce mineralization rates of organic matter in soils. From 10°C to nearly 35°C, mineralization will at least double for every 10°C increment increase in temperature ($Q_{10} = 2$). Raich and Schlesinger (1992) reviewed *in situ* measurements and concluded that a Q_{10} value of 2.4 adequately described temperature effects on CO_2 flux. Bergström et al. (1991) used a log function to describe temperature effects on CO_2 flux. Mineralization of composts in the field are best predicted when both temperature and moisture conditions are part of the equation.

C. Salinity

Salinity of compost-soil mixtures can affect N mineralization. Tester and Parr (1983) amended soil with biosolids compost at rates of 112 and 224 Mg ha^{-1} and monitored net N mineralization. Mineralization was nearly 5 times greater when the mixtures were leached with water as compared to unleached. Salinity level in the unleached 224 Mg ha^{-1} mixture was 4.62 dS m^{-1} and 1.01 dS m^{-1} in the leached mixture. Accordingly, decomposition as recorded by CO_2 evolution was similarly affected.

Salinity can also affect plant growth when composts are used as an amendment in plant growth media. Relative amendment rates are much greater than in most soils at nearly one third of the media by volume and plant growth may be affected by salinity. Mineralization of N may be reduced in high salt-containing media. However, the amount of total and mineralizable N in the media is large and therefore should not be a limiting factor for plant growth, even though N mineralization will be reduced.

D. pH

Irrespective of the fact that feedstocks used to make composts may vary widely in pH and as such influence the composting process, final pH is normally near neutrality (Gray et al., 1973). Lime-stabilized biosolids have a pH greater than 8 and, when composted by the static-aerated pile method, typically the compost has a final pH of 7.2 (McCoy et al, 1986). Soil pH probably has more effect on compost N mineralization than compost pH. Tester et al. (1977) demonstrated that adjusting soil from pH 5 to approximately 7 resulted in a substantial increase in N mineralization. Mineralization of organic N to NH_4^+ is less sensitive to pH changes than nitrification or the change of NH_4^+ to NO_3^-. This step is pivotal because the predominant form of N taken up by plants is NO_3^-. Therefore, the constituents or characteristics of the compost which affect nitrification would most influence the N mineralization rate of composts.

E. Forms of Organic Matter in Composts Influence Nitrogen Mineralization of Composts

Carbon to nitrogen ratio is regarded as the single most informative measurement of N mineralization capability of composts. Barbarika et al. (1985) found that C:N ratio was one of the primary factors controlling biosolids N mineralization. In field applications where plant growth is several weeks long, C:N ratio is a reasonably accurate predictor of N mineralization. In plant growth studies that occur over shorter periods (a few weeks as for some horticultural plants), other means for determining N mineralization such as short-term incubation would be required (Gilmour, 1998).

Sequential screening of biosolids compost resulted in greater N mineralization in the smaller sized fractions (Table 14.1). Compost material that passed through a 1 mm screen had a mineralization rate 3 times greater than compost that passed a 10 mm screen. The C:N ratio of the 10 mm pass compost was 14.7 and 10.4 for the 1 mm pass compost. A greater amount of woodchips (the bulking agent in the biosolids composting process) was removed by the 1 mm screen, which reduced the C:N ratio. These data suggest screening of composts will produce a number of fractions that mineralize differently and can be marketed for various purposes.

Table 14.1 Effects of Screening on N Mineralization of Biosolids Compost

Compost Fraction	Total N Content ($g\ kg^{-1}$)	C:N Ratio	N Mineralization (% of Total N)
10 mm pass	12.3	14.7	3.4
1 mm pass	13.5	10.4	10.7
1 mm retained	11.6	18.6	2.5

From Tester, C.F. et al., 1980. Effects of screening on compost properties, p. 126–135. In: *Proceedings of the National Conference on Municipal and Industrial Sludge Composting.* Information Transfer, Inc. Silver Spring, Maryland. With permission.

Yard trimmings composting operations have multiplied in urban areas. The sole N source for the compost is grass clippings that are only available in large quantities in the spring (Sikora and Sullivan, 2000). Municipal solid waste (MSW) composting facilities produce a compost that is deficient in plant-available N because the feedstocks are largely paper products and the resulting compost often has a high C:N ratio (Mamo et al., 1998). Therefore, MSW compost is largely ineffective as a source of fertilizer N. Both Hortenstine and Rothwell (1973) and Mays et al. (1973) report that plant yields increase only when N fertilizer is added along with MSW compost. Combining manures with these carbonaceous wastes in a proper ratio resulted in a mixture with optimum C:N ratio for composting and will produce a compost product that has significantly more plant-available N than MSW compost. Also, combining dairy manure and MSW or MSW compost may reduce volatile N and odor generation during the composting of manures (Martins and Dewes, 1992; Sikora, 1999). From these data the N fertilizer value of a compost is directly related to its C:N ratio.

III. NITROGEN MINERALIZATION OF DIFFERENT COMPOSTS

Mineralization of compost N is recorded as the "net" available or soluble N after an incubation that mimics a growing season. Net mineralization is the sum of mineralization and immobilization (the opposite of mineralization) that results from the compost-amended product. Table 14.2 contains mineralization percentages of composts made from different feedstocks. Note that mineralization ranges from 0 to 28% of the total organic N.

Recent evidence shows that blends of compost and inorganic fertilizer provide more N to crops than is predicted from adding the N equivalents from fertilizer and compost mineralization (Sikora and Enkiri, 1999). As thc percentage of N from compost in the blends increases and fertilizer (inorganic) N percentage decreases, yields and uptake of N by plants become greater than the equivalent N fertilizer alone (Figure 14.1). The reason for the increase appears to be that fertilizer stimulates compost N mineralization in soil, which provides more N to the plant (Sikora and Enkiri, 2000). Using ^{15}N-labeled fertilizer, Sikora and Enkiri (2000) demonstrated that compost could be the only source of extra N taken up by fescue (*Festuca arundinacea* L.) from a blend of NH_4NO_3 fertilizer and biosolids compost.

Table 14.2 Nitrogen Mineralization Percentages of Composts Made from Various Feedstocks

Primary Feedstock	Co-composted Feedstock	Test Location	N Mineralization (%) and Basis	Reference
Biosolids	Woodchips	Laboratory	7% of total N	Tester et al., 1977
Livestock manure	None	Field	10.5% of organic N	Wen et al., 1995
Municipal refuse	Biosolids	Laboratory	10.7% of total N to NO_3^-	Mattingly, 1956
Biosolids	Wheat straw	Laboratory	8.3% of total N to NO_3^-	Mattingly, 1956
Biosolids	Cotton trash	Laboratory	8.3% of total N to NO_3^-	Mattingly, 1956
Chicken manure	Not described	Laboratory	28% of total N	Castellanos and Pratt, 1981
Dairy manure	Not described	Laboratory	5% of total N	Castellanos and Pratt, 1981
Municipal refuse	None	Laboratory	0% of total N	Beloso et al., 1993
Sheep manure	Straw	Laboratory	13.4% of total N	Herbert et al., 1991
Cow manure	Straw	Laboratory	14.2% of total N	Herbert et al., 1991
Hog slurry	Sawdust	Laboratory	0% of total N	Herbert et al., 1991
Biosolids	Straw	Field	20% compared to urea	Baldoni et al., 1996
Food residuals	Yard trimmings	Field	10.6% of total N	Sullivan et al., 1998
Food residuals	Yard trimmings and paper	Field	8.1% of total N	Sullivan et al., 1998
Food residuals	Wood chips and sawdust	Field	7.3% of total N	Sullivan et al., 1998

Accurate compost mineralization determination is necessary for users to comply with nutrient management plans. For years, N was the only nutrient of concern because phosphorus (P) was considered essentially bound to the soil matrix. However, organic P forms are generally governed by different chemistry rules than inorganic P forms, and so organic P may leach deeper into the soil profile than inorganic P. One of the few field studies following the P mineralization from compost was performed by McCoy et al. (1986). They found that the P mineralization rate of biosolids was controlled by the treatment method for the biosolids and not by the compost process. Biosolids were a product of tertiary treatment to remove P from the wastewater. The final forms of P were iron- and calcium-P and these forms did not change significantly during composting or after application of compost to soils. Mineralization rate of P estimated from the 2-year field study was only 2 to 4% of the total P present.

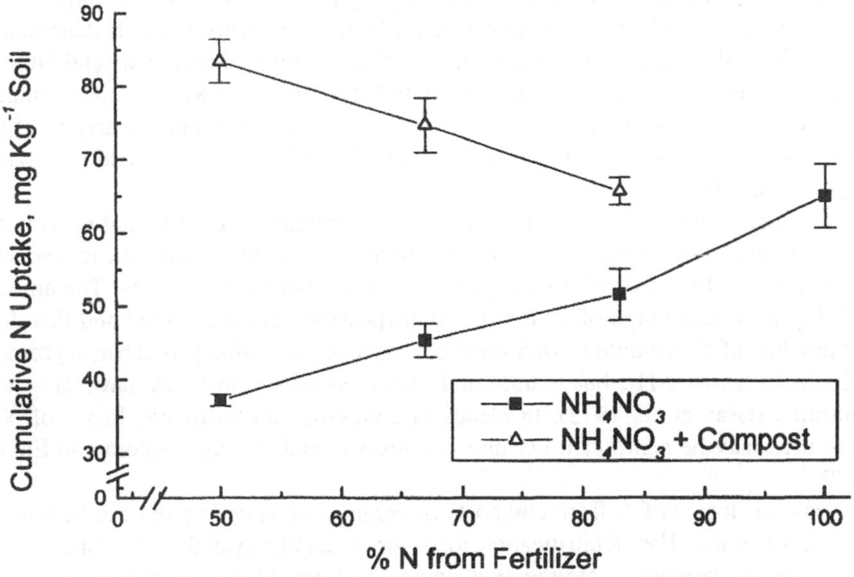

Figure 14.1 Cumulative N uptake by fescue from increasing amounts of NH_4NO_3 fertilizer or from blends of NH_4NO_3 fertilizer and biosolids compost that equal 100% NH_4NO_3 fertilizer application. Greater N uptake was recorded in pots containing increasing proportions of total available N as compost.

IV. NITROGEN LOSSES FROM COMPOSTS OR COMPOST-AMENDED MEDIA

Nitrogen losses from composts can be by volatilization of ammonia (NH_3), gaseous losses of N_2 or NO_x gases, or leaching of soluble compounds. These losses are much more difficult to predict than N mineralization and nearly all are influenced by environment of the compost or compost-amended medium. A study comparing fresh manure and composted manure indicated that yields were not different but N losses from volatilization or leaching were over twice as much from fresh manure than from compost (Brinton, 1985). Focusing only on mature and stable composts, we will discuss briefly mechanisms of N losses.

Most of the N in mature and stable composts is in the organic N form (Castellanos and Pratt, 1981). The small portion of total N that is inorganic N can be a mixture of NH_4^+ and NO_3^- and the proportion of each in a compost depends upon curing time and, in some instances, the feedstocks. Nitrate formation is sensitive to temperature and NO_3^- will not be present in composts until temperatures are near ambient and the curing process is well underway. At 45°C, practically no nitrification occurs. Nitrification is inhibited on the upper part of the mesophilic range while ammonification is vigorous in the thermophilic range (Harmsen and Kolenbrander, 1965). Several studies report that NO_3^- is a final product of composts and NO_3^- is sometimes used as an indicator of stability (Chefetz et al., 1998). Biosolids compost mineralization studies indicated that a 7-week-old compost that had nearly all of the inorganic N in the NH_4^+ form was quickly transformed to NO_3^- after addition to soil (Tester et al., 1977).

Organic N forms in composts or compost-amended media including soil are generally not lost because only small portions are soluble and susceptible to leaching. Therefore, the losses of N from composts involve inorganic N forms. The amount of NH_4^+ as a percentage of total N is small (possibly less than 10%) and therefore volatile loss of NH_3 would be of a small consequence. Alkaline pHs can also promote volatile losses of NH_3, but mature and stable composts normally have pHs near neutrality (Gray et al., 1973). In blends of composts and fertilizers, losses of NH_3 from urea may be significant because composts contain urease (Sikora and Enkiri, 1999).

Gaseous losses of N from composts or compost-amended media can be a result of denitrification. For denitrification to occur, a readily available C source, NO_3^-, and anaerobic conditions are required. In compost-amended soils, potting media, or beds, an available C source is normally present. But Beauchamp (1986) measured higher denitrification potential in soil amended with fresh or liquid manures that in soil amended with compost. These data suggest that denitrification is more prevalent when amending with manures than with composts because they have greater amounts of available C. Nitrate is rapidly produced when composts are added to soils, providing the benefit to plants seen when crops are grown in compost-amended soils (Sikora and Enkiri, 1999; Tester and Parr, 1983). Finally, denitrification requires anaerobic conditions. Anaerobic conditions can occur when the porosity of the compost-amended medium is low and gas exchange is limited. Composts having small particle size may have low porosity and anaerobic conditions may prevail. Excess moisture will also lead to anaerobic conditions and the possible loss of N via denitrification. In summary, losses of N from composts and compost-amended soil is dictated by environmental conditions and losses would most likely be caused by conditions that result in denitrification.

V. COMPOSTS AS GROWING MEDIA

A. General Considerations

In much of the literature there is confusion between composted material and so-called compost. The latter is often used to describe any soilless growing medium, such as that based on peat. Growing media is the preferred term and can be employed in conjuction with the source material description, e.g., composted growing media (Bragg, 1995).

For many horticultural applications, the use of soilless growing media is relatively straight forward. That is, the physical requirements of the substrate reflect the production system and all plant nutrients are selected and added to the mix as inorganic fertilizer. As an example, for most proprietary peat-based materials, such as containerized plant mixes, it is assumed that peat contributes no significant nutrients but is important in terms of providing the correct balance of air porosity and water availability, measured as air-filled porosity (AFP) (Bragg and Chambers, 1988; Bunt, 1988).

In hydroponic systems, the soilless substrate may be inert, such as perlite or rockwool, or absent altogether, for instance in the case of cultivation using Nutrient Film Technique (NFT) (Adams, 1991; Atherton and Rudich, 1986).

These technological developments of inert substrates coupled to supply of plant nutrients in inorganic form are in complete contrast to use of compost. While composted materials may look similar to peat, and have comparable physical attributes to peat, nutritional availability follows a different pattern. Because of the hidden differences, together with the inadvertent confusion over nomenclature, it often comes as a surprise to growers that the recommendations for use of peat-based growing media may not be identical to those for composted materials. This has resulted in a lack of confidence in use of composted materials as growing media, despite evidence that by careful use they can perform well and even, under some circumstances, perform better than available peat-based media (Gouin, 1998).

Any composted material considered as a candidate growing medium includes, by definition, a complex blend of organic and inorganic plant nutrients. The medium may have these nutrients at acceptable levels or it may have a toxic excess. Because composting is a dynamic biological process, the microbial activity of composted material may be high and therefore the potential for change remains, even in mature composts. If compost does not immediately meet the needs of the grower, specific changes may be made by adding organic or inorganic materials or by removing soluble salts, for instance by leaching. However, because the system is active, any such change is likely to have additional, possibly unexpected consequences, such as reactivating the microflora or changing the N balance in the material. Equally, material changes naturally over time and so nutrient availability may alter during prolonged storage.

The processes involved in mineralization of organic-N in compost (Figure 14.2) involve enzymic degradation, particularly of microbial protein to free amino acids (I) followed by ammonification. Ammonia may combine with acidic components to form NH_4^+ ions. If the substrate is neutral or alkaline then gaseous NH_3 may be released (II). However, providing the substrate is well aerated, nitrification (III) follows as a result of bacterial action. The sequence releases H^+ ions which will give a consequential acidification in the medium. Major shifts in pH will affect availability of other plant nutrients, particularly manganese (Mn) and boron (B) which are largely unavailable to plants at values around 7.5, and P, with principal unavailability at around pH 8.0 (Bunt, 1988). Therefore, changes in substrate pH will result in shifts to excess or inadequate mineral supply and appearance of induced deficiency symptoms in growing plants (Bould et al.,1983; Bragg, 1995). Where plant substrates have excessive levels of some ions, induced deficiency of others due to poor uptake may result (Bunt, 1988). This may be particularly true of composted growing media where complex interactions may result, particularly in terms of Mn, iron (Fe), and zinc (Zn) uptake. Raja Harun and McKenna (1998) studied the alleviation of chlorotic leaf symptoms by the manipulation of micronutrients. High Mn levels depressed magnesium (Mg) uptake, with levels half that in leaves of comparable peat-based controls. However, symptoms of leaf chlorosis were attributed to Fe deficiency, not Mn toxicity, which was ameliorated by incorporation of Fe into the growing medium. The suppression of Fe uptake by high Mn was also considered to result in high uptake of Zn. Addition of Fe reduced the severity of Mn toxicity and may be a method of improving acceptability of commercial composted growing media.

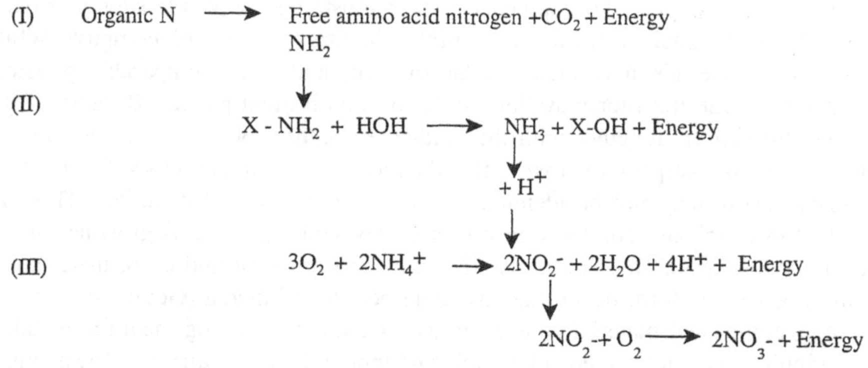

Figure 14.2 Mineralization route of organic-N in compost.

In some potential feedstocks, N may not be readily available. Examples include chitinaceous materials such as crab waste (Black, 1997). Degradation of chitin [a β (1–4) homopolymer of N-acetyl-glucosamine] is complex and specific to certain bacteria and fungi (Boyer, 1986; Chen, 1987). Consequently, incorporation of such materials may not result in an anticipated release of available N (Kuo, 1995) or N release may be unpredictably delayed during the course of composting and maturation, possibly resulting in secondary heating. This problem of secondary composting on release of N is significant as it can influence the available choices in constituting composts of a particular nutritional balance, either as fertilizers or as growing media. Despite this, such compost additives may have other benefits such as contributing to the control of plant parasitic nematodes by providing a slow-release source of nutrients for antagonistic microorganisms (Spiegel et al., 1987). A two-phase composting process may harness the benefits of secondary heating and compost stabilization as a result of use of chitinaceous materials and result in benefits of enhanced pest suppression (Roy et al., 1997).

B. Mushroom Substrates

Issues of source, availability, and mineralization of N in mushroom composts are somewhat different from other applications. Nevertheless, developments in this sector can demonstrate what is possible when improving process control for product quality of all composts. For mushrooms, compost must function in a very specific biological and biochemical way to meet the needs of a crop fungus. The predominant mushroom species grown worldwide is *Agaricus bisporus* (Lange) Imbach. This fungus is cultivated on a straw-based compost. Although a wide range of other species, such as Shiitake (*Lentinus edodes* [Berkley] Pegler) and Oyster mushrooms (*Pleurotus ostreatus* Jacq ex Fr.) are also grown, these are not typically grown on true composts. Rather, they may be grown on part-fermented substrates such as sawdust or straw respectively. Furthermore, their distribution is not as widespread as *A. bisporus*. Consequently, when considering compost-related issues for mushroom cultivation, *A. bisporus* is inevitably the species of choice.

There are two major nutritional issues in mushroom compost. First, the way in which compost is formulated must meet the requirements of the crop. Second, the way in which the crop utilizes compost subsequently influences the attributes of the spent mushroom substrate (SMS), which may be used either directly as a composted material or re-composted prior to use.

C. Mushroom Composts

Historically, the production of mushrooms is attributed to French horticulturists. Arisings of mushrooms, most probably *Agaricus campestris* Fries, were noted in so-called hot beds where natural heat generated from manure heaps was used as a means of protecting tender crops, such as cucumber (*Cucumis sativus* L.) and melon (*Cucumis melo* L.), from frost. The partly fermented manure, often with a high straw content, was also used as a crop substrate. The conditions under which mushrooms arose were replicated and were enhanced by the transfer of material from the source of mushrooms to a new pile. This practice, begun in the late 17th century, was developed until the industrial production of mushrooms was a commercial reality in the early 19th century. In time, these various practices of compost manufacture and inoculation were scientifically investigated and methodically developed (Atkins, 1966).

The manufacture of compost for mushroom production remained unpredictable until the positive management of manure and straw blends, initially by manual aeration. From this came the establishment of the French mushroom industry, centered on the caves of the Loire Valley in which the crops were grown. The chronology of mushroom cultivation was reviewed by van Griensven (1988).

Modern mushroom compost is predominantly based upon wheat (*Triticum aestivum* L.) straw with N supplied from manures of various forms. Traditionally, manure has come from horse stables and, most preferably, from situations with a high ratio of bedding straw to manure. This results in a proportion of straw having been physically abraded by hooves and a reduced risk of anaerobic conditions prior to controlled composting. In the 1970s there was a move towards reducing the proportion of horse manure and sourcing of N from poultry manure. Typically this is as sawdust/manure blends rather than slurries. Composts containing 100% N from non-horse manure sources were referred in some texts as "synthetic" composts (Randle, 1974), a term which is rather ambiguous and is best avoided. Nonetheless, researchers at the Glasshouse Crops Research Institute in the U.K. introduced the concept of replacing manures with alternative N sources, either from poultry manures or from inorganic sources such as urea at around 4 to 6% and 46% available N, respectively to achieve similar results (Flegg, 1983).

Fraser and Fujikawa (1958) showed that *A. bisporus* is incapable of utilizing NO_3^--N. While gaseous NH_3 is harmful to *A. bisporus*, NH_4^+-N can be readily utilized, as can urea and various amino acids (Flegg et al., 1985). Bacterial protein may also be used as a nutrient source by *A. bisporus* (Fermor and Wood, 1981). *A. bisporus* shares the ability to utilize microbial biomass as a nutrient source with a number of other fungi (Fermor and Wood, 1991; Grant et al., 1986). This is as a result of extracellular protease which may be produced both *in vivo* and *in vitro.* Composting has been shown to result in N being bound in the N-rich lignin-humus complex and in bacterial protein as the microbial biomass flourishes and is sequentially replaced through the different phases of compost processing (van Griensven, 1988; Figure 14.3). This microbial fixation is therefore of protein, peptide N as part of the lignin fraction of the substrate and of microbial protein (Gerrits, 1969, 1988) and is likely to be comparable with many composting processes. Under the specific conditions employed by mushroom growers, at the point at which composting is typically stopped and the fungus inoculated to the mass, the resulting compost is highly selective for growth of *A. bisporus.*

In comparison to mature materials such as biosolid-derived compost, mushroom compost is immature. However, investigations into nutritional dynamics for mushrooms show how nutrients, particularly N, may be transformed into slow-release organic forms (Flegg et al., 1985). Within compost, the level of NH_4^+ within the water fraction may be related to the level of free NH_3 (Miller et al., 1991). Savioe and Minvielle (1995) showed that the style of composting may influence N content and release. For instance, contained compost treatments that result in relatively high microbial biomass prior to high temperatures being achieved are likely to release higher levels of N than in other situations. Source of this may be from NH_4^+ immobilized in bacterial cells and released by lysis. Savioe and Minvielle (1995) also proposed that *de novo* production by thermophilic microorganisms by ammonification of N substrates released from microbial compounds is likely to be a major contributor. This is particularly important where facility design allows specific control, for instance of temperature. Such process control allows preservation of N and maximization of microbial biomass as a slow release source of N in subsequent use.

Mushroom yield is linked to the "efficiency" of the compost and therefore any composting process that precedes cropping. In order to enhance yield, it has become common practice to supplement compost with N-rich materials. This was first proposed by Schisler and Sinden (1963). Such procedures were developed to the point where they are now standard practice in industry. Incorporation of N-rich supplements may result in secondary reheating and generation of NH_3 (Flegg et al., 1984) and so delayed, or slow-release, materials such as those based on formaldehyde-treated plant protein are typically used.

D. Spent Mushroom Substrates (SMS)

The complex biology of *A. bisporus* requires for production of sporophores the combination of compost colonized by the crop-fungus with an inert over-layer,

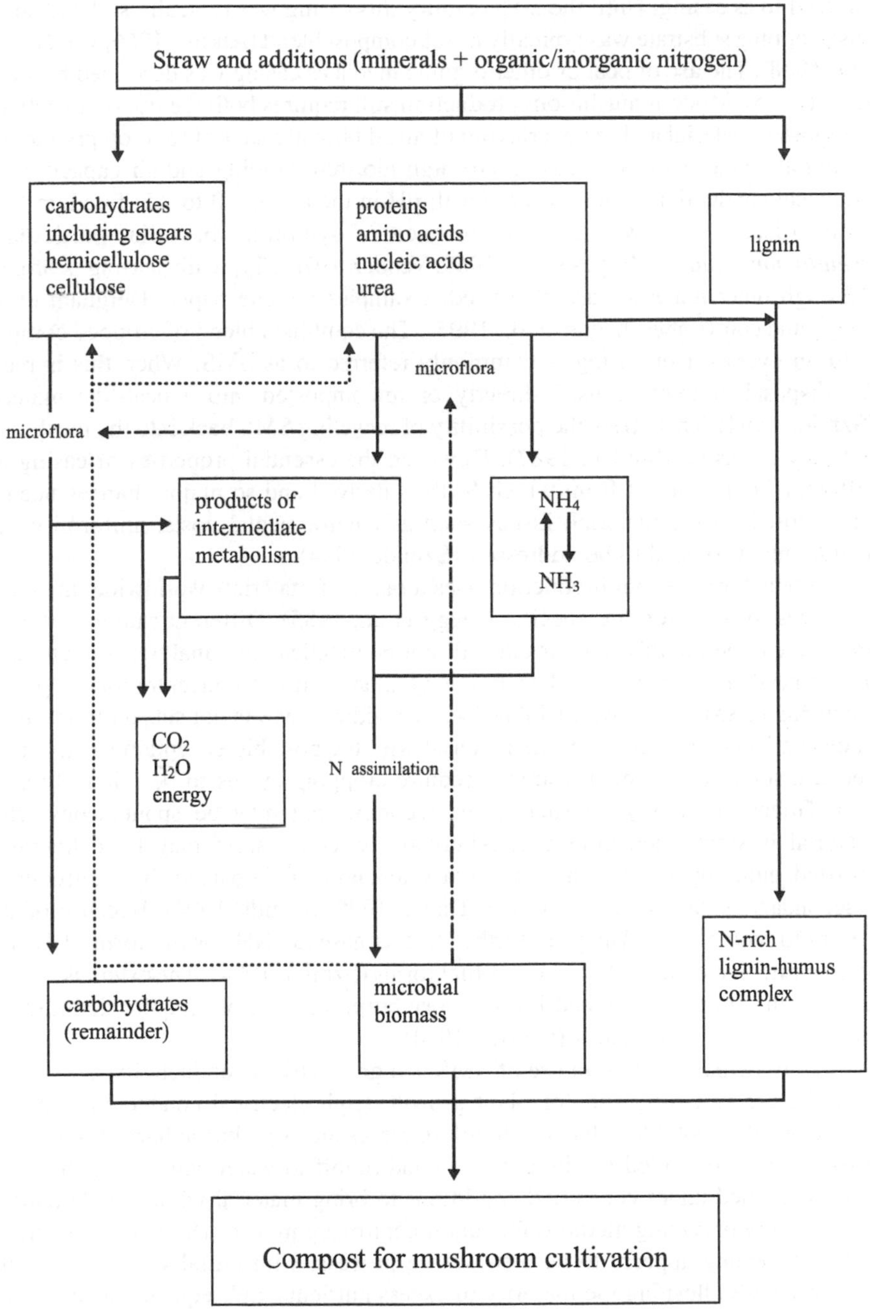

Figure 14.3 Flow chart of the mushroom composting process showing microflora which decomposes (–·–·–) and microflora which is decomposed (-----). (From van Griensven, L.J.L.D. (ed.). 1988. *The Cultivation of Mushrooms*. Darlington Mushroom Laboratories, Rustington, United Kingdom. With permission.)

referred to as casing. Until the 20th century this casing was typically soil and so the postcropping substrate was typically a soil/compost blend (Atkins, 1966; van Griensven, 1988). The use of peat or other organic media as casing was described by Flegg et al. (1985). Modern mushroom production still requires both the use of an actively composted material and the overlaying of an additional material for crop production. Essential characteristics of casing are high moisture holding and air capacity; low electrical conductivity and nutrient level; pH in the range 7.0 to 7.5; freedom from pests, disease, and weeds; and presence of symbiotic bacteria, particularly *Pseudomonas putida* (Hayes et al., 1969; Vedder, 1978). Typically, casing is of peat, although other materials are also used. Examples include paper (Dergham et al., 1991) and coir (Labuschagne et al., 1995). The combined blend of cropped compost with an overlayer of casing is commonly referred to as SMS. When this is ready for disposal, it may be used directly or re-composted into a peat-like material (Szmidt, 1994). This offers the possibility of recycling SMS back into the mushroom industry as casing (Sinden, 1987). However, the essential properties of casing are different from compost from which SMS is derived and so major changes need to be imposed on the material. Issues such as environmental costs, uniformity, and availability also need to be addressed (Szmidt, 1994).

Although mushroom production uses a range of materials worldwide, the target attributes for the crop are specific (Flegg et al., 1985). Different samples of SMS are likely to be broadly similar but will not be identical and analysis of material is recommended prior to use. The range of plant-nutrient concentrations typically occurring in SMS is shown in Table 14.3. Providing SMS is not subject to leaching, plant nutrients will remain in the material with the possible exception of N. This is because some management routines require cropping cycles to be relatively short (van Griensven, 1988). In such cases recomposting may be spontaneous when material is stock-piled after use. Alternatively, composting may be deliberately assisted, either by aeration or, if necessary, amending C:N balance by reintroducing a secondary N source (Levanon and Danai, 1995; Szmidt, 1994). Because of this ability to recompost, N may be further lost as gaseous NH_3 or the form of N may alter, particularly from NH_4^+ to microbial forms (Szmidt, 1994). The extent is unclear as most information collected for statutory purposes does not require definition of form of N, only total values (Gerrits, 1994).

The principal problem in use of SMS is high electrical conductivity. This is true both for use as casing and for plant growth. High electrical conductivity implies high content of soluble salts and therefore represents a pollution hazard in outdoor use due to uncontrolled leaching by rain and runoff to water courses. SMS can be readily leached under controlled conditions to bring material within the bounds of acceptability providing methods for runoff control are in place (Szmidt and Conway, 1995). Leaching appears to be a two-stage process, with initial volumes of water (1:2 to 1:8 v/v) flushing the majority of excess nutrients. Subsequent leaching (1:16 v/v) resulted in a second peak of conductivity in the elutent. It is likely that such high volumes of leachate release low-solubility components such as calcium sulfate (gypsum) and calcium oxalate subsequent to initial leaching of K^+, Ca^{2+}, SO_4^{2-}, Cl^-, and Na^+ ions (Chong and Rinker, 1994; Szmidt and Chong, 1995).

Table 14.3 Maximum and Minimum Plant Nutrient Concentrations and Other Key Attributes Recorded in Fresh Spent Mushroom Substrate (SMS)

Factor	Maximum	Minimum
pH	8.2	7.3
Electrical conductivity (μS cm^{-1})	2380	1410
C:N ratio	15:1	9:1
Dry matter (%)	55.3	53.1
Nitrate nitrogen	5.8	<1
Ammonium nitrogen	2.8	<1
Total available nitrogen[z]	9	4
Phosphorus	11.2	2.9
Potassium	18.2	3.3
Calcium	83.4	27
Magnesium	3.8	2.3
Sodium	4.1	1.4
Chloride	7.5	1.5
Zinc	0.97	0.44
Copper	0.37	—

Note: Units for plant nutrient concentrations are g kg^{-1}.

[z] Combined nitrate- and ammonium-N in any one sample, excluding microbial or organic N.

From: Szmidt, R.A.K. and C. Chong, 1995. Uniformity of spent mushroom substrate (SMS) and factors in applying recommendations for use. *Compost Science and Utilization* 3(1):64–71. With permission.

VI. CONCLUSIONS

Composts are categorized as slow-release N fertilizers because they release or mineralize only a fraction of their total N content. Reliable or predictable mineralization rate of a variety of composts is not always available to users. With increased emphasis on managing and recording nutrients applications to soils, the necessity to understand, measure, record, and account for N available in composts is important. Nitrogen mineralization from composts is affected by the same factors that affect the N mineralization of organic N in soils. Physical factors include moisture and temperature. Chemical factors include pH, salts, and the presence of toxic quantities of inorganic or organic compounds. Although composted materials may look similar to peat, and have comparable physical attributes to peat, nutritional availability follows a different pattern. Any composted material considered as a candidate growing medium includes, by definition, a complex blend of organic and inorganic plant nutrients. The medium may have these nutrients at acceptable levels or it may have a toxic excess. Because composting is a dynamic biological process, the microbial activity of composted material may be high and therefore the potential for change remains, even in mature composts. Mushroom compost is formulated to meet the requirements of the crop. The way in which the crop utilizes compost subsequently influences the attributes of the SMS which may be used either directly as a composted material or recomposted prior to use. Different samples of SMS are likely to be broadly similar but will not be identical. Composts are sources of fertilizer N to varying degrees, and understanding factors that control mineralization will make composts more valuable for agricultural and horticultural uses.

REFERENCES

Adams, P. 1991. Hydroponic systems for winter vegetables. *Acta Horticulturae* 287:181–190.

Atherton, J.G. and J. Rudich. 1986. *The Tomato Crop: A Scientific Basis for Improvement.* Chapman and Hall Ltd., London, United Kingdom, p. 661.

Atkins, F.C. 1966. *Mushroom Growing Today.* Faber and Faber Ltd., London, United Kingdom, p. 188.

Baldoni, G., L. Cortellini, and L. Dal Re. 1996. The influence of compost and sewage sludge on agricultural crops, p. 431–438. In: M. DeBertoldi, P. Sequi, B. Lemmes, and T. Papi (eds.). *The Science of Composting.* Blackie Academic and Professional, London, United Kingdom.

Barbarika, A., L.J. Sikora, and D. Colacicco. 1985. Factors affecting the mineralization of nitrogen in sewage sludge amended soils. *Soil Science Society of America Journal* 49:1403–1406.

Beauchamp, E.G. 1986. Availability of nitrogen from three manures to corn in the field. *Canadian Journal of Soil Science* 66:713–720.

Beloso, M.C., M.C. Villar, A. Cabaneiro, M. Carballas, S.J. Gonzalez-Prieto, and T. Carballas. 1993. Carbon and nitrogen mineralization in an acid soil fertilized with composted urban refuses. *Bioresource Technology* 45:123–129.

Bergström, L., H. Johnnson, and G. Torstensson. 1991. Simulation of soil nitrogen dynamics using the SOIL N model. *Fertilizer Research* 27:181–188.

Black, D. 1997. Crab composter gets claws on new feedstocks. *BioCycle* 38(8):30–33.

Bould, C., E.J. Hewitt, and P. Needham. 1983. *Diagnosis of Mineral Disorders in Plants.* Her Majesty's Stationery Office, London, United Kingdom, p. 170.

Boyer, J.N. 1986. End products of anaerobic chitin degradation by salt marsh bacteria as substrates for simulatory sulfate reduction and methanogenesis. *Applied and Environmental Microbiology* 52(6):1415–1418.

Bragg, N.C. 1995. *Growing Media.* Grower Books, Swanley, United Kingdom, p. 106.

Bragg, N.C. and B.J. Chambers. 1988. Interpretation and advisory applications of compost air-filled porosity of potting substrates. *Acta Horticulturae* 294:183–190.

Brinton, W.F., Jr. 1985. Nitrogen response of maize to fresh and composted manure. *Biological Agriculture and Horticulture* 3:55–94.

Bunt, A.C. 1988. *Media and Mixes for Container Grown Plants.* Unwin Hyman, London, United Kingdom, p. 309.

Castellanos, J.Z. and P.F. Pratt. 1981. Mineralization of manure nitrogen — correlation with laboratory indexes. *Soil Science Society of America Journal* 45:354–357

Chefetz, B., Y. Chen, and Y. Hadar. 1998. Water-extractable components released during composting of municipal solid waste. *Acta Horticulturae* 469:111–118.

Chen, A.A. 1987. Chitin metabolism. *Archives of Insect Biochemistry and Physiology* 6(4):267–277. (Abstract).

Cheneby, D. and B. Nicolardot. 1992. Use of mineralization kinetics to estimate the agricultural value of organic fertilizers, p. 188–196. In: D.V. Jackson, J.-M. Merillot, and P. L'Hermite (eds.). *Composting and Compost Quality Assurance Criteria*. Commission of the European Communities, Brussels, Belgium.

Chong, C. and D.L. Rinker. 1994. Use of spent mushroom substrate for growing containerized woody ornamentals: an overview. *Compost Science and Utilization* 2(3):45–53.

Dergham, Y., J. Lelley, and A.A. Ernst. 1991. Waste paper as a substitute for peat in the mushroom (*Agaricus bisporus*) casing soil production. *Mushroom Science* 13:263–267.

Fermor, T.R. and D.A. Wood. 1981. Degradation of bacteria by *Agaricus bisporus* and other fungi. *Journal of General Microbiology* 126:377–387.

Fermor, T.R. and D.A. Wood. 1991. Mushroom compost microbial biomass: a review. *Mushroom Science* 13:191–199.

Flegg, P.B. 1983. Rapidly prepared mushroom compost, p. 78–79. In: *Annual Report of The Glasshouse Crops Research Institute for 1981*. Little Hampton, United Kingdom.

Flegg, P.B., P.E. Randle, and J.F. Smith. 1984. Mushroom compost supplementation, p. 80–82. In: *Annual Report of The Glasshouse Crops Research Institute for 1982*. Little Hampton, United Kingdom.

Flegg, P.B., D.M. Spencer, and D.A. Wood. 1985. *The Biology and Technology of the Cultivated Mushroom*. John Wiley & Sons, Chichester, United Kingdom, p. 347.

Fraser, I.M. and B.S. Fujikawa. 1958. The growth promoting effect of several amino acids on the common cultivated mushroom, *A. bispordes*. *Mycology* 50:538–549.

Gerrits, J.P.G. 1969. Organic compost constituents and water utilized by the cultivated mushroom during spawn run and cropping. *Mushroom Science* 7:111–126.

Gerrits, J.P.G. 1988. Nutrition and compost, p. 29–72. In: L.J.L. van Griensven (ed.). *The Cultivation of Mushrooms*. Darlington Mushroom Laboratories, Rustington, United Kingdom..

Gerrits, J.P.G. 1994. Composition, use and legislation of spent mushroom substrate in the Netherlands. *Compost Science and Utilization* 2(3):24–30.

Gilmour, J.T. 1998. Carbon and nitrogen mineralization during co-utilization of biosolids and composts, p. 89–112. In: S. Brown., J.S. Angle, and L. Jacobs (eds.). *Beneficial Co-utilization of Agricultural, Municipal and Industrial By-products*. Kluwer, Dordrecht, Netherlands.

Gouin, F. 1998. Using compost in the ornamental horticulture industry, p. 131–138. In: S. Brown, J.S. Angle, and L. Jacobs (eds.). *Beneficial Co-utilization of Agricultural, Municipal and Industrial By-products*. Kluwer, Dordrecht, Netherlands.

Grant, W.D., L.L. Rhodes, B.A. Prosser, and R.A. Asher. 1986. Production of bacteriolytic enzymes and degradation of bacterial by filamentous fungi. *Journal of General Microbiology* 132:2353–2358.

Gray, K.R., K. Sherman, and A.N. Biddlestone. 1973. A review of composting — Process and products — Part 3. *Process Biochemistry* 8:11–16.

Harmsen, G.W. and G.J. Kolenbrander. 1965. Soil inorganic nitrogen, p. 43–92. In: W.V. Bartholomew and F.E. Clark (eds.). *Soil Nitrogen*. American Society of Agronomy Monograph Series No. 10. American Society of Agronomy, Madison, Wisconsin.

Hauck, R.D. 1980. Mode of action of nitrification inhibitors, p. 19–32. In:. J.J. Meisinger, G.W. Randall, and M.L. Vitosh (eds). *Nitrification Inhibitors — Potentials and Limitations*. ASA Special Publication 38. American Society of Agronomy (ASA), Madison, Wisconsin.

Hayes, W.A., P.E. Randle, and F.T. Last, 1969. The nature of the microbial stimulus affecting sporophore formation in *Agaricus bisporus*. *Annals of Applied Biology* 64:177–187.

Herbert, M., A. Karam, and L.E. Parent. 1991. Mineralization of nitrogen and carbon in soils amended with composted manure. *Biological Agriculture and Horticulture* 7:349–361.

Hortenstine, C.C. and D.F. Rothwell. 1973. Pelletized municipal refuse compost as a soil amendment and nutrient source for sorghum. *Journal of Environmental Quality* 2:343–344.

Howard, D.M. and P.J.A. Howard. 1993. Relationship between CO_2 evolution, moisture content and temperature for a range of soil types. *Soil Biology and Biochemistry* 25:1537–1546.

Krauss, G.D. and A.L. Page. 1997. Wastewater, sludge and food crops. *BioCycle* 38(2):74–75, 78–80, 82.

Kuo, S. 1995. Nitrogen and phosphorus availability in groundfish waste and chitin-sludge co-composts. *Compost Science and Utilization* 3(1):19–29.

Labuschagne, P., A. Eicker, and M. van Geuning. 1995. Casing mediums for *Agaricus bisporus* cultivation in South Africa: A preliminary report. *Mushroom Science* 14:339–344.

Levanon, D. and O. Danai. 1995. Chemical, physical and microbiological considerations in recycling spent mushroom substrate. *Compost Science and Utilization* 3(1):72–79.

Linn, D.M. and J.W. Doran. 1984. Effect of water-filled pore space on carbon dioxide and nitrous oxide production in tilled and nontilled soils. *Soil Science Society of America Journal* 48:1267–1272.

Mamo, M,C., J. Rosen, T.R. Halbach, and J.F. Moncrief. 1998. Corn yield and nitrogen uptake in sandy soils amended with municipal solid waste compost. *Journal of Production Agriculture* 11:469–475.

Martins, O. and T. Dewes. 1992. Loss of nitrogenous compounds during composting of animal manures. *Bioresource Technology* 42:103–111.

Mattingly, G.E.G. 1956. Studies on composts prepared from waste materials. III. Nitrification in soil. *Journal of the Science of Food and Agriculture* 7:601–605.

Mays, D.A., G.L. Terman, and J.C. Duggan. 1973. Municipal compost: effects on crop yields and soil properties. *Journal of Environmental Quality* 2:89–92.

McCoy, J.L., L.J. Sikora, and R.R.Weil. 1986. Plant availability of phosphorus in sewage sludge compost. *Journal of Environmental Quality* 15:403–409.

Miller, F.C., E.R. Harper, and B.J. Macauley. 1991. Investigation of various gases, pH and redox potential in mushroom composting phase I stacks. *Australian Journal of Experimental Agriculture* 31:415–425.

O'Keefe, B.E., A. Axley, and J. J. Meisinger. 1986. Evaluation of nitrogen availability indexes for a sludge compost amended soil. *Journal of Environmental Quality* 15:121–128.

Raich, J.W. and W.H. Schlesinger. 1992. The global carbon dioxide flux in soil respiration and its relationship to vegetation and climate. *Tellus* 44(B):81–99.

Raja Harun, R.M. and C. McKenna, 1998. Alleviation of interveinal yellowing in leaves of petunia 'Rose Frost' in composted spruce bark. *Acta Horticulturae* 469:235–244.

Randle, P.E. 1974. Compost, p. 82–84. In: *Annual Report of The Glasshouse Crops Research Institute for 1973.* Little Hampton, United Kingdom.

Roy, S., P. Leclerc, F. Auger, G. Soucy, C. Moresoli, L. Cote, D. Potevin, C. Beaulieu, and R. Brzezinski. 1997. A novel two-phase composting process using shrimp shells as an amendment to partly composted biomass. *Compost Science and Utilization* 5(4):52–64.

Savioe, J.-M. and N. Minvielle. 1995. Changes in N availability and effects of ammonia during composting. *Mushroom Science* 14:275–282.

Schisler, L.C. and J.W. Sinden. 1963. Nutrient supplementation of mushroom compost at spawning. *Mushroom Science* 5:150–164.

Sikora, L.J. 1998. Nitrogen availability from composts and blends of composts and fertilizers. *Acta Horticulturae* 469:343–351.

Sikora, L.J. 1999. MSW compost reduces nitrogen volatilization during dairy manure composting. *Compost Science and Utilization* 7(4):34–41.

Sikora, L.J., and N.K. Enkiri. 1999. Growth of tall fescue in compost/fertilizer blends. *Soil Science* 164:62–69.

Sikora, L.J. and N K. Enkiri. 2000. Efficiency of compost-fertilizer blends as compared to nitrogen fertilizer alone. *Soil Science*. In press.

Sikora, L.J. and W. J. Rawls. 2000. *In situ* respiration determination as a tool for classifying soils according to soil organic matter content. *Communications in Soil Science and Plant Analysis* 31: In press.

Sikora, L.J. and D.M. Sullivan. 2000. Case studies of municipal and on-farm composting in the United States of America, p. 605–623. In: J.F. Power and others (eds.). *Land Application of Agricultural, Industrial, and Municipal By-Products*. Soil Science Society of America Book Series No. 6. Soil Science Society of America, Madison, Wisconsin. In press.

Sinden, J. 1987. The American mushroom industry. *Mushroom Journal* 173:165–169.

Spiegel, Y., I. Chet, and E. Cohn. 1987. Use of chitin for controlling plant parasitic nematodes II: mode of action. *Plant and Soil* 98:337–345.

Sullivan, D.M., S.C. Fransen, A.I. Bary, and C.G. Cogger. 1998. Fertilizer nitrogen replacement value of food residuals composted with yard trimmings, paper or wood wastes. *Compost Science and Utilization* 6(1):6–18.

Szmidt, R.A.K. 1994. Recycling of spent mushroom substrates by aerobic composting to produce novel horticultural substrates. *Compost Science and Utilization* 2(3):63–72.

Szmidt, R.A.K. and C. Chong, 1995. Uniformity of spent mushroom substrate (SMS) and factors in applying recommendations for use. *Compost Science and Utilization* 3(1):64–71.

Szmidt, R.A K. and P.A. Conway. 1995. Leaching of recomposted spent mushroom substrates (SMS). *Mushroom Science* 14:901–905.

Tester, C.F., L.J. Sikora, J.M. Taylor, and J.F. Parr. 1977. Decomposition of sewage sludge compost in soil. I. Carbon and nitrogen transformations. *Journal of Environmental Quality* 6:459–463.

Tester, C.F., J.F. Parr, and P. Paolini. 1980. Effects of screening on compost properties, p. 126–135. In: *Proceedings of the National Conference on Municipal and Industrial Sludge Composting*. Organized by the Hazardous Materials Control Research Institute, Silver Spring, Maryland. Information Transfer, Inc., Silver Spring, Maryland.

Tester, C.F. and J.F. Parr. 1983. Decomposition of sewage sludge compost in soil: IV. Effect of indigenous salinity. *Journal of Environmental Quality* 12:123–126.

van Griensven, L.J.L.D. (ed.). 1988. *The Cultivation of Mushrooms*. Darlington Mushroom Laboratories, Rustington, United Kingdom, p. 515.

Vedder, P.J.C. 1978. *Modern Mushroom Growing*. Educaboek B.V., Culemborg, Netherlands, p. 420.

Wen, G., T.E. Bates, and R.P. Voroney. 1995. Evaluation of nitrogen availability in irradiated sewage sludge, sludge compost and manure compost. *Journal of Environmental Quality* 24:527–534.

Yakovchenko, V., L.J. Sikora, and D.D. Kaufman. 1996. A biologically based indicator of soil quality. *Biology and Fertility of Soils* 21:245–251.

CHAPTER 15

Plant Nutrition Benefits of Phosphorus, Potassium, Calcium, Magnesium, and Micronutrients from Compost Utilization

Zhenli He, Xiaoe Yang, Brian A. Kahn, Peter J. Stoffella, and David V. Calvert

CONTENTS

I. INTRODUCTION

The annual production of organic wastes in the U.S. exceeds 1 billion megagrams (Mg) (Karen et al., 1995). About one third of these materials are processed into composts. The benefits of compost utilization in agriculture have been well documented. Applications of composts have been reported to improve physical, chemical, and biological properties of soils, resulting in higher crop yield and quality (Giusquiani et al., 1995; Li et al., 2000; Stoffella et al., 1997). Among other beneficial effects, supply of nutrients to plants is one of the major considerations for utilizing

1-56670-460-X/01/$0.00+$.50

compost to improve soil productivity (Dick and McCoy, 1993). Nitrogen (N) nutrition benefits to plants from compost have been reviewed in chapter 14, by Sikora and Szmidt. Although most of the research on composts as a source of plant nutrients has focused on N (Hue et al., 1994), composts have much to offer as sources of other elements as well. This chapter focuses on plant nutrition benefits of phosphorus (P), potassium (K), calcium (Ca), magnesium (Mg), iron (Fe), manganese (Mn), copper (Cu), zinc (Zn), and boron (B) from compost utilization in agriculture, with an emphasis on horticultural crops.

II. ELEMENTAL CONCENTRATIONS IN COMPOST

A. Macronutrients (Other Than Nitrogen)

Phosphorus concentrations in composts generally range from <0.4 to >23 $g{\cdot}kg^{-1}$ (He et al., 1995; Vogtmann et al., 1993a), depending on sources. Biosolids generally contain a greater P concentration than most feedstocks. However, a varying proportion of P in biosolids is in organic forms, which are less available to plants (He et al., 2000).

Most municipal solid waste (MSW) composts from the U.S. and European countries have P concentrations of 2 to 6 $g{\cdot}kg^{-1}$ with a mean of 3.3 $g{\cdot}kg^{-1}$ (Table 15.1). This P concentration is about 2 to 10 times greater than the total P (0.2 to 2.0 $g{\cdot}kg^{-1}$) in most agricultural soils (Tisdale and Nelson, 1996). In addition, plant available P is usually <1% of the total soil P, whereas as much as 20 to 40% of the P in compost can be available to plants (Vogtmann et al., 1993b). Therefore, application of composts can increase plant available P in the soil.

Table 15.1 Concentrations of Phosphorus, Potassium, Calcium, and Magnesium in Various Composts in Selected Countries

Countries	Composts[z]	Organic C	P	K	Ca	Mg	C/P	Reference
U.S.	MSW	252	3.4	4.3	27.8	2.8	74	He et al., 1995
Germany	MSW	192	2.7	8.4	28.2	4.8	71	Vogtmann et al., 1993a
China	PMC	190	2.4	12.5	—	—	79	Yang, 1996
Spain	MSW	284	6.0	7.0	75.0	5.0	47	Gonzalez-Vila et al., 1982
France	MSW	—	2.6	2.5	40.0	3.0	—	De Haan, 1981
Italy	MSW	395	2.7	0.7	—	—	146	Petruzzelli et al., 1985
Netherlands	MSW	—	3.3	2.7	21.4	1.7	—	De Hann, 1981
Mean		263	3.3	5.4	38.5	3.5	85	

Note: Values for organic C, P, K, Ca, and Mg are given in $g{\cdot}kg^{-1}$ (oven dry basis).
[z] MSW = municipal solid waste; PMC = pig manure compost.

The concentrations of K in composts vary from 0.7 to >12 $g{\cdot}kg^{-1}$, with a mean of 5.4 $g{\cdot}kg^{-1}$ (Table 15.1). This value is lower than the K concentration in plant tissues of most crops (8 to 35 $g{\cdot}kg^{-1}$). Potassium is highly mobile in plants at all levels within individual cells and within tissues (Marschner, 1995). Therefore, part of the K in plant materials may be lost during composting. Most agricultural soils contain total K concentrations from 4 to 25 g·kg^{-1} (Tisdale and Nelson, 1996), but with <1% of the total K available to plants. Composts can be an alternative source of K for crops.

Calcium concentrations in composts vary from 21 to 75 g·kg^{-1}, with a mean of 39 g·kg^{-1} (Table 15.1). The forms and availability of Ca depend on the source and feedstock composition of the compost. Most neutral to alkaline soils contain sufficient available Ca for plant growth. However, Ca deficiency can be a problem for acid soils and crop quality usually is affected if Ca is not sufficient. Therefore, for acid soils, especially acid, sandy soils, composts can increase Ca availability for crop growth.

Composts contain Mg concentrations in the range of 1 to 5 g·kg^{-1}, with a mean of 3.5 g·kg^{-1} (Table 15.1). Magnesium is needed for crops on soils with an insufficient supply of Mg or an imbalance of Ca and Mg. Compost application can contribute to soil-available Mg for most agricultural soils.

B. Micronutrients

Iron (Fe), Mn, Cu, Zn, B, and molybdenum (Mo) are essential elements for crop production and food quality (Marschner, 1995). A long-term diet containing low concentrations of Fe, Mn, Cu, and Zn has been reported to cause human malnutrition (Yang et al., 2000). The availability of Fe, Cu, and Zn in calcareous soils is generally low, and an external source of these nutrients is needed for improved crop yield and food quality (Martens and Westermann, 1991). Composts contain variable amounts of micronutrients (Table 15.2).

Table 15.2 Mean Concentrations of Micronutrients in Municipal Solid Waste Composts in the U.S. and European Countries

Countries	Fe	B	Cu	Mn	Mo	Zn	Reference
U.S.	16400	54.1	250	431	7.2	609	He et al., 1995
Germany	—	—	43.2	—	—	211	Vogtmann et al., 1993a
Spain	2200	3.0	200	500	—	700	Gonzalez-Vila et al., 1982
France	—	60.0	250	600	—	1000	De Hann, 1981
Italy	—	—	422	—	—	857	Petruzzelli et al., 1985
Netherlands	—	60.0	630	400	—	1650	De Haan, 1981
Mean	9300	44.3	299	483	7.2	838	

Note: Values for micronutrients are given in mg·kg^{-1} (dry weight basis).

Total Fe concentration in composts is from 2000 to 16,000 mg·kg^{-1}. Biosolids have a relatively high concentration of Fe, but water solubility of Fe is low (insoluble forms of Fe are dominant) (He et al., 2000). Composts made from MSW generally contain less Fe than those made from biosolids.

Boron concentration in composts ranges from 3 to 60 mg·kg^{-1} (Table 15.2), which is quite comparable with the range for plant tissue (5 to 70 mg·kg^{-1}) (Romheld and Marschner, 1991). Boron deficiency is relatively common in plants as compared with a deficiency of the other micronutrients, especially when plants are grown in acid, sandy soils with a low supply of total B. Composts may be a source of available B for crops grown on B-deficient soils.

Copper concentrations in composts range from 43 to 630 mg·kg^{-1} based on data from the U.S. and some European countries (Table 15.2). This concentration range is about 4 to 50 times greater than the mean value of total Cu in the world soil (12 mg·kg^{-1}) (Berrow and Reaves, 1984). However, Cu in composts is generally chelated by organic matter and may not be available to plants before mineralization of the organic fractions.

Mean Mn concentration ranges from 400 to 600 mg·kg^{-1} for composts from various countries (Table 15.2), which is comparable to the average total Mn in soils (450 mg·kg^{-1}) (Berrow and Reaves, 1984). Manganese in composts may be more bioavailable than soil Mn.

Molybdenum in composts from the U.S. ranges from 1 to 12.8 mg·kg^{-1}, with a mean of 7.2 mg·kg^{-1} (Table 15.2), as compared with 1.5 mg·kg^{-1} total Mo in the world soil (Berrow and Reaves, 1984). Deficiencies of Mo in the U.S. occur on acid, sandy soils. Although composts can be a potential source of Mo for deficiency correction, the use of composts as Mo fertilizers has not been widely investigated.

Composts generally contain more Zn than Cu, with a concentration range from 211 to 1650 mg·kg^{-1} (Table 15.2), which is 5 to 40 times greater than the mean total Zn in the world soil (40 mg·kg^{-1}) (Berrow and Reaves, 1984). Zinc deficiencies are commonly found in calcareous soils and sandy soils (Martens and Westermann, 1991). Applications of composts to such soils can increase Zn availability and improve Zn nutrition of plants.

III. PLANT NUTRIENT AVAILABILITY AND MINERALIZATION EFFECTS

Organic amendments, such as yard waste compost and biosolids, can enrich the soil P status by their direct contribution and by alteration of the soil P sorption capacity (Hue et al., 1994). Phosphorus in plant tissue is mainly incorporated into organic fractions as nucleic acids, phospholipids, and phytin, with a small proportion in free inorganic forms. Organic P in composts from plant materials is readily decomposed to release ortho-phosphate, which is available to plants. The availability of P in composts ranges from 20 to 40% of the total P content (Vogtmann et al., 1993a). He et al. (2000) reported that the availability of P in MSW compost and yard waste was greater than that in biosolids. Biosolids contained a larger amount of P, but the P was primarily in the inorganic form and associated with Ca, Fe, or aluminum (Al) (Table 15.3). Mineralization significantly increased P availability in the MSW and yard waste compost based on Mehlich 3 extraction, but had minimal effect on the extractability of P from biosolids. For instance, field incubation for 180 days increased extractable P from 3 to 17% and from 4 to 26% of the total P in the MSW and yard waste composts, respectively, but the increase was only 0.1% for biosolids (Table 15.3).

Plant availability of K in composts can be more than 85% of the total K content (Vogtmann et al., 1993a). This emphasizes the importance of compost as a potential source of K for crop production. Varying contents of K in compost can be attributed to different sources of feedstocks. However, the composting process may also have a substantial influence on K availability. Due to the high water solubility of K,

Table 15.3 Mineralization Effect on Extractability of Macro- and Micronutrients in Composts by Mehlich 3 Reagent

	Biosolids			MSW Compost			Yard Waste		
Nutrients	0 d[z]	180 d	360 d	0 d	180d	360 d	0 d	180 d	360 d
P	0.9	1.0	1.3	3.3	17.3	18.4	3.7	25.5	27.0
K	38.7	27.6	23.8	33.9	71.1	57.7	73.4	35.7	28.5
Fe	0.7	0.6	0.4	6.8	8.5	9.5	13.6	14.0	16.8
Mn	6.9	15.9	21.1	24.5	38.8	62.7	47.1	51.9	57.6
Zn	7.0	19.0	32.4	59.0	67.3	71.9	67.7	76.6	80.0
Cu	1.9	3.7	5.7	13.3	16.4	16.8	33.8	50.0	63.8

Note: Values are total nutrient extractability (%).

[z] Days after field incubation in the University of Florida research farm, Fort Pierce, Florida.

leaching losses can occur during the composting or mineralization processes if the compost is exposed to rainfall. Extractability of K in composts by the Mehlich 3 reagent can be as high as 73% of the total K (Table 15.3). Mineralization decreased extractable K from 73 to 29% for yard waste and from 39 to 24% for biosolids. For MSW compost, K extractability increased during the first 6-month field incubation, followed by a subsequent decrease (Table 15.3). The decrease in extractable K due to mineralization was probably due to leaching of K from composts (He et al., 2000).

Mineralization increased plant availability of Fe in MSW and yard waste compost, as estimated by the Mehlich 3 extraction method (Table 15.3). The availability of Fe is very low in biosolids and the amount of Fe extractable by the Mehlich 3 reagent accounts for less than 1%. In contrast to MSW and yard waste composts, mineralization decreased extractability of Fe in biosolids, probably because of differences in Fe forms.

Water solubility of Mn in composts is much higher than Fe (Table 15.3), with 7 to 47% of the total Mn being extractable by the Mehlich 3 reagent. Mineralization increased Mn availability, especially in the MSW composts (Table 15.3).

As with Mn, Zn in composts is highly water soluble and readily available to plants. Feedstock sources and properties of composts greatly influenced the percentage of total Zn extractable by the Mehlich 3 reagent, with values ranging from 7% for biosolids to 68% for yard waste (Table 15.3). Mineralization further increased Zn availability or extractability. The increased Zn availability due to mineralization may be beneficial on Zn-deficient soils, but also raises the possibility of toxicity to the soil microbial community.

Copper in composts is less water soluble than Zn, but water solubility of Cu varied greatly among different composts (Table 15.3). Mineralization generally increased the amount of Cu extractable by the Mehlich 3 reagent, and the increase was more significant for yard waste compost than for biosolids (Table 15.3).

Microbial biomass carbon (C), N, sulfur (S), and P generally increase immediately after compost addition because of an increased supply of organic C (Smith, 1991). This stimulating effect can last a few months, depending on the quantity and quality of the amended compost and nutrient availability in the soil (Perucci, 1990). Similarly, activities of several enzymes in the soil are significantly increased by compost amendments (Giusquiani et al., 1995; Perucci, 1990). Microbial biomass is considered the most active fraction of soil organic matter and represents a significant source of plant-available nutrients (Smith and Paul, 1991). The increased total microbial biomass and enzyme activity due to compost amendments can improve soil fertility over a long time.

IV. PLANT NUTRITION EFFECTS FROM VARIOUS COMPOSTS

A. Macronutrients (Other Than Nitrogen)

In a classic study of data from the 1940s, Bunting (1963) reported that sewage sludge (biosolids) was a source of N and P but not of K. Biosolids contained 0.3 to 0.4% K_2O on a dry matter basis, compared to 2% K_2O for farmyard manure.

A review of older literature on the use of waste materials (including MSW composts and biosolids) as sources of P was provided by Sommers and Sutton (1980).

Wastewater treatment method controls the P mineralization rate of biosolids, and greatly impacts the usefulness of biosolids composts as P fertilizers (McCoy et al., 1986). In the McCoy et al. (1986) study, extractable soil P increased with increasing biosolids compost applications, but corn (*Zea mays* L.) uptake of P did not increase similarly. The forms of P extracted from the compost-amended soil were not available for plant uptake. Only 3 to 5% of the P in composts made from biosolids precipitated with Fe or Al was in an organic form, so microbial mineralization had little effect on P availability.

Buchanan and Gliessman (1990) concluded that organic matter additions by compost, combined with inorganic N or P fertilizer, may enhance P use efficiency compared to fertilization with acid-forming N fertilizer plus inorganic P and no compost. Composts may improve the soil environment for root extension and P absorption, due at least in part to a pH buffering effect. Total P accumulation and yield of broccoli (*Brassica oleracea* L., Italica Group) were highest in this study in treatments incorporating only a spent mushroom bedding-horse manure-crop residue compost or a horse manure vermicompost at 30 $Mg \cdot ha^{-1}$.

In a greenhouse study, an MSW compost added to a calcareous loam soil at a long-term rate equivalent to about 150 $Mg \cdot ha^{-1}$ resulted in increased total P in fruits of eggplant (*Solanum melongena* L.) and tomato (*Lycopersicon esculentum* Mill.) compared to an unfertilized control and a treatment given inorganic fertilizer (Cabrera et al., 1991). With this soil type, the continuous use of MSW compost increased the levels of organic- and available-P in the soil, and produced a consequent increase in P uptake by the test plants.

James and Aschmann (1992) concluded that a co-compost produced from lime-stabilized, anaerobically digested biosolids and wood chips may have been a better source of plant-available P than uncomposted, aerobically digested biosolids. This was true even though the compost contained primarily insoluble, organic P and the biosolids contained higher levels of soluble ortho-phosphate.

Four MSW composts (one amended with $CaCO_3$) were analyzed as potential fertilizers by Villar et al. (1993). The four composts could supply all the macronutrients recommended for plant growth, although not in high amounts. Most of the Ca, about half the K and Mg, and small proportions of N and P would be quickly available to plants shortly after these composts were applied to soils. However, except for Ca, macronutrient concentrations were much lower than those found in other organic wastes.

Warren and Fonteno (1993) reported that adding 20% composted poultry litter (by volume) to a loamy sand soil in 3.8-L containers raised the available P and exchangeable K, Ca, and Mg to levels within a recommended range for landscape plants.

Hue et al. (1994) conducted a pot study in which a yard waste compost with 5.7% Ca was blended at 0, 25, 50, 75, and 100% with a low-Ca Ultisol. Although as little as 25% compost greatly increased soil extractable Ca, corn shoot composition of Ca was not increased by compost amendments. Corn shoot P, in contrast, increased with increasing percent compost, but for best growth, at least 75% compost was needed.

The effects of MSW compost on fruit quality of 'Pope' orange (*Citrus sinensis* [L.] Osbeck.) on sour orange (*Citrus aurantium* L.) rootstock were evaluated in a commercial grove by Stoffella et al. (1996). MSW compost (256 $Mg \cdot ha^{-1}$) was surface applied, disc-incorporated to a depth of 10 cm, or not applied within the tree canopy area of 42-year-old trees. Surface soil C, N, K, and Fe contents in the disced plots, 5 months after MSW application, were higher than in the surface applied or unamended plots. Fruit quality characteristics such as percent juice and sugar/acid ratio did not differ among treatments. However, plots receiving a disc-incorporation of compost had a higher percentage of large fruit than control plots.

In a pot experiment, corncob compost was superior to liming for improving growth of lettuce (*Lactuca sativa* L.), pea (*Pisum sativum* L.), and corn on an acid Oxisol with a pH of 4.07 (Chung and Wu, 1997). Although the compost did little to neutralize soil pH, it substantially increased P uptake in all three species. The improved P uptake was attributed primarily to alleviation of Al and Mn toxicity by the compost, possibly due to the complexing of these elements by the organic matter in the compost.

A biosolids/sawdust compost applied to an eutric sandy Cambisol at 7.5, 15, 22.5, and 30 $g \cdot kg^{-1}$ had little effect on organic P and on labile forms of P in a study by Coutinho et al. (1997). The compost significantly increased hydroxide-extractable inorganic P and nonextractable soil P fractions. Chemical rather than biological reactions lead to the redistribution of compost-borne P to more firmly held forms after its application to the soil. Their conclusion echoed that of McCoy et al. (1986): The effectiveness of biosolids as a P fertilizer will vary with their treatment for stabilization, with Fe and Al salts limiting P effectiveness.

Although biosolids are relatively P-rich wastes, P uptake by lettuce, bean (*Phaseolus vulgaris* L.) and petunia (*Petunia hybrida* Vilm.) was not improved by addition of biosolids, whether composted or not, compared to a treatment receiving no P in a study by Wen et al. (1997). Extractable soil P also did not increase with biosolids versus the no-P treatment. The percentage of total P that was extractable by 0.5 *M* $NaHCO_3$ was 30 to 70% for composted livestock manure, versus 2.5 to 5% for composted biosolids. In contrast to the biosolids treatments, P uptake by lettuce increased with rate of P application with composted livestock manure. This was another study where Fe and Al salts added during wastewater treatment resulted in low availability of P from biosolids.

Bittenbender et al. (1998) reported that a macadamia husk-cattle manure compost had minimal effects on leaf nutrient composition of macadamia (*Macadamia integrifolia* Maiden & Betche) trees compared to a chemically fertilized control, except that slightly lower N and B concentrations occurred at two orchard sites with the compost treatment. Compost applications slightly increased soil K, Ca, and Mg due to (1) a direct contribution of these nutrients from the compost itself, and (2) the increased cation exchange capacity of the soil due to compost addition, which enhanced K, Ca, and Mg retention. Although effective, compost fertilization was not profitable under the experimental conditions.

Amendment with an immature MSW compost or a vermicompost made from the same MSW to the B horizon of a Cambisol was studied by Sainz et al. (1998). Amendment with 10 or 50% vermicompost increased dry matter yields of red clover (*Trifolium pratense* L.) and cucumber (*Cucumis sativus* L.) compared to soil alone. Vermicompost amendment also increased shoot P, Ca, and Mg concentrations, but may have reduced the activity of arbuscular mycorrhizal fungi in the soil.

Fishwaste compost was tested as a P source in soilless growing media by Kuo et al. (1999). The compost contained 3.6 $g \cdot kg^{-1}$ of P on a dry weight basis and had high P solubility. The compost was a viable potential P source for container-grown crops in soilless media if the P leaching potential was minimized. Kuo et al. (1999) recommended that the compost comprise no more than 25% of a growing medium, or that it be used in conjunction with controlled drip irrigation rather than with overhead sprinklers.

In a field experiment, applications of a manure-sawdust co-compost increased tissue nutrient concentrations of P, K, Ca, Mg, and Zn in pak choi (*Brassica chinensis* L.) and corn compared to an unamended soil (Wong et al., 1999). The recommended rates of amendment were 25 $Mg \cdot ha^{-1}$ for pak choi and 50 $Mg \cdot ha^{-1}$ for corn.

From these studies, it is apparent that differing feedstocks affect the potential of composts to be sources of plant macronutrients. Even within a given type of feedstock, such as biosolids, results vary due to factors such as treatment for stabilization and compost pH. An important point which both researchers and compost users must keep in mind is that increases in an extractable macronutrient from a growing medium do not necessarily translate into increased plant uptake of that element (e.g., McCoy et al., 1986).

B. Micronutrients

Media amended with MSW compost produced higher quality chrysanthemum (*Dendranthema* sp.) plants with higher foliar K, Cu, Zn, and B compared to unamended media in a study by Gogue and Sanderson (1975). Although plants were not Fe deficient, the MSW compost was not a good source of Fe. A slight foliar necrosis was observed on older leaves, which may have been due to B toxicity from the compost.

In a study with marigold (*Tagetes erecta* L.), biosolids compost addition to a soilless growing medium provided all the required trace elements for plant growth and several times more Fe than that supplied by a fritted trace element fertilizer (Chaney et al., 1980). The compost used in this study was high in Fe due to Fe compounds used to remove P during wastewater treatment. They stated that biosolids composts are likely to contain adequate Fe for plant growth, but should not cause Fe toxicity due to their microelement balance and neutral pH.

Broschat (1991) stated that biosolids and manure composts would not be recommended for use in media for ornamental plants susceptible to Mn deficiency, such as queen palm (*Syagrus romanzoffiana* [Chamisso] Glassman]), since these composts tend to bind Mn. Further, he noted that Mn binding appears to be much less of a problem with MSW composts, while yard trash composts usually have such low levels of total Fe and Mn that supplemental micronutrient fertilization is needed when these composts are used in potting media.

Leaves of tomato plants grown in soil amended with MSW compost showed decreased Mn and Cu contents compared to leaves from plants grown in unamended soil in a study by Stilwell (1993). These results were attributed to reduced availability of Mn and Cu in the compost-amended soil due to increases in pH and organic matter content.

Villar et al. (1993) (see previous section) also studied effects of four MSW composts as potential micronutrient fertilizers. Useful quantities of Fe, Mn, Cu, and Zn all would be quickly available to plants shortly after these composts were applied to soils. The Fe was mainly in non-available forms and thus was unlikely to be phytotoxic. Correct dosage of composts would be essential to avoid potential phytotoxicity from Mn, Cu, and Zn because these elements were mainly in available forms.

Certain composts may cause micronutrient deficiencies even after they have aged for several weeks. Baca et al. (1995) grew sunflower (*Helianthus annuus* L.) in 1:1 v/v mixtures of a Fluvisol and three kinds of compost made from sugarcane (*Saccharum officinarum* L.) bagasse plus other agricultural wastes. A progressive immobilization of N, P, and micronutrients as composts aged resulted in reduced sunflower yield and severe nutrient deficiencies of Fe, Zn, and B, even in composts matured for over 60 days.

Brown et al. (1997) reported on methods to correct limed-biosolids-induced Mn deficiency under field conditions. Soil amendment with 200 $kg \cdot ha^{-1}$ $MgSO_4$ prevented Mn deficiency in biosolids-treated plots in this study. Soil acidulation with S was suggested as an alternative to prevent Mn deficiency when lime-stabilized biosolids are used.

Incorporating an MSW compost at 48 $Mg \cdot ha^{-1}$ or an MSW-biosolids co-compost at 24 $Mg \cdot ha^{-1}$ into a calcareous limestone soil increased concentrations of soil-extractable metals, but caused no significant changes in tomato and squash (*Cucurbita pepo* L.) fruit concentrations of Cu and Zn compared to an unamended control (Ozores-Hampton et al., 1997).

Wilkinson (1997) studied the potential use of poultry litter that included recycled paper treated with boric acid as a fertilizer for tall fescue (*Festuca arundinacea* Schreb.). Several composts based on chicken manure were tested. B in the composts was as available to tall fescue as was B from boric acid, but B toxicity was not observed with normal agronomic rates of compost application based on the N requirements of the tall fescue.

In an experiment on potatoes (*Solanum tuberosum* L.) in 4-L pots, increasing rates of MSW compost amendment (0, 10, 20, or 30 g per pot) resulted in increased shoot and tuber levels of Fe, Mn, Cu, and Zn (Karam et al., 1998). Shoots contained phytotoxic levels (31.7 $mg \cdot kg^{-1}$) of Cu at the 30 g per pot rate in one of two studies.

Sainz et al. (1998) (see previous section) also studied effects of soil amendment with MSW-derived compost or vermicompost on micronutrient concentrations in red clover and cucumber. Vermicompost amendment increased shoot Mn, Cu, and Zn concentrations.

These studies show that complex interactions among plant nutrient elements may occur in composted growing media, especially for uptake of micronutrients such as Fe, Mn, and Zn. For example, with Fe and Mn, a large supply of one element alone may worsen the deficiency of the other element (Olsen, 1972). Enrichment of compost amendments with microelements may need to be considered if an imbalance is likely. As one example, incorporation of Fe into composted spruce (*Picea* sp.) bark alleviated Mn toxicity to petunia brought about by high Mn levels in the compost (Raja Harun and McKenna, 1998). The additional Fe reduced the uptake of Mn and Zn by the petunia plants.

Another consideration is that compost feedstock composition may change over time, so that results from past decades may not apply to the current situation. Chaney and Ryan (1993) noted that MSW from the 1970s often contained phytotoxic levels of B (see Purves and MacKenzie, 1974), but most current MSW composts no longer contain high B. The potential for further changes in B levels in MSW composts was demonstrated by Cook et al. (1994).

V. SUMMARY

Relatively little attention has been paid to the potential plant nutrition benefits of compost application for elements other than N and P. Interest will surely grow in the future as more organic wastes are composted and as inorganic sources of plant nutrient elements become scarcer and more expensive. Complicating the situation is the great variation in plant nutrient element concentrations which are possible both between and within feedstocks, as well as variations in elemental availability. Other considerations in using composts as fertilizers have been discussed elsewhere in this book. Possible trends for the future may include standardization of compost products where possible (for example, biosolids compost produced by a particular wastewater treatment plant or system), faster and more reliable testing of finished compost products so that producers have better knowledge of the nutrients being supplied, and "tailor-made" composts designed for specific applications, as proposed in chapter 16 in this book.

REFERENCES

Baca, M.T., I.C. Delgado, M. De Nobili, E. Esteban, and A.J. Sanchez-Raya. 1995. Influence of compost maturity on nutrient status of sunflower. *Communications in Soil Science and Plant Analysis* 26(1–2):169–181.

Berrow, M.L. and G.A. Reaves. 1984. Background levels of trace elements in soils, p. 333–340. In: Proceedings of the First International Conference on Environmental Contamination. CEP Consultants, Edinburgh, Scotland.

Bittenbender, H.C., N.V. Hue, K. Fleming, and H. Brown. 1998. Sustainability of organic fertilization of macadamia with macadamia husk-manure compost. *Communications in Soil Science and Plant Analysis* 29(3–4):409–419.

Broschat, T.K. 1991. Manganese binding by municipal waste composts used as potting media. *Journal of Environmental Horticulture* 9:97–100.

Brown, S., J.S. Angle, and R.L. Chaney. 1997. Correction of limed-biosolid induced manganese deficiency on a long-term field experiment. *Journal of Environmental Quality* 26:1375–1384.

Buchanan, M.A. and S.R. Gliessman. 1990. The influence of conventional and compost fertilization on phosphorus use efficiency by broccoli in a phosphorus deficient soil. *American Journal of Alternative Agriculture* 5:38–46.

Bunting, A.H. 1963. Experiments on organic manures, 1942–49. *Journal of Agricultural Science (Cambridge)* 60:121–140.

Cabrera, F., J.M. Murillo, R. Lopez, and J.M. Hernandez. 1991. Fate of phosphorus added with urban compost to a calcareous soil. *Journal of Environmental Science and Health, Part B: Pesticides, Food Contaminants, and Agricultural Wastes* 26:83–97.

Chaney, R.L. and J.A. Ryan. 1993. Heavy metals and toxic organic pollutants in MSW-composts: research results on phytoavailability, bioavailability, etc., p. 451–506. In: H.A.J. Hoitink and H.M. Keener (eds.). *Science and Engineering of Composting: Design, Environmental, Microbiological and Utilization Aspects*. Renaissance Publications, Worthington, Ohio.

Chaney, R.L., J.B. Munns, and H.M. Cathey. 1980. Effectiveness of digested sewage sludge compost in supplying nutrients for soilless potting media. *Journal of the American Society for Horticultural Science* 105:485–492.

Chung, R.S. and S.H. Wu. 1997. Effect of corncob compost on plant growth in an acid red soil. *Communications in Soil Science and Plant Analysis* 28(9–10):673–683.

Cook, B.D., T.R. Halbach, C.J. Rosen, and J.F. Moncrief. 1994. Effect of a waste stream component on the agronomic properties of municipal solid waste compost. *Compost Science and Utilization* 2(2):75–87.

Coutinho, J., M. Arrobas, and O. Rodrigues. 1997. Effect of composted sewage sludge amendment on soil nitrogen and phosphorus availability. *Communications in Soil Science and Plant Analysis* 28(19–20):1845–1857.

De Haan, S. 1981. Results of municipal compost research over more than fifty years at the institute for soil fertility at Haren Groningen, The Netherlands. *Netherlands Journal of Agricultural Science* 29:49–61.

Dick, W.A. and E.L. McCoy. 1993. Enhancing soil fertility by addition of compost, p. 622–644. In: H.A.J. Hoitink and H.M. Keener (eds.). *Science and Engineering of Composting: Design, Environmental, Microbiological and Utilization Aspects*. Renaissance Publications, Worthington, Ohio.

Giusquiani, P.L., M. Pagliai, G. Gigliotti, D. Businelli, and A. Benetti. 1995. Urban waste compost: effects on physical, chemical, and biochemical soil properties. *Journal of Environmental Quality* 24:175–182.

Gogue, G.J. and K.C. Sanderson. 1975. Municipal compost as a medium amendment for chrysanthemum culture. *Journal of the American Society for Horticultural Science* 100:213–216.

Gonzalez-Vila, F.J., C. Saiz-Jimenez, and F. Martin. 1982. Identification of free organic chemicals found in composted municipal refuse. *Journal of Environmental Quality* 11:251–254.

He, X., T.J. Logan, and S.J. Traina. 1995. Physical and chemical characteristics of selected U.S. municipal solid waste composts. *Journal of Environmental Quality* 24:543–552.

He, Z.L., A.K. Alva, P. Yan, Y.C. Li, D.V. Calvert, P.J. Stoffella, and D.J. Banks. 2000. Nutrient availability and changes in microbial biomass in composts during field incubation. *Compost Science and Utilization* In press.

Hue, N.V., H. Ikawa, and J.A. Silva. 1994. Increasing plant-available phosphorus in an ultisol with a yard-waste compost. *Communications in Soil Science and Plant Analysis* 25(19–20):3291–3303.

James, B.R. and S.G. Aschmann. 1992. Soluble phosphorus in a forest soil Ap horizon amended with municipal wastewater sludge or compost. *Communications in Soil Science and Plant Analysis* 23(7–8):861–875.

Karam, N.S., K.I. Ereifej, R.A. Shibli, H. AbuKudais, A. Alkofahi, and Y. Malkawi. 1998. Metal concentrations, growth, and yield of potato produced from *in vitro* plantlets or microtubers and grown in municipal solid-waste-amended substrates. *Journal of Plant Nutrition* 21:725–739.

Karen, D.L., R.J. Wright, and W.D. Kemper (eds.). 1995. *Agricultural Utilization of Urban and Industrial By-Products*. American Society of Agronomy, Crop Science Society of America, and Soil Science Society of America, Madison, Wisconsin.

Kuo, S., R.L. Hummel, E.J. Jellum, and D. Winters. 1999. Solubility and leachability of fishwaste compost phosphorus in soilless growing media. *Journal of Environmental Quality* 28:164–169.

Li, Y.C., P.J. Stoffella, and H.H. Bryan. 2000. Management of organic amendments in vegetable crop production systems in Florida. *Soil and Crop Science Society of Florida Proceedings 59*: in press.

Marschner, H. 1995. *Mineral Nutrition of Higher Plants*. 2nd edition. Academic Press, New York.

Martens, D.C. and D.T. Westermann. 1991. Fertilizer applications for correcting micronutrient deficiencies, p. 549–592. In: J.J. Morvedt, F.R. Cox, L.M. Shuman, and R.M. Welch (eds.). *Micronutrients in Agriculture*. 2nd edition. Soil Science Society of America, Madison, Wisconsin.

McCoy, J.L., L.J. Sikora, and R.R. Weil. 1986. Plant availability of phosphorus in sewage sludge compost. *Journal of Environmental Quality* 15:403–409.

Olsen, S.R. 1972. Micronutrient interactions, p. 243–264. In: J.J. Mortvedt, P.M. Giordano, and W.L. Lindsay (eds.). *Micronutrients in Agriculture*. Soil Science Society of America, Madison, Wisconsin.

Ozores-Hampton, M., E. Hanlon, H. Bryan, and B. Schaffer. 1997. Cadmium, copper, lead, nickel and zinc concentrations in tomato and squash grown in MSW compost-amended calcareous soils. *Compost Science and Utilization* 5(4):40–45.

Perucci, P. 1990. Effect of the addition of municipal solid waste compost on microbial biomass and enzyme activities in soils. *Biology and Fertility of Soils* 10:221–226.

Petruzzelli, G., L. Lubrano, and G. Guidi. 1985. Heavy metal extractability. *BioCycle* 26(8):46–48.

Purves, D. and E.J. MacKenzie. 1974. Phytotoxicity due to boron in municipal compost. *Plant and Soil* 40:231–235.

Raja Harun, R.M. and C. McKenna. 1998. Alleviation of interveinal yellowing in leaves of petunia "Rose Frost" in composted spruce bark. *Acta Horticulturae* 469:235–244.

Romheld, V. and H. Marschner. 1991. Function of micronutrients in plants, p. 297–328. In: J.J. Mortvedt, F.R. Cox, L.M. Shuman, and R.M. Welch (eds.). *Micronutrients in Agriculture*. 2nd edition. Soil Science Society of America, Madison, Wisconsin.

Sainz, M.J., M.T. Taboada-Castro, and A. Vilarino. 1998. Growth, mineral nutrition and mycorrhizal colonization of red clover and cucumber plants grown in a soil amended with composted urban wastes. *Plant and Soil* 205:85–92.

Smith, J.L. and E.A. Paul. 1991. The significance of soil microbial biomass estimations, p. 359–396. In: J.J. Bollag and G. Stotzky (eds.). *Soil Biochemistry*. Volume 6. Marcel Dekker, New York.

Smith, R.S. 1991. Effects of sewage sludge application on soil microbial processes and soil fertility, p. 191–212. In: B.A. Stewart (ed.). *Advances in Soil Science*. Volume 16. Springer-Verlag, New York.

Sommers, L.E. and A.L. Sutton. 1980. Use of waste materials as sources of phosphorus, p. 515–544. In: F.E. Khasawneh, E.C. Sample, and E.J. Kamprath (eds.). *The Role of Phosphorus in Agriculture*. American Society of Agronomy, Crop Science Society of America, and Soil Science Society of America, Madison, Wisconsin.

Stilwell, D.E. 1993. Evaluating the suitability of MSW compost as a soil amendment in field grown tomatoes: Part B: Elemental analysis. *Compost Science and Utilization* 1(3):66–72.

Stoffella, P.J., D.V. Calvert, Y.C. Li, and D.H. Hubbell. 1996. Municipal soil waste (MSW) influence on citrus leaf nutrition, yield, and fruit quality. *Proceedings of the InterAmerican Society of Tropical Horticulture* 40:157–160.

Stoffella, P.J., Y.C. Li, N.E. Roe, M. Ozores-Hampton, and D.A. Graetz. 1997. Utilization of organic waste composts in vegetable crop production systems, p. 253–269. In: R.A. Morris (ed.). *Managing Soil Fertility for Intensive Vegetable Production Systems in Asia*. Asian Vegetable Research and Development Center, Tainan, Taiwan.

Tisdale, S.L. and W.L. Nelson. 1996. *Soil Fertility and Fertilizers*. Macmillan, New York.

Villar, M.C., M.C. Beloso, M.J. Acea, A. Cabaneiro, S.J. Gonzalez-Prieto, M. Carballas, M. Diaz-Ravina, and T. Carballas. 1993. Physical and chemical characterization of four composted urban refuses. *Bioresource Technology* 45:105–113.

Vogtmann, H., K. Fricke, and T. Turk. 1993a. Quality, physical characteristics, nutrient content, heavy metals, and organic chemicals in biogeneic waste compost. *Compost Science and Utilization* 1(1):69–87.

Vogtmann, H., K. Matthies, B. Kehres, and A. Meier-Ploeger. 1993b. Enhanced food quality induced by compost application, p. 645–667. In: H.A.J. Hoitink and H.M. Keener (eds.). *Science and Engineering of Composting: Design, Environmental, Microbiological and Utilization Aspects*. Renaissance Publications, Worthington, Ohio.

Warren, S.L. and W.C. Fonteno. 1993. Changes in physical and chemical properties of a loamy sand soil when amended with composted poultry litter. *Journal of Environmental Horticulture* 11:186–190.

Wen, G., T.E. Bates, R.P. Voroney, J.P. Winter, and M.P. Schellenbert. 1997. Comparison of phosphorus availability with application of sewage sludge, sludge compost, and manure compost. *Communications in Soil Science and Plant Analysis* 28(17–18):1481–1497.

Wilkinson, S.R. 1997. Response of Kentucky-31 tall fescue to broiler litter and composts made from broiler litter. *Communications in Soil Science and Plant Analysis* 28(3–5):281–299.

Wong, J.W.C., K.K. Ma, K.M. Fang, and C. Cheung. 1999. Utilization of a manure compost for organic farming in Hong Kong. *Bioresource Technology* 67:43–46.

Yang, X., Z.L. He, and Z.Q. Ye. 2000. Improving human nutrition through agriculture and plant nutrition. In: *Proceedings of the International Conference on Improving Human Nutrition through Agriculture*. Institute of Food Production Research, Washington, D.C. In press.
Yang, Y. 1996. Composts, p. 49–50. In: X. Sun (ed.). *Encyclopedia of China Agriculture*. Agricultural Science Press, Beijing, China.

Section IV

Potential Hazards, Precautions, and Regulations of Compost Production and Utilization

CHAPTER 16

Heavy Metal Aspects of Compost Use

Rufus L. Chaney, James A. Ryan, Urszula Kukier, Sally L. Brown, Grzegorz Siebielec, Minnie Malik, and J. Scott Angle

CONTENTS

1-56670-460-X/01/$0.00+$.50

I. INTRODUCTION

Composts prepared from municipal solid waste (MSW), biosolids (municipal sewage sludge), food processing wastes, manures, yard debris, and agricultural byproducts and residues are increasingly available for agricultural use. Utilization of composts as fertilizers and soil conditioners provides benefits from nutrients, from organic matter, from biodegradation of organic matter, and from organisms in the composts. Remarkable benefits have been identified in new approaches for control of plant diseases by use of composts in media or in field plantings, and in revegetation of disturbed soils and mine wastes. Production of composts provides an important cost saving to cities, industries, and agricultural users, and allows recycling for beneficial use of more of society's discards. Although some compost products continue to be poorly manufactured, scientists have discovered improved manufacturing methods, methods to monitor or evaluate composts inexpensively, and criteria for quality control of compost products. Because production of composts will offer large cost savings to both urban and agricultural areas, such composts will be available at relatively low cost to horticultural industries for use as fertilizers, soil conditioners, and when prepared properly, potting media components.

Although many benefits are possible from use of composts, these products must be safe for sustainable agriculture for their use to be permitted by governments. These products also must reliably supply nutrient and organic matter benefits to become competitive in the marketplace. The potential presence of pathogenic organisms, heavy metals/trace elements, potentially toxic synthetic organic compounds (compounds that are not normally biosynthesized are referred to as "xenobiotic" compounds), and possible element imbalance in composts have caused concern to some potential compost users. Some believe that because the concentration of zinc (Zn) or copper (Cu) in composts is higher than found in background soils, these materials must not be utilized on soils. However, practicing horticulturists and researchers have used high-quality organic matter/compost products for decades without adverse effects (Andersson, 1983; Chaney and Ryan, 1993; de Haan, 1981; Mays and Giordano, 1989; Sanderson, 1980; Woodbury, 1992). Boron (B) phytotoxicity was observed when high rates of MSW-compost were used in media in the 1970s, but changes in glue formulations removed this possible adverse effect of MSW-composts (Chaney and Ryan, 1993; Sanderson, 1980). How could such high benefits be observed so often if the metals and other constituents were so dangerous? In brief, the logical flaw, in presuming that metals in composts must cause adverse effects in the future, is the focus on total concentrations, when phytoavailability of microelements is well known to vary as a function of source. Similarly, biosolids, pet excreta in yard debris, and manures contain pathogenic organisms, but proper composting generates products that comprise no human or plant pathogen risk. It is not biosolids and composts that should cause concern, but whether these products meet enforceable standards of acceptable quality composting technologies, maturity of composts, and composition of composts that are to enter the marketplace of organic amendments and media components.

In 1970, when modern interest in the safety of utilizing organic byproducts and composts on cropland began its rapid increase, biosolids were often highly contaminated with metals and xenobiotics. Few MSW composts were available for use, and only a few of these were mature composts ready for use in crop production. Substantial efforts were undertaken in many countries to conduct research to characterize the potential for adverse effects from use of composts and biosolids so that regulations could be developed to protect soil fertility and food-chain safety.

In the U.S., these efforts culminated in the development of the U.S. Environmental Protection Agency (U.S. EPA) Clean Water Act Section 503 Rule on land application of biosolids (U.S. EPA, 1989a, 1993), hereafter called the 503 Rule. Such U.S. rules are "proposed" for public comment, to allow errors and omissions to be identified, and other data to be provided to the U.S. EPA to improve the scientific basis of the rule. Errors were found and questions were raised after the U.S. EPA prepared the first Proposed Draft 503 Rule (U.S. EPA, 1989b). Therefore, the scientific community thoroughly evaluated data from many experiments to develop improved risk assessment models to protect soil fertility and food-chain safety during use of composts and biosolids. Development of the final corrected 503 Rule is discussed later.

A. Industrial Pretreatment Improves Quality of Biosolids and Composts

Fortunately, pretreatment of industrial wastewaters has allowed most municipalities to produce biosolids and composts with low concentrations of metals and synthetic organic compounds (Table 16.1), reducing the potential for adverse effects. The median concentrations of metals in biosolids have fallen substantially over the last 25 years. When pretreatment of industrial sources is complete, biosolids still contain significant levels of Zn, Cu, and some other elements because such elements are in foods (hence in human wastes) and food wastes, or are leached from the pipes which carry water to and in our homes. Interestingly, the need to keep lead (Pb) in drinking water at low concentration at the home tap to protect children is requiring many municipalities to treat their water to reduce the corrosion of water pipes. This improvement in drinking water treatment to reduce risks from Pb in water transmission systems has reduced Zn, Cu, Pb, and cadmium (Cd) levels in biosolids formed during treatment of domestic wastewaters.

These management options have made it possible to attain biosolids and composts with reduced concentrations of metals and xenobiotics. Some have argued that only products which are as low in contaminants as possible should be allowed to be used in agriculture. Although common sense dictates that avoidable metals should be avoided, costs involved with avoidance of metals in biosolids or composts become an issue. At some point, the increase in benefits from lower concentrations are less than the costs associated with reduction. It is especially difficult to define concentration limits in composts for the elements that are naturally present in all soils and foods; it is even more difficult to do so for those metals that are micronutrients for

Table 16.1 Range of Contaminant Concentrations Reported for Biosolids Before Pretreatment Enforcement

	Historic Reported Range			MSW-Compost (Median)					
Element	Min.	Max.	1990 Survey Median	Mixed	Separated	"Green" MSWC	German Richtwert	NOAEL Biosolids	Attainable Biosolids
Zn	101	49,000	725	563	353	255	400	2800	<1500
Cd	<1	3,410	7	3.5	1.9	0.5	2.0	21	<10
Cd/Zn, %	0.01	110	0.8	0.61	0.47	—	—	1.5	<1.0
Cu	84	17,000	463	194	72	40	100	1500	<500
Ni	2	8,330	29	29	24	17	50	290	<100
Pb	13	26,000	106	261	152	86	150	300	<100
As	—	—		3.7	—	—	—	54	<25
Hg	0.6	110	2	1.5	1.3	0.17	1.5	17[z]	<5
Cr	10	99,000	40	43	34	—	100	—	<500
PCBs[y]	<0.1	>1,000	0.02	—	—	0.14	—	2.2	<0.5

Note: 1990 median concentrations from National Sewage Sludge Survey (U.S. EPA, 1990), typical "mixed" or "separated" MSW-composts (MSWC) (Epstein et al., 1992), "Green" MSW-composts (Fricke et al., 1989), and the German MSW-Compost Richtwert (Guide Value), compared to the "No Observed Adverse Effect Level" (NOAEL) biosolids and "Attainable Quality" biosolids products with effective pretreatment and control of water corrosivity. All values are mg per kg dry weight.

[z] Appropriate limits for all biosolids/compost uses except mushroom production.

[y] PCBs = polychlorinated biphenyls.

plants and animals, but can become phytotoxic or zootoxic when present in excess under soil management conditions that maximize phytoavailability. These questions can be addressed through field experiments to learn what the response to different quality composts and biosolids may be when these products are used in a range of soils for production of a wide assortment of crops, considering long-term high cumulative applications. Data are usually available as needed for such quantitative risk assessment for soil or compost trace element limits.

For MSW composts, "pretreatment" to separate compostable and noncompostable materials can significantly reduce the levels of metals and xenobiotics in the compost (see Table 16.1) (Cook and Beyea, 1998; Epstein et al., 1992; Richards and Woodbury, 1992). Here pretreatment is more complex because there are at least two major ways to obtain this pretreatment. One is requiring all citizens to separate home solid wastes into different fractions to allow recycling of glass, plastics, clean paper and metals; composting of biodegradable organic materials; and safely disposing of the noncompostable materials. Depending on one's view, this is either having government encourage the lower cost better solution, or the imposition of intrusive government into one's kitchen. The alternative to home separation is development of machines to separate the recyclable materials from those that are best used in producing composts. Charging a higher rate for MSW that has not been preseparated in the home might increase compliance, but with human fallibility, such home separation cannot be perfect. Separation at the MSW handling facility can be done before or after shredding, and the completeness of separation of recyclable and compostable materials is variable when different technologies are used at different facilities. None of these separation technologies are as effective as home separation by responsible citizens who believe that such separation is part of better environmental protection (see Cook and Beyea, 1998).

High-quality composts should contain low levels of glass, plastics, and pieces of metals from the noncompostable materials in the initial MSW. Glass is silica similar to sand; but some glass had Pb, Cd, and other metals added for color or for function (x-ray shielding in TV tubes). Cans are commonly welded steel today rather than soldered with Pb, but most are galvanized with Zn. The higher value for these metals is recycling rather than shredding until they are inseparable from compost organic particles. Other materials commonly placed in MSW at homes include pesticide wastes, batteries, and such household hazardous wastes as "pressure-treated lumber" containing high levels of chromium (Cr), Cu, and arsenic (As) or pentachlorophenol (PCP) with contaminant dioxins. Although very effective demonstration programs for home separation have been tested in several cities by the Audubon Society and their cooperators (Cook and Beyea, 1998), few cities have provided the educational resources needed to achieve such effective home separation of recyclable and hazardous constituents from compostable materials in household solid wastes. Inclusion of treated lumber wastes with yard debris causes even the normally uncontaminated yard debris to be enriched in metals and xenobiotics. Shredding companies need to aggressively exclude chromated copper arsenate (CCA)- and PCP-treated wood wastes from mulch or composting feedstocks.

B. Nutrient Supply from Biosolids and Composts

Organic soil amendments such as biosolids and composts are used as fertilizers and soil conditioners (see guidance for beneficial use in Hornick et al., 1983; Wright, 1999; and other chapters in this book). Different products have different rates of nitrogen (N) mineralization, different phosphate availability, etc., related to their composition and method of manufacture. Most users want to apply the amendments at N or phosphorus (P) fertilizer rates, but it has been difficult to measure the N-mineralization rate of different products in different soils. Composting of biosolids and manures with cellulosic materials lowers the mineralization rate due to a change in carbon to nitrogen (C:N) ratio and changes in chemical forms of N present. Improved methods of estimating N-mineralization rate from a wide range of organic amendments have been reported by Gilmour (1998) and Gilmour and Skinner (1999).

Although N and P supply are always a consideration during utilization of organic amendments such as biosolids, composts, and manures, biosolids-containing products can also serve as potassium (K), sulfur (S), calcium (Ca), magnesium (Mg), and microelement fertilizers except for B (Chaney et al., 1980). The use of biosolids-compost potting media for horticultural crops was evaluated by Sterrett et al. (1982, 1983). They observed effective transplant production with such media, and found no change in Cd or Pb in edible crop tissues at maturity. Such media can provide savings for producers without threatening food safety because composts of high-quality were used to prepare the media. In later tests, they showed that adding composted biosolids to metal-rich urban garden soils could reduce concentrations of Pb and Cd in lettuce (*Lactuca sativa* L.) (Sterrett et al., 1996), illustrating soil metal remediation principles later characterized by Brown et al. (1998b).

C. Defining High-Quality Biosolids and Composts for Sustainable Use

Because biosolids and composts cannot be as low in metals as background soils, some method is needed to evaluate whether high-quality organic resources are low enough in metals and xenobiotics that they may be safely used in agriculture. "Quantitative risk assessment" is the method developed to make these determinations, a method using scientific data from field research studies with amendments, soils, crops, and animals to determine whether harm is possible or likely when different quality organic resources are used on cropland. Actually, many of the most limiting pathways for risk assessment are in horticultural use. For example, home gardens allow individuals to be exposed to elements absorbed by garden crops from the amended soils, or allow children to ingest amended soils or even commercial products purchased for use as a mulch or soil conditioner for edible and ornamental plants.

This chapter cannot be comprehensive, but it does provide a summary and references to the research and risk assessment conducted on land application of biosolids and composts. Although MSW and yard debris composts are generally less contaminated with metals and xenobiotics than biosolids, they also contain lower levels of adsorbent iron (Fe) and manganese (Mn) oxides. Less research has been conducted with these composts than with biosolids. Further, because composts have lower concentrations of metals than biosolids, the data from study of biosolids have been more important in developing the limits for regulations for several organic amendments. Because contaminant applications in compost remain low, adverse effects of compost use are seldom, if ever, observed. Application of the 503 limits developed for biosolids are often used for other organic amendments; but if the adsorbents in the compost are lower than in biosolids, protection against metal uptake may also be lower. Further research is needed to determine whether the limits developed for biosolids-applied metals should be applied to other organic amendments.

II. RISK ASSESSMENT METHODOLOGY FOR CONTAMINANTS IN BENEFICIALLY USED BIOSOLIDS AND COMPOSTS

A. Pathway Risk Assessment Used for the U.S. EPA Section 503 Rule

This section summarizes risk assessment for trace elements and current regulations needed to keep composts and biosolids utilization in agriculture an environmentally acceptable practice, and future directions in markets and research for these products. In the U.S., work has gone on since about 1978 to develop regulations for biosolids utilization. During 1989, a proposed rule was published for public comments (U.S. EPA, 1989b). Unfortunately, the U.S. EPA did not seek peer review of the science in the proposed rule before publication for public comment. But the proposed rule did not properly use the available science and data sets, and consequently, it was discarded after public review comments showed these flaws (U.S. EPA, 1989b). U.S. EPA staff and contractors had decided to use only data that showed adverse effects to establish limits. For example, if they could find no phytotoxicity in the field from high-quality composts and biosolids, then they looked at greenhouse data. If metals in biosolids did not injure plants, they looked at data for added metal salts. They made numerous errors in simply collecting data from published research results (see Page et al., 1989). Because inappropriate data were used to estimate limits needed to protect the environment from contaminants in biosolids, EPA calculated such low allowed cumulative application rates that publically owned treatment works (POTWs) would have been forced to use landfills or incinerators rather than using biosolids beneficially on farmland or for remediation of disturbed land.

These errors were fundamental scientific mistakes that could be corrected. EPA subsequently selected an expert workgroup to identify and correct the mistakes in the 1989 proposed rule. This team worked part time for 2 years to carefully check the algorithms used to calculate limits, and to review the data collected by EPA and add other relevant data in the literature, and correct these algorithms and data. The expert workgroup examined data from experiments from around the world, and a number of very important principles were identified (discussed later).

The Clean Water Act Part 503 Rule evaluates 14 Pathways (Table 16.2) by which an applied contaminant in biosolids/compost could cause risk to highly exposed individuals (HEIs). The different pathways protect the HEIs from adverse effects of contaminants that might be transferred to humans via the general agricultural food chain, or by growing a substantial fraction of their annual food supply on home gardens that had received a huge maximum cumulative biosolids application, 1000 $Mg \cdot ha^{-1}$ on a dry weight basis. The rule protects livestock, and humans who consumed meats and organ meats from livestock that were maximally exposed to biosolids. Pathway 8 protects plants by calculating limits to prevent phytotoxicity. Pathway 9 protects soil organisms (e.g., earthworms, bacteria, fungi), and Pathway 10 protects predators of soil organisms. The remaining pathways involved surface and groundwater (important for nitrate), volatilization to air, and inhaled suspended dust, which are generally not limiting for trace elements.

Each pathway was constructed to protect organisms identified as the HEIs — humans, plants, or animals — at the 95th to 98th percentile of exposure. These are much more protective than it seems at first glance. Rather than the 95th to 98th percentile of the whole U.S. population, it is this level of exposure among the subset of the population which were actually significantly exposed to contaminants from biosolids. Only a small part of the U.S. population lives on a farm or eats (for their lifetime) crops grown on a garden that has high regular compost or biosolids application. Even in these cases, transfers were linearly extrapolated from applications in experiments to the assumed 1000 $Mg \cdot ha^{-1}$ cumulative loading rate, considerably over-estimating transfers.

Although potentially toxic organic compounds were deleted from the 503 Rule during the correction of the calculations, a thorough risk assessment for polychlorinated biphenyls (PCBs) applied via biosolids or compost was prepared by Chaney et al. (1996). For organics, the most limiting pathway is the ingestion of soil or biosolids by grazing livestock, because the livestock bioconcentrate PCBs from their diets, and there is very little transfer of PCBs from soil to plants (no uptake, only

Table 16.2 Pathways for Risk Assessment for Potential Transfer of Biosolids-Applied Trace Contaminants to Humans, Livestock, or the Environment, and the Highly Exposed Individuals to be Protected by a Regulation Board on the Pathway Analysis

	Pathway	Highly Exposed Individuals
1	Biosolids→Soil→Plant→Human	Individuals with 2.5% of all food produced on amended soils with 1000 $Mg{\cdot}ha^{-1}$.
2	Biosolids→Soil→Plant→Human	Home gardeners with 1000 $Mg{\cdot}ha^{-1}$; 60% garden foods for lifetime.
3	Biosolids→Human	Ingested biosolids product; 200 mg per day.
4	Biosolids→Soil→Plant→Animal→Human	Farms; 45% of "homegrown" meat; 1000 $Mg{\cdot}ha^{-1}$; lifetime.
5	Biosolids→Soil→Animal→Human	Farms; 45% of "homegrown" meat; lifetime.
6	Biosolids→Soil→Plant→Animal	Livestock feeds; 100% grown on amended land with 1000 $Mg{\cdot}ha^{-1}$.
7	Biosolids→Soil→Animal	Grazing livestock; 1.5% biosolids in diet.
8	Biosolids→Soil→Plant	"Crops"; strongly acidic soil with 1000 $Mg{\cdot}ha^{-1}$.
9	Biosolids→Soil→Soil Biota	Earthworms, microbes, in soil with 1000 $Mg{\cdot}ha^{-1}$.
10	Biosolids→Soil→Soil Biota→Predator	Shrews; 33% earthworms diet, living on site with 1000 $Mg{\cdot}ha^{-1}$.
11	Biosolids→Soil→Airborne Dust→Human	Tractor operator; dusty tillage; 1000 $Mg{\cdot}ha^{-1}$.
12	Biosolids→Soil→Surface Water→Human	Subsistence fishers where erosion moved products from fields to lakes.
13	Biosolids→Soil→Air→Human	Farm households.
14	Biosolids→Soil→Groundwater→Human	Well water on farms; 100% of lifetime supply.

Note: Each Pathway presumes 1000 Mg dry biosolids per ha and/or annual application of biosolids as N fertilizer.

From Chaney and Ryan, 1994; U.S. EPA, 1989a, 1992, 1993. With permission.

volatile transport). Production and use of PCBs has been prohibited for several decades, and PCB concentrations in biosolids/composts are now very low (Table 16.1). Even the highly exposed farm family consuming meat and dairy products produced on fields with biosolids or compost amendments are highly protected from the traces present today.

B. U.S. Limits on Contaminants in Biosolids and Composts

Table 16.3 shows the limits established for regulated potentially toxic elements by the final EPA 503 Rule, and the pathway that was most limiting (Column 2) for each metal using the final EPA calculations (U.S. EPA, 1993). The Ceiling Limit (Column 3) is a maximum concentration of metals in biosolids which was established because (1) if only cumulative loading limits were imposed, pretreatment effectiveness might be threatened; and (2) if more highly contaminated biosolids were land applied, because biosolids metals are more phytoavailable the higher the total metal concentration, phytotoxicity might occur at lower cumulative metals applications than estimated using data obtained from studies with better quality biosolids. If no ceiling limit were required, highly contaminated biosolids in which the metals have higher phytoavailability could be used, which could cause the cumulative loading

Table 16.3 Comparison of the Ceiling and Cumulative Limits for Biosolids Contaminents Under the Final CWA-503 Limits vs. No Observed Adverse Effect Level Limits

	Limits Under the EPA 503 Rule			Limits Under the NOAEL Approach			
Element	Limiting Pathway	Ceiling 99th (mg·kg⁻¹)	Cumulative (kg·ha⁻¹) = APL (mg·kg⁻¹)	Limiting Pathway	Limit (mg·kg⁻¹)	Percentile of NOAEL	Attainable Quality
1	2	3	4	5	6	7	8
As	3	75	41	3	54	98	<25
Cd	3	85	39	2	21	91	<5–10
Cd/Zn	—	—	—	2	1.5%	87	1.0%
Pb	3	840	300	3	300	90	<100
Hg[z]	3	57	17	12	17	93	<5
Mo[y]	6	75	35	6	54	98	<50
Se	6	100	36	6	28	98	<15
Cr	8	3000	1300	—[x]	—	—	—
Cu	8	4300	1500	8	1500	89	<500–750
Ni	8	420	420	8	290	98	<100
Zn	8	7500	2800	8	2800	91	<1500–2000

Note: Ceiling (Column 3, based on the lower of the risk-based Pathway Limit or the 99th percentile of biosolids contaminant concentrations from the 1990 National Sewage Sludge Survey [U.S. EPA, 1990]). Cumulative (Column 4, Pathway Risk Assessment based) limits for biosolids contaminants under the Final CWA-503 Limits (Feb. 19, 1993). No Observed Adverse Effect Level (NOAEL) limits (column 6) estimated by Chaney (1993a) and Chaney and Ryan (1994). Columns 2 and 5 refer to Pathways listed in Table 2. Biosolids which meet the Alternative Pollutant Limit (APL) (Column 4) or NOAEL (Column 6) quality limits may be applied at up to 1000 $Mg \cdot ha^{-1}$ before reaching the Cumulative Limit, and still protect Highly Exposed Individuals according to the Technical Support Document for the 503 Rule. For the APL and NOAEL biosolids, adsorption of contaminants by biosolids constituents lowers the potential for risk sufficiently to allow general marketing and continuing use in sustainable agriculture. The percentile of the NOAEL (Chaney,1993a) concentration in the National Sewage Sludge Survey (US-EPA, 1990) is shown for comparison (Column 7). Column 8 shows the metal limits that we believe are "attainable" by publically owned treatment works (POTWs) that enforce industrial pretreatment standards; the corrosivity of drinking water may need to be controlled in some cities to achieve NOAEL levels of Pb, Cu, Zn, and Cd.

[z] Valid for all biosolids uses except mushroom production.

[y] Mo limit corrected by adding appropriate data, and omitting data from a biosolids containing Mo at 1500 $mg \cdot kg^{-1}$.

[x] Cr limits for biosolids were deleted from the 503 Rule after Court-required review; no adverse effects of Cr(III) present in biosolids or composts have been identified despite intense investigations (see Chaney et al., 1997).

limit to be unprotective. Column 4 shows the Cumulative Application Limit for the metals (in $kg \cdot ha^{-1}$), the most limiting pathway for each element of all the risk assessment pathway calculations for the element.

In order to develop a new method of biosolids regulation (the alternative pollutant limit [APL]), the Cumulative Application Limit was assumed to be applied by 1000 $Mg \cdot ha^{-1}$ of biosolids, and expressed on a $mg \cdot kg^{-1}$ basis. Because limiting the concentration of metals in biosolids or compost can provide lower percent bioavailability and hence greater protection than simply limiting the cumulative application of metals, the expert workgroup recommended that EPA provide a regulatory mechanism which reduced the regulatory burden for higher quality biosolids (which met the limits in Column 4). These APL biosolids may be marketed for general use without cumulative site loadings for the regulated metals, if treated to kill pathogens (e.g., composted, heat dried, or equivalent pathogen reduction). EPA has further described APL-quality biosolids that have been aerobically stabilized and received an effective pathogen reduction treatment as "exceptional quality" biosolids. These products may be marketed as commercial products for lawns, and for use in home gardens.

Column 6 is the so-called No Observed Adverse Effect Level quality of biosolids as recommended by Chaney (1993a). For a number of the elements, U.S. Department of Agriculture (USDA) review of the final EPA 503 Rule indicated that policy decisions had led to limits which USDA concluded were less appropriate than needed for such a rule. For several elements (As, Cd, mercury [Hg]), EPA presumed 100% bioavailability of metals in soil or biosolids ingested by children; this caused limits to be lower than needed to protect HEI individual children (e.g., As); further, EPA used the 99th percentile as the ceiling limit instead of the 98th percentile used in the 1989 proposed rule. Raising this ceiling seemed unwarranted (Chaney, 1993a).

C. Hidden Safety Factors in Pathway Calculations

Ryan and Chaney (1993, 1997) provide a detailed discussion of the impact of protecting HEIs on the level of protection actually achieved by the complex algorithms of the 503 Rule methodology (U.S. EPA, 1989b). In the garden pathway, many factors are multiplied together to make the final calculation. The whole population for this pathway is those individuals who consume a high fraction (approximately 60%) of their lifetime garden foods from crops grown on soil in home gardens amended with very high levels of biosolids, on the order of 1000 $Mg{\cdot}ha^{-1}$. These individuals are assumed to be exposed for 70 years for cancer endpoints, and 50 years for Cd injury to the kidney. The soil is presumed to be acidic (geometric mean soil pH by EPA, and pH $\leq$6.0 by Chaney) for the whole lifetime. Transfer of Cd from the soil is estimated by the linear regression uptake slope for the crop and soil times the amount of a food ingested per day (lifetime average), times the fraction of diet presumed to be produced on the biosolids-amended home garden (37% of potatoes [*Solanum tuberosum* L.] and 60% of other garden foods).

Further, the uptake slope used in the 503 Rule calculation of food-chain transfer is not the increment reached at the plateau in the usually observed long-term relationship between soil Cd and crop Cd (Chaney and Ryan, 1994; Corey et al., 1987), but the linear regression for the data. This linear regression approach gives a much higher predicted plant Cd concentration at 1000 $Mg{\cdot}ha^{-1}$ biosolids than observed in long-term field studies (see later). The smaller the cumulative application rate for the actual data used in the regression, the larger the error of over-prediction. Also, the risk reference dose (RfD) which may not be exceeded (e.g., for Cd, 1 µg Cd per kg body weight per day) is a conservative estimate of the intake of Cd that over a lifetime causes the first sign of mild kidney disease (Chaney et al., 1999).

So the final 503 algorithms were revised to include some calculation factors that are central-tendency rather than all being worst case. The EPA calculation of the mean uptake slope for valid field data used the geometric mean of all Cd uptake slope data, not just the soil with pH ≤6.0. Because this geometric mean increased the allowed soil Cd for garden soils to 120 kg·ha^{-1}, the USDA (1993) advised U.S. EPA that they should use the data from acidic soils, and consider using arithmetic means of the valid field data for plant uptake slope. If the arithmetic mean of the whole field data set was used, the maximum soil Cd allowed would have been 12 kg·ha^{-1}. However, this data set included three field studies using highly contaminated biosolids. In general, the higher the biosolids Cd concentration, the higher the uptake slope regardless of the cumulative applied Cd (Jing and Logan, 1992). Since the rule would be limiting maximum biosolids Cd to relatively lower levels than those found in the three studies with highly contaminated biosolids, USDA reasoned that the results from these three studies should not be included in the EPA calculations (Chaney and Ryan, 1994). When the data from the three highly contaminated biosolids field studies were omitted from the dataset used to make the calculation, the estimated allowed cumulative application of biosolids Cd was 21 kg·ha^{-1} (Table 16.4). Stern (1993) was also concerned about the garden food pathway and conducted a Monte Carlo calculation of risks from biosolids-applied Cd, as noted by Chaney and Ryan (1994) for the complete dataset. Stern's calculation estimated the same 12 kg·ha^{-1} of Cd when all data were used. But he also did not evaluate the effect of deleting the uptake slopes from studies that used very high Cd biosolids. In the end, USDA recommended that biosolids to be used on farmland should contain no higher than 21 mg·kg^{-1} of Cd (USDA,1993), and noted that median quality biosolids contained only 7 mg·kg^{-1} of Cd and domestic source biosolids commonly contain <2 mg·kg^{-1} of Cd. When these lower levels are so attainable today, there is no need to allow biosolids containing Cd at 120 mg·kg^{-1} to be used on cropland.

As noted above, the 503 limit for an element was the lowest risk based limit for all pathways. For Cd, it had been evident for 30 years that the garden foods pathway comprised higher potential risk than the soil ingestion pathway. In the U.S. EPA Rule (U.S. EPA, 1993), soil ingestion was the limiting pathway, 39 kg·ha^{-1} of Cd. This would not have been the outcome if the algorithm noted above with acidic soils had been used, or if soil ingestion had been corrected for the low bioavailability of Cd in ingested soil. Table 16.5 shows the calculated limit for each pathway for Cd.

Even with the revised algorithms of the final 503 Rule, it is more likely that regulators erred on the conservative side (making low estimates of allowed cumulative applications) than on the high side. Some of the hidden safety factors that remain in the risk assessment include

1. Individuals cannot harvest a mixture of high-uptake slope leafy vegetables grown in a single garden for the whole year due to climate limitations on crop growth; thus they cannot practically ingest 60% of their annual intake of leafy vegetables grown on the model home garden. This is the food group which transfers most soil Cd into garden foods — therefore, the estimated risk is higher than the maximum potential risk.
2. Individuals are very unlikely to consume garden foods from a garden with 1000 Mg·ha^{-1} biosolids for 50 years.

Table 16.4 Calculated Increased Cd in Garden Food Groups Due to Biosolids Use According to the USDA (1993) Recommendation

Food Group	UC_i	DC_i	FC_i	$UC_i \cdot DC_i \cdot FC_i$	%
Potatoes	0.008	15.60	0.37	0.0462	1.8
Leafy vegetables	1.719	1.97	0.59	1.995	79.9
Non-dry legumes	0.004	3.22	0.59	0.0076	0.3
Root vegetables	0.094	1.60	0.59	0.0885	3.5
Garden fruits	0.113	4.15	0.59	0.277	11.1
Sweet corn	0.097	1.60	0.59	0.0814	3.3
All Garden Foods				2.496	100

Calculation algorithm:

$$RP_C = \frac{\underset{\text{WHO limit}}{70\ \mu g/day} - \underset{\text{Background intake}}{16.1\ \mu g/day}}{\underset{\Sigma(UC_i \cdot DC_i \cdot FC_i)}{2.496}} = \frac{\underset{\text{Allowed increase}}{53.9\ \mu g/day}}{2.496} = \underset{\text{Pathway 2 limit}}{21.5\ kg \cdot ha^{-1}}$$

Note: Arithmetic means of Cd uptake slopes by leafy vegetables were used rather than geometric means; mean for leafy vegetables calculated only for acid soils (pH <6) and biosolids with Cd <150 mg·kg^{-1}. The lower panel shows the full detail of the calculation algorithm. UC_i = Cd uptake slope for the i th food group (mg Cd per kg dry food group per cumulative kg biosolids-Cd applied per ha); DC_i = lifetime (chronic) average i th food group ingestion rate (dry grams per day); and FC_i = fraction of the i th food group supplied by the home garden where biosolids were applied at 1000 Mg·ha^{-1}. RP_C = "Reference Pollutant Application Rate" in kg·ha^{-1}.

3. Use of linear regression slopes for the uptake of metals to edible portions of crops is a high estimate of the increase when the plateau is reached (about three- to ten-fold error; see below).
4. Cd in crops grown on recommended quality biosolids/compost (APL) has low bioavailability (Chaney et al., 1978b) because of Zn also accumulated by the crop.

Multiplying the combination of central tendency and worst case variables together, one may still be estimating exposures greater than the most highly exposed individual for their lifetime, and thus calculating excessively low allowed cumulative loadings. Many of the most limiting pathways for a contaminant were those that calculated a lower estimate than needed to provide full lifetime protection to the HEIs. One cannot estimate the actual percentile of the HEI in the final rule due to the lack of data on measured intake of vegetables grown on biosolids or compost amended soils, and the cumulative application on these soils. It may be several hundred years before an appreciable number of individuals could build garden biosolids and composts applications to high enough levels to meet the limits of the 503 Rule for cumulative application rate (see Ryan and Chaney [1993] for discussion of time to reach 1000 Mg·ha^{-1} while following the 503 Rule).

An important source of significant conservatism in the final 503 Rule is the failure to adjust ingested metals for fractional bioavailability. The Pathway 3 limit (soil ingestion) shown in Table 16.4 is 39 kg·ha^{-1} for the U.S. EPA calculation. Table 16.5 lists the Pathway 3 corrected limit, 183 kg·ha^{-1} of Cd, when bioavailability is

Table 16.5 Pathway Limits for Biosolids Cd Calculated Using Final 503 Rule Methodology

	Pathway	EPA Calculation (mg Cd per kg)	Corrected Limit (mg Cd per kg)	Further Information
1	Food crops		>300	Default to Pathway 2
2	Garden crops	120	21	Most limiting Pathway
3	Soil Ingestion	39	182	Corrected for bioavailability
4	Soil→Crop→Livestock→Human	1,600	>1,000	Zn phytotoxicity prevents Cd transfer and risk to livestock or humans
5	Soil→Livestock→Human	68,000	>1,000	Pathway 7 is more limiting
6	Soil→Crop→Livestock	140	>1,000	Zn phytotoxicity prevents Cd risk
7	Soil→Livestock	650	>>300	Low bioavailability of soil Cd
8	Phytotoxicity	—	>>21	Food chain much more sensitive; Zn phytotoxicity limits at 1:100
9	Soil organisms/earthworms	—	>>300	Pathway 10 more limiting
10	Earthworms→Shrews	53	296	Earthworms bioconcentrate Cd
11	Dust to farm worker	8,000	—	Higher still if bioavailability included
12	Surface water	63,000	—	
13	Air volatiles	—	—	Not applicable
14	Groundwater	Unlimited	>1000	

Note: Corrected limits reflect corrections for calculations for Pathways 2 and 3 reported in Chaney and Ryan (1994) and discussed in this chapter.

included in the calculation (see Chaney and Ryan, 1994). Feeding studies have indicated that when Cd is incorporated into foods, and especially when Zn is present at the usual ratio of 100 μg Zn per 1 μg Cd in biosolids, soils, and foods, the food Cd has very low bioavailability. This was illustrated in studies of Swiss chard (*Beta vulgaris* L., Cicla group) grown on biosolids-amended soils fed to Guinea pigs (Chaney et al., 1978b) and Romaine lettuce fed to mice (Chaney et al., 1978a). When Cd to Zn in biosolids and crops were normal (1:100), no increase in kidney or liver Cd was observed. But when the biosolids had high Cd:Zn (pre-1980 Milorganite™), chard and lettuce accumulated much higher concentrations of Cd (without Zn phytotoxicity), and kidney and liver tissues were significantly increased in Cd. Soil Zn can inhibit crop uptake of Cd, leaf Zn can inhibit translocation of Cd to storage tissues, and food Zn can inhibit absorption of Cd in the intestine of consumers. Further, when crop Zn reaches phytotoxic levels, crop Cd remains at low concentrations when Cd to Zn is low. Crop Cd cannot exceed about 5 $mg \cdot kg^{-1}$ in crops which appear healthy, but leaf Cd can be substantially increased before Zn phytotoxicity reduces yield when Cd to Zn is high. Similar low bioavailability of biosolids-applied Cd to livestock has been repeatedly observed when high-quality biosolids were used to produce feeds, or directly fed to livestock to test the element transfer (Decker et al., 1980; Smith et al., 1985).

D. Phytotoxicity of Trace Elements

The most limiting pathway for Zn, Cu, and nickel (Ni) applied in biosolids/composts is phytotoxicity. Reactions of these elements in soils, and phytotoxicity of these elements to plants in relation to the concentration of the element in plant shoots, make it highly unlikely that any animal will suffer toxicity if the forage crops they chronically ingest are already experiencing Zn, Cu, or Ni phytotoxicity. This protection has been called the "soil-plant barrier" (Chaney, 1983). That plants are harmed by excessive soil Zn, Cu, or Ni before other HEIs in other pathways are harmed is a valuable protection of humans, livestock, and wildlife. But farmers do not want yield reduction when they expect beneficial response of crops to applied biosolids and composts. We believe that the 503 Rule provides adequate protection against future yield reduction under recommended farming pH management based on long-term studies. Shifting to the APL approach to regulate biosolids quality makes phytotoxicity even less likely to occur.

There have been many misunderstandings about the protection against phytotoxicity afforded by the 503 Rule. The Technical Support Document (U.S. EPA, 1992) presents the evidence. An article by Chang et al. (1992) summarizes the logic and data used as part of the development of the Pathway 8 (phytotoxicity) limits. The authors searched the literature to find the concentration of Zn, Cu, or Ni in seedlings that caused reduction in shoot weight in pot or nutrient solution tests with one added metal in the tests. This resulted in a tabulation of plant tissue concentration vs. growth retardation in immature plants (2 to 6 week growth period) without an understanding of the cause of the inhibition of growth or its impact on the growth of the mature plant. Utilizing this information to develop a phytotoxicity value requires an assumption that short-term reduction in shoot growth translates to yield reduction at maturity. As the scientific literature does not adequately address the validity of this assumption, a 50% growth retardation (phytotoxicity threshold; PT_{50}) was used as the threshold. Then, the field data from many studies were examined to search for the probability that the plant tissue metal concentration associated with the PT_{50} was exceeded in the field, with a 1% probability used to set soil metal limits (approach 1). A 1% chance of exceeding the plant tissue concentration associated with the PT_{50} concentration is quite protective, especially considering the observed probability for all recorded studies, far less than 1%. In none of the field studies identified by EPA have soils reached sufficiently high rates of biosolids metals (Zn, Cu, or Ni) to produce a yield reduction except for highly contaminated biosolids prohibited under the 503 Rule or strongly acidic soils that promote metal uptake and phytotoxicity. Therefore the biosolids cumulative application limits were set at a probability of 0.001% for Zn and Cu, and for Ni at 0.005%, or set at the 99th percentile of biosolids composition (the ceiling limit).

A second approach (approach 2) to calculate allowable loading rates to estimate Zn, Cu, and Ni limits used plant tissue concentration associated with potential phytotoxicity in sensitive crops in the field (obtained from the literature) and the plant response curve to biosolids-applied metals in acidic fields, subtracting the background concentration of the element in leafy vegetables, to calculate an allowable loading rate. The lower value from these two approaches was used as the final

503 Rule cumulative limit. Approach 2 using lettuce with tissue concentration from Logan and Chaney (1983) produced a soil loading with Zn of 2800 $kg \cdot ha^{-1}$, whereas approach 1 yielded a probability of 0.001% at a loading with Zn of 2500 to 3000 $kg \cdot ha^{-1}$. In the case of the other two metals (Cu and Ni), approach 2 yielded substantially higher loading rates than approach 1 (2500 vs. 1500 $kg \cdot ha^{-1}$ for Cu, and 2400 vs. 425 $kg \cdot ha^{-1}$ for Ni). As an additional observation that these values are nonphytotoxic, Mahler et al. (1987) have reported that the yield of Swiss chard and corn were not different between biosolids-amended and control soils in growth chamber pot experiments using soil samples from high metal loadings experiments. Also, Hinesly et al. (1984) reported only yield increases in corn grain at these high levels of biosolids metals applications. Further discussion of the limited potential for metal phytotoxicity when high-quality biosolids are used on cropland is reported by Smith (1996).

Some readers of the Chang et al. (1992) article and the Technical Support Document for the 503 Rule (EPA, 1992) have expressed great concern about use of the PT_{50}. A careful examination of the article shows the use of alternative phytotoxicity threshold Zn levels for 8, 10, and 25%, as well as for 50%. The cumulative Zn limit is not practically altered by using the PT_{25} Zn concentration.

Scientists experienced with the study of metals in soils and plants are also familiar with many possible errors in the kinds of studies that made up the PT_{50} databases. In nutrient solution and sand culture tests, researchers commonly used Fe-chelates to supply plant available Fe. When nonselective chelators such as ethylenediamine-tetraacetate (EDTA) are used to supply Fe, added Zn, Cu, or Ni displaces Fe from Fe-EDTA, causing the Fe to be precipitated and much less phytoavailable. Metal-induced Fe-deficiency has often been an artifact of Fe-EDTA chemistry in nutrient solutions and sand culture. Using Fe-EDTA strongly confounded a study of corn (*Zea mays* L.) and other Poaceae that used phytosiderophores to dissolve and absorb soil Fe, but also confounded studies with other plant families (Parker et al., 1995). Part of the added Zn, Cu, or Ni is chelated with the EDTA. Although EDTA is not usually important in uptake of these elements, when high levels of Fe-EDTA were supplied, and thus high levels of Zn-, Cu-, or Ni-EDTA chelates were formed in the test solutions, some direct leakage of metal chelate into the roots occurred. When Fe is supplied at 100 $\mu mol \cdot L^{-1}$, uptake of chelated metal can be appreciable, thereby confounding the goal of the tolerance vs. residue test. In a highly regarded set of studies by Beckett, Davis, and colleagues (e.g., Davis and Beckett, 1978), barley (*Hordeum vulgare* L.) was studied with Fe-EDTA as the Fe source in nutrient solutions applied to sand cultures. For at least Ni and Cu, the phytotoxicity diagnostic concentrations found in these sand culture studies are appreciably lower than found in many soil studies. If one finds such disagreement between basic research sand culture or nutrient solution studies compared to soil studies while conducting risk assessment research, one should give much greater weight to results of soil studies.

If the Pathway 8 limits had relied on common phytotoxicity diagnostic concentrations of these elements based on soil studies in pots and fields, the outcome of the 503 Rule risk assessment for metal phytotoxicity would not have been practically changed. Chaney and Oliver (1996) made such a list of diagnostic concentrations, taking into account errors in research methods such as those discussed previously. The phytotoxicity diagnostic concentrations in diagnostic foliar plant tissues they listed were Zn, 500 mg·kg^{-1} dry leaves; and Cu, 30 mg·kg^{-1} dry leaves (Chaney and Oliver, 1996). More recent evaluation of Ni phytotoxicity strongly supports diagnosis of possible Ni phytotoxicity-induced yield reduction at 25 to 50 mg·kg^{-1} dry shoots for most species, while some species have little or no yield reduction from Ni at 100 mg·kg^{-1} shoots (Chambers et al., 1998). Chang et al. (1992) discussed other plant species having PT_{50} of 50 to 100 mg·kg^{-1} leaves, but identified a PT_{50} for corn at 3 mg·kg^{-1} from the reports they relied on. This PT_{50} was in substantial error as subsequently shown by the study of L'Huillier et al. (1996).

Although some have questioned using PT_{50}, it is evident that appropriate protection of the environment and of crop yields is obtained by the 503 Rule cumulative Zn, Cu, and Ni limits and APLs based on an "holistic" evaluation of metal phytotoxicity from land-applied biosolids. There are reports in the literature of metal phytotoxicity that resulted from biosolids-applied metals; but if one examines the factors which contributed to the metal phytotoxicity, it is evident that these reports are not a valid criticism of the 503 limits. Sensitive crops (such as lettuce) have suffered yield reduction at pH ≥6.0 with foliar metal concentrations above phytotoxicity diagnostic levels in studies with extremely contaminated biosolids (Marks et al., 1980). Adding such high metal amendments allows one to apply a lot of metals at one time to examine potential phytotoxicity, but because the biosolids matrix alters metal phytoavailability, results from such studies are relevant only to highly contaminated biosolids. Indeed, such tests are strong evidence that highly contaminated biosolids or composts are not acceptable for use in horticulture and strongly support the requirement for industrial pretreatment. A second category involves tests with biosolids containing more typical concentrations of Zn, Cu, and Ni in which soil pH was allowed to drift to very low levels where sensitive crops suffered phytotoxicity (Brallier et al., 1996; King and Morris, 1972; Lutrick et al., 1982). Allowing pH of test plots to drop over time while one observes potential development of phytotoxicity is a time-honored method to conduct basic research studies, but such data are not an appropriate basis for development of regulations. When soil pH has dropped to 4.6, soil aluminum (Al) and Mn contribute to any observed phytotoxicity, and such pH levels are not considered reasonable for farm management because of the predictable yield loss. In studies with very strongly acidic soils in which crop yield reductions were evident on soils amended with APL quality biosolids, full yield was regained by simply applying the limestone needed for normal production of the sensitive crops. For these reasons, EPA decided to set the Zn, Cu, and Ni cumulative metal application limits for Pathway 8 based on field research with soils at about pH 5.2 to 5.5, where natural soil Al and Mn can cause natural phytotoxicity to the same sensitive crops, and at higher pH.

Another category of field-observed metal phytotoxicity is sites where pesticide metal accumulation or industrial contamination occurred, and crops have suffered yield reduction. Although data from such studies were considered, again the role of the biosolids matrix reducing metal phytoavailability indicated that such data are not relevant to regulations for metals applied in modern high-quality biosolids or composts.

Another aspect of metal phytotoxicity in the field should be considered. In the field, roots quickly grow through the tilled soil depth where the applied metals accumulate, and then are much less likely to be harmed by the metals. We have seen high metal biosolids field plots which caused visible metal-induced Fe-deficiency-chlorosis during early seedling growth of sensitive species, viz., soybean (*Glycine max* [L.] Merrill) or lettuce, but where the plants fully recovered and had no significant reduction in yield. In pot studies, roots are constrained to the treated soil, and once toxicity begins, it usually worsens. Metal concentrations are higher in plants grown in pots than in the same plants grown in the field where the same soil and biosolids mixtures were used for the pot test or large lysimeters in which the biosolids were incorporated in the surface 10 cm depth (deVries and Tiller, 1978). In practice, when high-quality biosolids or compost were land applied, phytotoxicity (usually from Zn) was observed only (1) when the soil pH was well below 5.0, typically near 4.5; and (2) the crop was one known to be highly sensitive to excessive soil metals such as peanut (*Arachis hypogaea* L.), lettuce, snap bean (*Phaseolus vulgaris* L.), spinach (*Spinacia oleracea* L.), etc., as discussed previously.

The role of Fe, Mn, and Al and adsorbed phosphate in reducing uptake and phytotoxicity of applied metals has also become more appreciated. With high-quality compost products, there is no reason to suspect that Zn, Cu, or Ni phytotoxicity will occur in the field unless extreme soil acidity is reached. Effective farmers do not tolerate such poor management of pH. Thus great concern about potential phytotoxicity of biosolids/compost-applied metals (Zn, Cu, Ni) is not supported by research tests or practical farm use of these products with modern regulations. Only older highly contaminated biosolids, or extremely acidic pH, have allowed phytotoxicity.

Another perspective on metal phytotoxicity may be helpful. Farmers with medium textured soils downwind of Zn smelters may have soils with 2000 to 3000 ppm Zn, but if they keep soil pH in the 6.5 range or higher, they profitably produce even sensitive legumes and vegetable crops. Also, when soils are made calcareous with biosolids or compost treatments to aid in revegetation, mine wastes and smelter contaminated soils with over 10,000 $mg \cdot kg^{-1}$ Zn have no adverse effect on grasses and legumes (Brown et al., 1998b; Li et al., 2000).

E. Phytoavailability of Applied Trace Elements over Time

Although concern about long-term phytoavailability of metals in soils amended with biosolids and compost has been part of the focus of research in this area of science for over 30 years (Beckett et al., 1979; Chaney, 1973; Page, 1974), some continue to express concern about the long-term risk from metals in biosolids and composts. Although we believe it is scientifically valid to criticize the final Pathway Analysis as being very conservative (see previous discussion), some scientists have suggested that the pathway calculations are not protective enough. For example, McBride (1995) challenged some of the conclusions of the expert workgroup, concluding that metals in land-applied biosolids/compost may threaten soil fertility and food chain safety and comprise a "time bomb" because applied organic matter is biodegradable.

The "time-bomb model" for the worst case risk from biosolids metals was published by Beckett et al. (1979); similar considerations were reported by Chaney (1973), Page (1974), and most other researchers who began research on biosolids utilization in the 1970s. Beckett et al. (1979) considered that organic matter must comprise the most important metal-adsorbing constituents in biosolids-amended soils, and because the added organic matter will eventually be oxidized to the level appropriate for the climate, texture, and cropping pattern of the soil in question, the added metals would become more plant available over time and eventually poison plants and animals. Many researchers had this concern in the 1970s, before extensive research conducted in several nations failed to support this model (Chaney and Ryan, 1994; Corey et al., 1987; Johnson et al., 1983). Interestingly, Beckett's team also reported that Zn in biosolids is bound with organic-iron oxide assemblages rather than simply bound to organic matter (Baldwin et al., 1983), and it is generally agreed that only Cu is predominantly bound to organic matter as the starting form in stabilized biosolids or composts.

McBride (1995) also challenged the concept of the plateau response (as illustrated by Chaney et al., 1982) that results from the biosolids-applied adsorbent materials (hydrous oxides of Fe, Mn, Al, etc.) that persist in biosolids-amended soils (see review in Corey et al., 1987). This argument was based on the presumed loss of organic-matter-specific metal binding sites in amended soils as organic matter was biodegraded.

Chaney, Ryan, and other scientists have examined this question by measuring Cd phytoavailability of soils from long-term biosolids field plots. An important set of studies by Mahler et al. (1987) and Mahler and Ryan (1988a, 1988b) involved additions of Cd salts to soils collected from farmer's fields that received high cumulative applications of biosolids and adjacent untreated fields of the same soil series. They grew Swiss chard, a spinach-like vegetable with high Cd uptake ability. This plant is tolerant of high foliar Cd, so it has a very wide range of linear response to phytoavailable Cd in soils. A careful examination of the full data from Mahler et al. (1987) and Mahler and Ryan (1988a, 1988b) shows that the uptake slopes for the unamended soils are in general higher than the slopes for the biosolids-amended soils. When the untreated and biosolids-amended soils were at the same pH, or taken to the same pH by addition of limestone, the slope was lower (= lower phytoavailability) for the amended soil. The reduction in Cd uptake slope was especially evident for those soils which had received high cumulative biosolids applications or biosolids-metals applications. High applications of biosolids would add higher amounts of specific metal binding strength to the amended soil, which would cause lower Cd uptake slopes compared to lower cumulative biosolids/compost application rates. Because soil pH strongly affects uptake of Cd, valid comparisons of soil differences should be made at equivalent soil pH levels.

Figure 16.1 shows models of the patterns of plant uptake of metals in relation to soil metal concentrations observed in studies of long-term biosolids-treated or metal-salt-treated soils. This figure is our summary of the response patterns found in long-term field studies; all lines start at the linear slope usually found for low levels of added Cd salts, and model equal Cd additions to one soil but in different forms. Curve A represents the linear response to small additions of salt Cd in nearly all studies in the literature which we reviewed; in curve B, the pattern has increasing

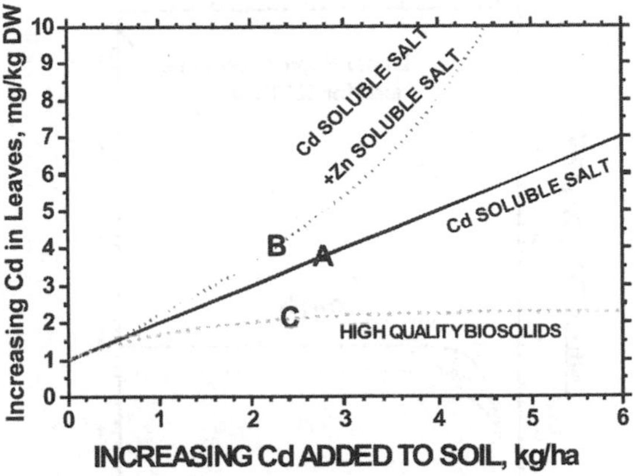

Figure 16.1 Corrected hypothetical models of plant Cd response to soil total Cd concentrations. (A) from addition of a soluble Cd salt; (B) from addition of a Cd soluble salt with 100 times more Zn as a soluble salt; and (C) from addition of NOAEL quality biosolids, after organic matter stabilization to background levels. (From Chaney, R.L. et al., 1998. Soil-root interface: ecosystem health and human food-chain protection, p. 279–311. In: P.M. Huang et al. (eds.). *Soil Chemistry and Ecosystem Health.* Soil Science Society of America, Madison, Wisconsin. With permission.)

slope with higher Cd applications because Zn is also added, at 100 times the Cd additions; the added Zn competes for the stronger Cd adsorption sites in the soil, increasing Cd phytoavailability. These first two patterns have been repeatedly observed in many studies around the world, and are illustrated well by the data in White and Chaney (1980).

In contrast with the response pattern of Cd salt additions, the model slope C in Figure 16.1 is for biosolids-applied Cd, in which the slope decreases at higher biosolids application rate toward a plateau with the x axis. Figure 16.2 shows the results for lettuce uptake of Cd on long-term biosolids field plots at Beltsville, MD, and shows the plateau regression response compared with linear regression to estimate the uptake slope for the plateau data (Brown et al., 1998a; Chaney et al., 1982). Using orthogonal contrast from an analysis of variance (ANOVA), they found that the $(\text{Rate})^2$ term was significant and negative. Using plateau regression, they found that the plateau converged, supporting that model over a simple Rate, Rate^2, and pH model. Clearly, the simple linear regression fails to model the plant accumulation of Cd and Zn appropriately. The slope is clearly controlled more by the adsorption chemistry of the biosolids-applied constituents than of the unamended soil. The Corey et al. (1987) workgroup reviewed the available field response data and concluded that this plateau response is commonly observed and is the theoretically expected pattern of response. Strong adsorption sites on hydrous Fe and Mn oxides and organic matter control adsorption of Cd in soils rather than precipitation as inorganic solids. For biosolids with basal low concentrations of adsorbent metal oxides, the pattern is more between linear and plateau responses (e.g., Chang et al., 1997).

Some field experiments have appeared to support the time bomb hypothesis, in that plant uptake of Cd from Cd-enriched, nonbiosolids-amended soils increased

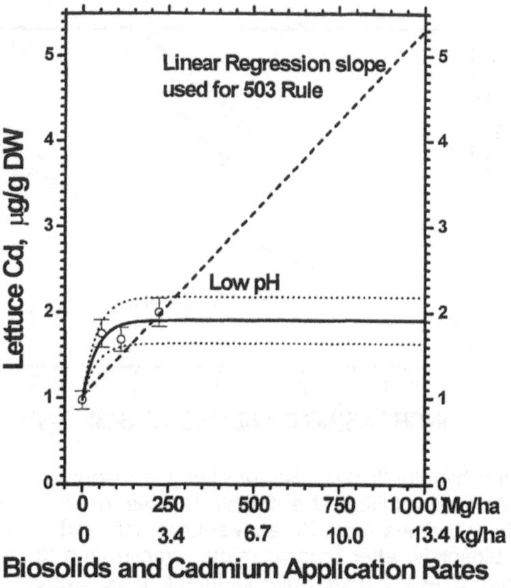

Figure 16.2 Linear vs. plateau regression analysis of lettuce uptake of Cd from acidic Christiana fine sandy loam amended with 0, 56, 112, or 224 Mg dry heat-treated sludge per ha, and pH uncontrolled (≤ 5.5 in 1983) (Low pH). Predicted response extrapolated to 1000 Mg·ha^{-1} to show implications of the model used. Results are average for 1976 to 1983 lettuce crops. Data points shown are arithmetic means ± one standard error; plateau regression shows predicted (dashed lines) with ± 95% confidence interval (dotted lines). Equation for linear regression (solid line) is: Lettuce Cd = 1.22 + 0.00390·Rate (low pH). Predicted increment in lettuce Cd at 1000 Mg·ha^{-1} for the low pH treatment is 0.89 mg·kg^{-1} for plateau regression vs. 4.28 mg·kg^{-1} for linear regression. Biosolids applied in 1976 contained 13.4 µg Cd, 1330 µg Zn, and 83 mg Fe per g dry weight. (Data originally reported in Chaney et al., 1982.)

over time as soil organic matter was biodegraded. Much of the organic carbon (OC) in biosolids also is lost due to biodegradation over time. If the OC disappeared and other biosolids components *did not* provide increased metal adsorption capacity to the soil, the response pattern for biosolids-amended soil should be increased phytoavailability of biosolids-applied Cd over time. On the other hand, having no change in plant Cd accumulation over time supports a negative conclusion about the time bomb hypothesis (Brown et al., 1998a; Chang et al., 1997; McGrath et al., 2000; Sloan et al., 1997). As is clear from Figures 16.1, 16.3, and 16.4 that show crop Cd vs. soil Cd, a plateau toward the x axis is the plateau of Chaney et al. (1982), Corey et al. (1987), and Mahler et al. (1987). For example, under highly controlled conditions, and equal pH between unamended and amended soils, Mahler et al. (1987) reported that for each test soil, the Cd response to added salt-Cd was highly linear with regression R^2 greater than 0.9 (two examples shown in Figure 16.3). But where the soil had a substantial application of biosolids such that possible changes in soil metal phytoavailability might be expected, the biosolids-amended soil had lower slopes than the unamended comparison/control soil. Clearly, the added biosolids caused a change in the Cd binding by the soils such that slopes were linear but reduced on the biosolids-amended soil compared to the nonamended soil.

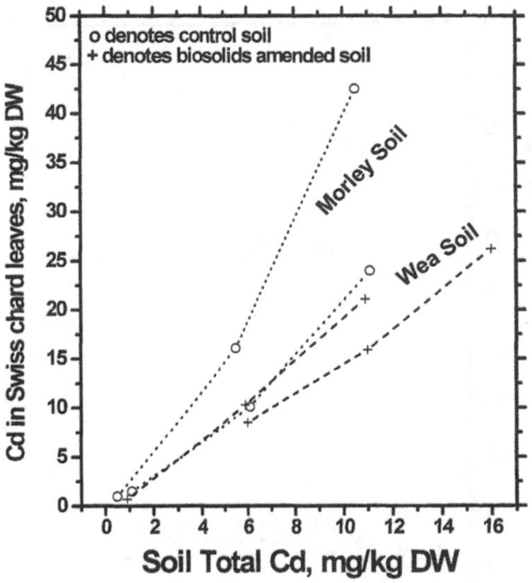

Figure 16.3 Relationship of total soil Cd and Cd accumulation in leaves of Swiss chard grown on control and long-term biosolids-amended soils with addition of 0–10 mg Cd per kg dry soil; soil series from two cities; all soils made calcareous to avoid unequal pH comparisons. (Based on data in Mahler et al., 1987.)

Brown et al. (1998a) recently examined the relationship of Cd to OC in soil and uptake of Cd by lettuce and other crops on field plots of different biosolids products that were established in 1976–1978 at Beltsville, MD. The plots were cropped again with Romaine lettuce and other garden crops in 1991–92 to examine the relative uptake of Cd by different garden foods to allow use of Cd uptake by lettuce to model Cd increase in all garden foods (Brown et al., 1996). Although McBride's model would require that plant uptake of biosolids-applied-Cd should be increased toward the salt Cd response line over time as the biosolids-added OC was lost, plant Cd was about the same or lower in 1992 than in 1978, even though soil pH had declined over time on the acidic biosolids-amended plots. In this field experiment, a Cd salt treatment at acidic and limed pH, at the same Cd rate (21 $kg{\cdot}ha^{-1}$ of Cd) as the Cd rich biosolids from Chicago (210 $mg{\cdot}kg^{-1}$ of Cd in 1978) was established. Figure 16.4 shows the Cd to OC for these treatments at the beginning of the study (1979–1981) and the 1991–1992 crops (Brown et al., 1998a). It is clear that loss of OC from the biosolids-amended soils did not increase Cd uptake by lettuce on these treatments. Thus the presence of metal adsorbent materials (hydrous oxides of Fe and Mn) in composts and biosolids can add to the safety of these products as well as to the plant nutrient value of the products.

We believe these results illustrate ways that manufacturers of biosolids and composts could increase the inherent safety of these products, and that such manufacturers should seek Fe and Mn rich wastes or purchase Fe and Mn ores to be included in the products to increase metal adsorption and fertility where these products are used in agriculture.

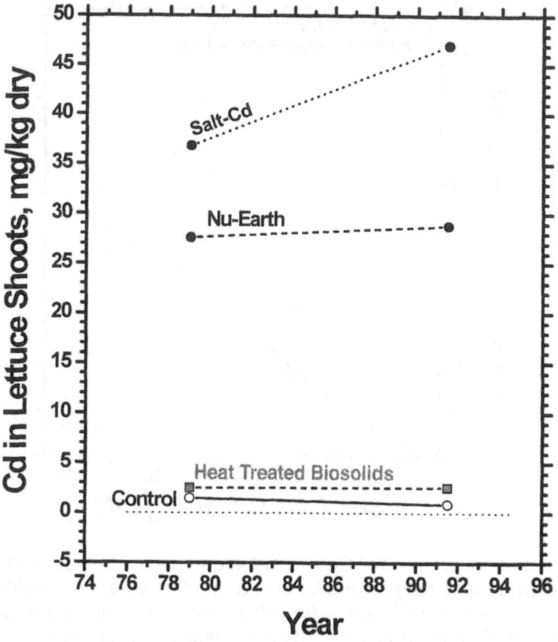

Figure 16.4 Effect of time after application of biosolids or salt-Cd to a low pH Christiana fine sandy loam at Beltsville, MD, on phytoavailability of soil Cd to Romaine lettuce. Cadmium at 21 $kg \cdot ha^{-1}$ was applied in 100 dry $Mg \cdot ha^{-1}$ of Nu-Earth biosolids (210 mg Cd per kg dry weight) or Cd-salt, while Cd at 3.0 $kg \cdot ha^{-1}$ was applied in 224 $Mg \cdot ha^{-1}$ heat-treated biosolids with 13 mg Cd and 83 g Fe per kg dry weight. Soil pH was allowed to drift downward over time. Biosolids-applied Cd did not become more phytoavailable despite loss of most of the organic carbon applied in the biosolids. (Based on Chaney et al., 1982 and Brown et al., 1998a.) Results are an average of 1976–1981 and 1991–1992 Romaine lettuce crops.

F. Labels May Confuse Risk Communication

Perhaps some of the unscientific fear of biosolids is related to the labels used to describe metals and xenobiotics in biosolids. U.S. EPA designated the chemicals in biosolids as "pollutants" rather than constituents or contaminants. We believe these labels should be used as described by Davies (1992) who argued that until an element or xenobiotic is present at levels that could cause adverse effects to highly exposed and sensitive individuals, it should not be called a pollutant. Rather, it is only a contaminant until it can cause some adverse effect under reasonable worst case conditions of soil management. Many contaminants have been emitted by industrial society, and PCBs have been dispersed around the world due to their volatility, as were Hg, Cd, Zn, and Pb, due to the transport of aerosols for long distances. Because of the extreme unlikelihood that individuals will meet all the criteria of the HEI simultaneously, the population is still highly protected at the point where the 503 Rule process has concluded the first adverse effect may occur; thus the soil would not be "polluted" on a practical basis. If such highly conservative protection factors are used in risk analysis, the most highly exposed and most sensitive individuals are still well protected. Thus, contaminated soils should still not be a source of special concern until the HEIs would achieve the lifetime dose mean of the 503 Rule pathway analysis. Failure to describe the magnitude of the over-protections provided against human Cd disease by the 503 Rule and misunderstandings of the risk assessment process are very likely the reasons why some individuals remain concerned about risk of Cd disease, or of metal phytotoxicity, in soils where high-quality composts or biosolids are applied to reach very high cumulative loadings. Certainly there are no field observations to support this concern.

G. Soil Cadmium Risk to Humans

The understanding of Cd risk to humans has improved since the 1980s. Chaney and Ryan (1994) presented a comprehensive review of Cd in soils, plants, foods, and humans, and concluded that Cd in rice (*Oryza sativa* L.) consumed by subsistence farm families is not a valid model for Cd risk in Western nations. Rice grown on flooded soil allows much higher maximum grain and diet Cd than possible with lettuce and other garden crops. Indeed, rice now appears to be in a Cd-risk class all by itself (Chaney et al., 1999).

An epidemiological study in Shipham, England showed that soils with average Cd concentrations of over 100 $mg{\cdot}kg^{-1}$ (50 times higher than required to cause health effects in subsistence rice consumers in Japan and China) caused no human disease to long-term gardening residents (Strehlow and Barltrop, 1988); here Zn accompanied the Cd in the gardens (from mine wastes that were dispersed several centuries before housing was constructed), and moved to the edible tissues of the garden crops. Two similar situations of persons growing gardens on contaminated soils without adverse effects other than Zn phytotoxicity to sensitive crops have been reported: one in a village at Stolberg, Germany developed on mine or smelter wastes after World War II (Ewers et al., 1993), and another in Palmerton, PA (Sarasua et al., 1995) where smelter emissions contaminated home gardens to 100 mg Cd and 10,000 mg Zn per kg of soil. Epidemiological studies on long-term populations in these communities found no evidence of Cd-induced renal tubular dysfunction in any of the older persons, even those who might have had high enough Cd ingestion to raise concern. On the other hand, in China numerous cases similar to the rice Cd poisoning of subsistence rice farm families in Japan were observed (Cai et al., 1990).

Thus, a combination of many sources of data now indicate that both the sociology and agronomy of soil Cd exposure to humans were more important to whether disease resulted than was the simple toxicology of Cd salts added to diets. Individuals in Japan and China, who consumed high amounts of rice locally grown on highly contaminated soils, suffered Cd disease because of a combination of soil chemistry of flooded soils, the microelement physiology of rice, the interactions between Zn and Cd at the plant membranes and in the intestine, and the Zn- and Fe-deficiency suffered by people who subsist on rice-based diets. Agronomy of Cd risk predominated over the toxicology of Cd risk in these cases. Epidemiologists who worked diligently to characterize the medical bases of Cd risk to individuals in Japan (e.g., Nogawa et al., 1987) or in Europe (e.g., Friberg et al, 1985) had little appreciation for the plant nutrition aspects of Cd disease, or for the human nutrition aspects of the disease. It now seems clear how they were led to the conclusion that soil Cd was dangerous to humans, because that is the pattern observed in Japan and China, where rice provides the bulk of the Cd exposure. But lettuce, wheat (*Triticum aestivum* L.), and potatoes are much more important sources of exposure to Cd from contaminated soils in the West, and soil Cd comprises far lower risk through these foods than through rice. As summarized in Chaney and Ryan (1994), Zn phytotoxicity is indeed a very strict limit on Cd in edible plant tissues, and the home garden cannot provide excessive bioavailable dietary Cd when Cd to Zn is 0.015 or lower. Cd and Zn in mine wastes and smelter emissions usually occur at the geological ratio, about 0.005 to 0.01 Cd to Zn ratio, while Cd in plating wastes, wastes from Cd-Ni batteries, and wastes from Cd pigments or plastic stabilizers containing Cd, have little accompanying Zn. These latter sources allow greater potential for flow of Cd in the food chain, and potential higher Cd bioavailability to humans who consume crops grown on soils strongly contaminated by these sources, especially so if the soils are acidic. Nearly all industrial injury to humans from Cd results from Cd unaccompanied by the 100:1 Zn in geological sources.

As agronomists who have worked for several decades to better understand the nature of soil Cd risk to humans, in order to set limits for Cd in biosolids, composts, soils, and crops, we find it especially ironic that agronomy has proved important in Cd risk. Toxicologists usually study the contaminant they are examining in a pure form, such as soluble Cd salts without accompanying Zn or other materials in Zn ores. They provide the Cd salts to animals that are often minimally adequate or deficient in Fe or Zn to observe maximal Cd retention. Moreover, Cd often is supplied to fasting individuals or by injection rather than by ingestion routes. Some of the potential adverse toxicological effects of Cd result from Cd-induced Zn deficiency, but the environmental relevance of these findings is restricted to sites where Cd contamination is not accompanied by Zn — the rare and very minor case for soil Cd enrichment.

A National Research Council Committee has reported on their review of the information used by U.S. EPA to develop the 503 Rule (National Research Council, 1996). They examined the risk assessment pathways, the data used to provide transfer coefficients, the data on pathogen reduction, etc., in the context of the safety of foods grown on soils amended with biosolids or composts, or where treated wastewater effluent was used for irrigation. The National Research Council committee concluded that there was no evidence that either growers or consumers were at any risk when the EPA regulations are followed, even at the extreme cumulative applications modeled by U.S. EPA.

H. Food Consumption Rates vs. Cadmium Risk Potential

Harrison et al. (1999) raised another concern about the 503 Rule Risk Assessment, noting that the USDA's Food and Nutrition Service presently recommends diets with higher daily intakes of vegetables, fruits and grains compared to the lifetime diets used in the 503 Rule Risk Assessment. This new approach to dietary recommendations, the Food Pyramid, may cause the recommended shift in diet patterns of U.S. citizens. A higher intake of leafy vegetables from acidic biosolids-amended gardens should increase total Cd ingested. Within the context of the 503 Rule Risk Assessment, higher Cd risk would be expected, but this is another artifact or error of the risk assessment by EPA. EPA made a policy decision that the bioavailability of elements in foods or soils would be assumed to be 100% that of soluble metal salts added to diets. This is simply untrue, and diet and food factors that affect element bioavailability should have been included in the original risk assessment. This is particularly important for Cd in foods where normally Zn is 100 times higher than Cd and inhibits Cd absorption (Chaney et al., 1978a, 1978b, 1999). Cadmium in leafy vegetable foods has very low or negligible bioavailability (based on feeding studies noted previously). Thus increased consumption of these foods will comprise no change in lifetime Cd risk of highly exposed gardeners. Zero bioavailability times some increased factor of food intake still gives zero bioavailable Cd from lettuce and chard, and no increase in risk even with "improved" diets.

I. Are Soil Microbes Protected by the U.S. EPA Section 503 Rule?

Some have questioned whether soil organisms are protected by the 503 Rule. Soil microbes are clearly part of Pathway 9, in which the soil organism most sensitive to an element in soil becomes the HEI to be protected. Considerable disagreement among different studies has been noted regarding rhizobium in soil amended with biosolids. A researcher in Europe found that the rhizobium for white clover (*Trifolium repens* L.) was "ineffective" on a set of long-term biosolids plots in the U.K. (McGrath et al., 1988); small nodules formed on the roots, but they did not fix atmospheric N. The soil was sandy, and the biosolids had high levels of some metals. When the soil was inoculated with an "effective" strain, the plants developed nodules which fixed N. In another study conducted in Germany, where metals were mixed with biosolids to make it "high metal", and soils were allowed to become strongly acidified, plants also did not fix N (Chaudri et al., 1993; Fließbach et al., 1994).

In a series of studies led by J.S. Angle, metal toxicity to rhizobium on long term biosolids plots, and in soils contaminated by a Zn smelter, were investigated. First, it was observed that rhizobium for soybean and alfalfa (*Medicago sativa* L.) were not harmed even when metals became phytotoxic to these crops (Angle, 1998). In a comparable study with white clover and alfalfa, in the field, and in chelator-buffered nutrient solutions, plants were more sensitive to metals than were the microbes or nodulated plants; and root hair growth and infection were the most sensitive processes in N fixation to excessive Zn or Cd. But at the levels of Zn and Cd that occurred near the Zn smelter (Palmerton, PA), or in long-term high Cd biosolids-amended plots at Beltsville, MD, only adverse effects of soil acidity, not soil metals, on survival and infection by rhizobium were found (Ibekwe et al., 1997a). Further, Ibekwe et al. (1997b) found that genetic diversity was improved and cell density was increased by high-quality biosolids, while cell density was not improved by the biosolids with higher metals than recommended. Still, acidity was the key factor that reduced genetic diversity and reduced cell density, as well as selection of ineffective strains of rhizobium. Although concern has been expressed about the need to protect soil rhizobium, we believe some perspective is needed regarding this goal. Rhizobium is long known to be sensitive to acidic soils. Farmers usually add limestone to soils to raise the pH to 6.5 to 7 before seeding legumes, and usually inoculate the seeds.

Both survival and infection of most rhizobia are greatly reduced at strongly acidic pH, and alfalfa and white clover cannot fix N in such acidic soils. About one-third of the arable soils of Earth are so acidic that soil Al or Mn limit root growth, and rhizobia survival and function also are inhibited at these levels of acidity. Thus, although at least one-third of Earth's surface has surface soil pH conditions where rhizobia cannot survive, there is no evidence that high-quality biosolids or composts endanger rhizobium unless soil pH is dropped to levels which harm the crop. We conclude that with biosolids and composts that meet or are superior to the APL quality standard, used under soil pH management appropriate for high yields of legumes, biosolids composts are favorable to legume production, not a risk to rhizobium and hence not a risk to legumes. Perspective is necessary for all risk assessment judgments, and clearly very important in the case of soil rhizobium.

J. Using the U.S. EPA Section 503 Biosolids Limits for Other Composts and Organic Amendments

Many jurisdictions have adopted the 503 limits (APL limits) for both biosolids and biosolids composts, and are using these limits for other kinds of composts (yard debris, manure, etc.). Research on potential risks from biosolids-applied elements has been extensive, and adequate data were available to develop the rule for biosolids materials. Other kinds of composts may not contain the high level of adsorbent materials found in biosolids, and thus higher phytoavailability could occur when these amendments are used with sensitive crops and strongly acidic soils. On the other hand, these other composts usually contain lower levels of Zn, Cu, Cd and other elements than found in biosolids. The lower supply of adsorbent surfaces in non-biosolids composts may require that APL-like limits for these other products should differ from the APL limits for biosolids. Only research on these other products can answer the question of whether using the 503 APL limit for other products is insufficiently protective of humans and the environment. As noted in this chapter, adding byproducts rich in Fe and Mn oxides can both increase the metal adsorption ability of composts, and increase their Fe and Mn fertilizer value. Improving many types of compost by including other byproducts with beneficial constituents should offer increased value products, and provide additional protections against adverse effects of metals in compost products.

III. SOME FUTURE DIRECTIONS FOR COMPOSTS AND BIOSOLIDS IN AGRICULTURE

As noted previously, and in Chaney and Ryan (1994), we believe the use of quality-based limits for metals will remove the potential for any adverse effects under reasonable farming conditions, including pH above about 5.2 to 5.4 where Al and Mn phytotoxicity become important to farm productivity. We believe that over time, research in other nations will corroborate the U.S. data and they will change to quality based regulations in place of soil metal concentration based limits, or cumulative application based limits. With time, others will come to understand the finding that biosolids rich in Fe and Mn add to the metal specific adsorption capacity of soils, and cause the plateau response rather than the linear response, or the time bomb model for risk. Phytoavailability and bioavailability will become key components of risk evaluation protocols and regulation development. Research at the fundamental level in soils, plants, and animal nutrition is likely to validate further the points we have raised, including the importance of agronomy and human nutrition in the potential for Cd risk.

A. Remediation of Metal Toxic Soils Using Composts and Biosolids

Another area of organic amendment science in which progress will be seen is the technical understanding that one can solve important societal problems using biosolids or composts. Because applying a combination of biosolids rich in Fe and limestone equivalent actually corrects or prevents Zn phytotoxicity in contaminated soils, Chaney has argued that we will see more use of biosolids to remediate contaminated soils (Chaney, 1993b). Our research on inactivation of soil Pb using biosolids has shown that added Fe-rich composts can reduce the bioavailability of soil Pb to rats (fed to simulate *pica* children [children with *pica* ingest nonfood items]) by 65% compared to an unamended control for the soil, or to 10% of the Pb bioavailability of food or water Pb (Brown et al., 1997b; Chaney and Ryan, 1994). Addition of Fe or Mn during manufacture of composts may reduce Pb bioavailability even further.

We have conducted or cooperated in conducting tests of the use of biosolids, composts, limestone, etc. in remediation of metal toxic soils where mine tailings were disposed, or where smelter emissions killed plants and destroyed ecosystems. Basic studies on metal adsorption by biosolids constituents showed that sorption plus phosphate plus limestone could aid in revegetation of Zn phytotoxic soils (containing 1.5% Zn, pH 6.3) at Palmerton, PA (Li and Chaney, 1998; Li et al., 2000; Siebielec and Chaney, 1999). With this evidence, demonstrations were conducted first in Silesia, Poland (Daniels et al., 1998; Stuczynski et al., 2000), and then in Kellogg, ID and Leadville, CO (Brown et al., 1998b), where mine wastes or smelter emissions left barren eroding soils rich in metals and severely infertile. Tailor-made mixtures can provide great benefits — these mixtures can restore soil fertility, remediate metal phytotoxicity, reduce soil Pb bioavailability, provide improved soil physical properties, increase organic matter, and provide soil microbial inoculum for dead soils. "One-shot" highly effective revegetation-remediation has been achieved by application of these mixtures of organic and inorganic byproducts to achieve improved soil properties. The sites become revegetated with plants that are safe for consumption by wildlife as 100% of their diets, and even inadvertent soil ingestion comprises lower risk due to the inactivation of soil Pb. Well-vegetated soils are unlikely to become erosion problems, and are hard to ingest because the plant cover limits animal access to the soil. Further, by making mixtures of biosolids, alkaline ash materials, and other organic and inorganic wastes and byproducts, considerable savings are achieved in wastes handling, and in reduced cost for effective revegetation of barren soils.

We conclude that the extensive studies on soil Pb bioavailability during the last few years indicate that composted biosolids can be manufactured to make products that are optimized to remediate soil metal contamination problems. Instead of biosolids being a heavy metal problem in the environment, biosolids can remediate toxic soils and heal harmed ecosystems because of the nutrient supply, nutrient balance, and persistent metal sorption capacity they provide. Extensive Pb contamination of urban soils occurred in most of the developed world (from automotive exhaust, and from paint used from the 1800s to 1976). Use of modern high-quality biosolids composts to cure problems with contaminated or infertile, disturbed soils (e.g., brownfields and general urban soils) will also win wider public support for the beneficial use of biosolids. By a careful program of pretreating industrial wastewaters to capture contaminants at the industrial source where they can be recycled economically, and monitoring the composition of the biosolids and composts produced for beneficial use, the biosolids industry can protect the biosolids marketplace from problems common in previous decades. Chaney has called this program of manufacturing improved biosolids products “tailor-made remediation biosolids and composts.” Within a community, by combining different wastes with beneficial constituents, and mixing them with the biosolids, the biosolids and composted biosolids can be made more beneficial and inherently safer than high-quality biosolids of today. Higher Fe and Mn in biosolids would both increase the phytoavailable Fe and Mn in amended soils, and increase specific metal adsorption by the soil-biosolids mixture. Inclusion of Fe or Mn byproducts, N, P, K, or $CaCO_3$-rich byproducts, etc. can reduce the need for landfills while reducing the potential for risk from the metals in soils or in biosolids. Manufacturing improved biosolids and compost products appears to be a valuable strategy to assure that these products will be of interest to purchasers in the marketplace (Brown and Chaney, 2000).

B. Lime-Induced Manganese Deficiency

In research on biosolids use in Maryland, Brown et al. (1997a) have found some instances in which application of limed-biosolids [$Ca(OH)_2$ added to reduce malodor and pathogens] on low Mn Coastal Plain soils caused Mn deficiency in susceptible plants such as wheat and soybeans. This deficiency can easily be avoided by limiting the amount of biosolids-applied $CaCO_3$ to the amount needed to reach pH 6.5 rather than using N supply as the limiting factor in annual applications. They have verified methods to cure Mn deficiency in the plants or in the soils to correct any limed-biosolids-induced Mn deficiency that is observed (Brown et al., 1997a). Further, they have tested the utility of adding Mn during biosolids manufacturing to prevent potential Mn deficiency where limed biosolids are land applied. So far, $MnSO_4$ added to dewatered limed biosolids can prevent Mn deficiency of wheat grown on susceptible soil at pH 7.5 (Brown and Chaney, 1998). U.S. Coastal Plain soils are often incubated under warm, anaerobic conditions, and have lost much or nearly all of their original Mn and Fe. Adding Mn-rich byproducts to biosolids to prevent lime-induced Mn deficiency is also part of the “tailor-made biosolids and composts” model because additions of wastes from other industries, or of ores for Fe or Mn, can increase the potential benefit from agricultural use of these products. The compost and biosolids research community should search for other ways to improve the beneficial aspects of biosolids and composts to maintain an adequate market to assure that treatment works can market their products at minimal costs. These materials can be manufactured to contain high levels of Fe or Mn, and when used with drip irrigation, can prevent Mn or Fe deficiency in crops where these problems are common (such as in the western U.S. with highly calcareous soils). These concepts extend to most organic matter and livestock waste resources, not just biosolids.

IV. SUMMARY

The levels of nutrients and metals in composts depend on the materials used to produce the composts. If composts have been appropriately matured, nutrients are in plant available forms for crop production, and the compost pH will be near neutral. After 25 years of research and development of regulations and advice for biosolids and compost utilization, pretreatment of industrial wastes allows biosolids composts, and composts prepared from biosolids mixed with MSW or yard debris, to contain levels of microelements needed for plant nutrition but not high levels that could cause phytotoxicity (except under extremely low soil pH not appropriate for crop production). Composts can supply N, P, K, S, Ca, Mg, Fe, Zn, Cu, Mn, B, molybdenum (Mo), and selenium (Se) required by plants or animals. When used in potting media, supplemental N fertilization is usually required depending on crop requirements. Use of compost can replace other forms of microelements used as fertilizers in media or fields.

Detailed evaluation of potential food-chain transfer of Cd, Pb, and other elements in composts clearly shows that consumption for 70 years (lifetime) of 60% of garden foods (an avid gardener) produced on pH 5.5 soils with 1000 Mg of compost per ha would not comprise risk to these highly exposed individuals, nor would ingesting the composts at 200 mg per day for 5 years. These risk assessment models are very conservative, showing that under normal use compost does not cause problems with soil fertility or food safety. Potentially toxic organic compounds are either destroyed during composting, or bound very strongly by the compost so that plant uptake is trivial. Phytoavailability and bioavailability of trace elements in compost-amended soils are low compared to assumptions of toxicologists. With the normal >100 g Zn per 1 g Cd present in composts, crop Zn inhibits intestinal absorption of the small increase in crop Cd that is possible at very acidic soil pH, preventing risk regardless of the fraction of diet grown on the amended soil. Regulations should reflect measured phytoavailability and bioavailability of compost-applied trace elements in the field such as used in the 40 CFR 503 Rule for biosolids, which covers composts including biosolids. Compost use can be a safe and wise choice for both home and commercial use to replace peat, uncomposted manures, etc. Many states have developed regulatory controls to assure that pathogenic organisms are killed during composting and that product quality standards are attained, which allow marketing for general use in the community.

Composts can have remarkable benefits when used to remediate disturbed land, mine wastes, and even hazardous soils. By combining many inorganic and organic waste materials from farms, cities, and industry; composting to produce a stable mature compost; and applying the compost to meet crop cultural goals, it is possible to inexpensively manufacture "tailor-made" composts for both commercial products and for solving environmental problems. The balanced plant nutrition, high microbial inoculum potential, and microbial energy source in a mature compost aids in rapid biodegradation of most xenobiotic organic compounds; thus composts can remediate soils containing hazardous levels of many xenobiotics. The slow-release N and high phosphate and metal adsorption capacity of Fe-rich composts plus limestone can remediate barren Zn/Pb mine wastes, allowing one-shot revegetation. Further, such composts can help protect urban children from high Pb soils by precipitating and adsorbing soil Pb, thereby reducing the bioavailability of soil Pb to children who ingest such soils. Strong turfgrass growth on these soils also reduces children's exposure to soil Pb. Well-manufactured modern composts do not comprise environmental risk even at massive cumulative applications, and they can be utilized to help solve important societal problems for which other cost-effective solutions have not been available.

REFERENCES

Andersson, A. 1983. Composted municipal refuse as fertilizer and soil conditioner. Effects on the contents of heavy metals in soil and plant, as compared to sewage sludge, manure and commercial fertilizers, p. 146–156. In: S. Berglund, R.D. Davis, and P. L'Hermite (eds.). *Utilization of Sewage Sludge on Land: Rates of Application and Long-Term Effects of Metals*. D. Reidel Publishers, Dordrecht, The Netherlands.

Angle, J.S. 1998. Impact of biosolids and co-utilization wastes on rhizobia, nitrogen fixation and growth of legumes, p. 235–245. In: S.L. Brown, J.S. Angle, and L.W. Jacobs (eds.). *Beneficial Co-Utilization of Agricultural, Municipal and Industrial Byproducts*. Kluwer Academic Publishers, Dordrecht, The Netherlands.

Baldwin, A., T.A. Brown, P.H.T. Beckett, and G.E.P. Elliott. 1983. The forms of combination of Cu and Zn in digested sewage sludge. *Water Research* 17:1935–1944.

Beckett, P.H.T., R.D. Davis, and P. Brindley. 1979. The disposal of sewage sludge onto farmland: the scope of the problems of toxic elements. *Water Pollution Control* 78:419–445.

Brallier, S., R.B. Harrison, C.L. Henry, and X. Dongsen. 1996. Liming effects on availability of Cd, Cu, Ni and Zn in a soil amended with sewage sludge 16 years previously. *Water, Air and Soil Pollution* 86:195–206.

Brown, S.L. and R.L. Chaney. 1998. Manganese deficiency induced by lime rich co-utilization products, p. 289–298. In: S.L. Brown, J.S. Angle, and L.W. Jacobs (eds.). *Beneficial Co-Utilization of Agricultural, Municipal and Industrial Byproducts*. Kluwer Academic Publishers, Dordrecht, The Netherlands.

Brown, S.L. and R.L. Chaney. 2000. Combining residuals to achieve specific soil amendment objectives, p. 343–360. In: W.A. Dick (ed.). *Land Application of Agricultural, Industrial and Municipal Byproducts*. Soil Science Society of America, Madison, Wisconsin.

Brown, S.L., R.L. Chaney, C.A. Lloyd, J.S. Angle, and J.A. Ryan. 1996. Relative uptake of cadmium by garden vegetables and fruits grown on long-term sewage sludge amended soils. *Environmental Science and Technology* 30:3508–3511.

Brown, S.L., J.S. Angle, and R.L. Chaney. 1997a. Correction of limed-biosolids induced manganese deficiency on a long-term field experiment. *Journal of Environmental Quality* 26:1375–1384.

Brown, S.L., Q. Xue, R.L. Chaney, and J.G. Hallfrisch. 1997b. Effect of biosolids processing on the bioavailability of Pb in urban soils, p. 43–54. In: *Biosolids Management — Innovative Treatment Technologies and Processes*. Proceedings of the Water Environment Research Foundation Workshop #104, Chicago, Illinois, 4 Oct. 1997. Water Environment Research Foundation, Alexandria, Virginia.

Brown, S.L., R.L. Chaney, J.S. Angle, and J.A. Ryan. 1998a. The phytoavailability of cadmium to lettuce in long-term biosolids-amended soils. *Journal of Environmental Quality* 27:1071–1078.

Brown, S.L., C.L. Henry, R.L. Chaney, and H. Compton. 1998b. Bunker Hill Superfund Site: ecological restoration program, p. 388–393. In: D. Throgmorton, J. Nawrot, J. Mead, J. Galetovic, and W. Joseph (eds.). *Proceedings of the 15th National Meeting of the American Society for Surface Mining and Reclamation*, St. Louis, Missouri, 17–21 May.

Cai, S., Y. Lin, H. Zhineng, Z. Xianzu, Y. Zhaolu, X. Huidong, L. Yuanrong, J. Rongdi, Z. Wenhau, and Z. Fangyuan. 1990. Cadmium exposure and health effects among residents in an irrigation area with ore dressing wastewater. *Science of the Total Environment* 90:67–73.

Chambers, D.B., R.L. Chaney, B.R. Conard, N.C. Garisto, U. Kukier, H.A. Phillips, and S. Fernandes. 1998. *Risk Assessment for Nickel in Soil, with a Critical Review of Soil–Nickel Phytotoxicity, and Annotated Bibliography on Nickel in Soil, Plants, and the Environment*. White Paper Report to Ontario, Canada Ministry of the Environment for Inco Ltd.

Chaney, R.L. 1973. Crop and food chain effects of toxic elements in sludges and effluents, p. 120–141. In: *Proceedings of the Joint Conference on Recycling Municipal Sludges and Effluents on Land*. National Association of State Universities and Land Grant Colleges, Washington, D.C.

Chaney, R.L. 1983. Potential effects of waste constituents on the food chain, p. 152–240. In: J.F. Parr, P.B. Marsh, and J.M. Kla (eds.). *Land Treatment of Hazardous Wastes*. Noyes Data Corporation, Park Ridge, New Jersey.

Chaney, R.L. 1993a. Risks associated with the use of sewage sludge in agriculture, p. 7–31. In: *Proceedings of the 15th Federal Convention of the Australian Water and Wastewater Association*, Gold Coast, Queensland, 18–23 Apr. Volume 1. Australian Water and Wastewater Association (Queensland Branch), Queensland, Australia.

Chaney, R.L. 1993b. Zinc phytotoxicity, p. 135–150. In: A.D. Robson (ed.). *Zinc in Soils and Plants*. Kluwer Academic Publishers, Dordrecht, The Netherlands.

Chaney, R.L. and J.A. Ryan. 1993. Heavy metals and toxic organic pollutants in MSW-composts: Research results on phytoavailability, bioavailability, etc., p. 451–506. In: H.A.J. Hoitink and H.M. Keener (eds.). *Science and Engineering of Composting: Design, Environmental, Microbiological and Utilization Aspects*. Renaissance Publications, Worthington, Ohio.

Chaney, R.L. and J.A. Ryan. 1994. *Risks Based Standards for Arsenic, Lead and Cadmium in Urban Soils*. DECHEMA, Frankfurt, Germany.

Chaney, R.L. and D.P. Oliver. 1996. Sources, potential adverse effects of and remediation of agricultural soil contaminants, p. 323–359. In: R. Naidu, R.S. Kookana, D.P. Oliver, S. Rogers, and M.J. McLaughlin (eds.). *Contaminants and the Soil Environment in the Australasia-Pacific Region*. Kluwer Academic Publishers, Dordrecht, The Netherlands.

Chaney, R.L., G.S. Stoewsand, C.A. Bache, and D.J. Lisk. 1978a. Cadmium deposition and hepatic microsomal induction in mice fed lettuce grown on municipal sludge-amended soil. *Journal of Agricultural and Food Chemistry* 26:992–994.

Chaney, R.L., G.S. Stoewsand, A.K. Furr, C.A. Bache, and D.J. Lisk. 1978b. Elemental content of tissues of guinea pigs fed Swiss chard grown on municipal sewage sludge-amended soil. *Journal of Agricultural and Food Chemistry* 26:994–997.

Chaney, R.L., J.B. Munns, and H.M. Cathey. 1980. Effectiveness of digested sludge compost in supplying nutrients for soilless potting media. *Journal of the American Society for Horticultural Science* 105:485–492.

Chaney, R.L., S.B. Sterrett, M.C. Morella, and C.A. Lloyd. 1982. Effect of sludge quality and rate, soil pH, and time on heavy metal residues in leafy vegetables, p. 444–458. In: *Proceedings of the Fifth Annual Madison Conference on Applied Research and Practice with Municipal and Industrial Waste*. University of Wisconsin–Extension, Madison, Wisconsin.

Chaney, R.L., J.A. Ryan, and G.A. O'Connor. 1996. Organic contaminants in municipal biosolids: Risk assessment, quantitative pathways analysis, and current research priorities. *Science of the Total Environment* 185:187–216.

Chaney, R.L., J.A. Ryan, and S.L. Brown. 1997. Development of the U.S. EPA limits for chromium in land-applied biosolids and applicability of these limits to tannery by-product derived fertilizers and other Cr-rich soil amendments, p. 229–295. In: S. Canali, F. Tittarelli, and P. Sequi (eds.). *Chromium Environmental Issues*. Franco Angeli, Milano, Italy.

Chaney, R.L., S.L. Brown, and J.S. Angle. 1998. Soil-root interface: ecosystem health and human food-chain protection, p. 279–311. In: P.M. Huang, D.C. Adriano, T.J. Logan, and R.T. Checkai (eds.). *Soil Chemistry and Ecosystem Health.* Soil Science Society of America (SSSA), Madison, Wisconsin. SSSA Special Publication Number 52.

Chaney, R.L., J.A. Ryan, Y.-M. Li, and S.L. Brown. 1999. Soil cadmium as a threat to human health, p. 219–256. In: M.J. McLaughlin and B.R. Singh (eds.). *Cadmium in Soils and Plants*. Kluwer Academic Publishers, Dordrecht, The Netherlands.

Chang, A.C., T.C. Granato, and A.L. Page. 1992. A methodology for establishing phytotoxicity criteria for chromium, copper, nickel, and zinc in agricultural land application of municipal sewage sludges. *Journal of Environmental Quality* 21:521–536.

Chang, A.C., H.N. Hyun, and A.L. Page. 1997. Cadmium uptake for Swiss chard grown on composted sewage sludge treated field plots: Plateau or time-bomb? *Journal of Environmental Quality* 26:11–19.

Chaudri, A.M., S.P. McGrath, K.E. Giller, E. Rietz, and D.R. Sauerbeck. 1993. Enumeration of indigenous *Rhizobium leguminosarum* biovar *trifolii* in soils previously treated with metal-contaminated sewage sludge. *Soil Biology and Biochemistry* 25:301–309.

Cook, J. and J. Beyea. 1998. Potential toxic and carcinogenic chemical contaminants in source-separated municipal solid waste composts: review of available data and recommendations. *Toxicology and Environmental Chemistry* 67:27–69.

Corey, R.B., L.D. King, C. Lue-Hing, D.S. Fanning, J.J. Street, and J.M. Walker. 1987. Effects of sludge properties on accumulation of trace elements by crops, p. 25–51. In: A.L. Page, T.J. Logan, and J.A. Ryan (eds.). *Land Application of Sludge — Food Chain Implications.* Lewis Publishers, Chelsea, Michigan.

Daniels, W.L., T. Stuczynski, K. Pantuck, R. Chaney, and F. Pistelok. 1998. Stabilization and revegetation of smelter wastes in Poland with biosolids, p. 267–273. In: *Proceedings of the 12th Annual Residuals and Biosolids Management Conference*, Bellevue, Washington, 12–15 July 1998. Water Environment Federation, Alexandria, Virginia.

Davies, B.E. 1992. Trace metals in the environment: Retrospect and prospect, p. 1–17. In: D.C. Adriano (ed.). *Biogeochemisty of Trace Metals*. Lewis Publishers, Boca Raton, Florida.

Davis, R.D. and P.H.T. Beckett. 1978. Upper critical levels of toxic elements in plants. II. Critical levels of copper in young barley, wheat, rape, lettuce and ryegrass, and of nickel and zinc in young barley and ryegrass. *New Phytologist* 80:23–32.

Decker, A.M., R.L. Chaney, J.P. Davidson, T.S. Rumsey, S.B. Mohanty, and R.C. Hammond. 1980. Animal performance on pastures topdressed with liquid sewage sludge and sludge compost, p 37–41. In: *Proceedings of the National Conference on Municipal and Industrial Sludge Utilization and Disposal*. Information Transfer, Inc., Silver Spring, Maryland.

de Haan, S. 1981. Results of municipal waste compost research over more than fifty years at the Institute for Soil Fertility at Haren/Groningen, The Netherlands. *Netherlands Journal of Agricultural Science* 29:49–61.

deVries, M.P.C. and K.G. Tiller. 1978. Sewage sludge as a soil amendment, with special reference to Cd, Cu, Mn, Ni, Pb, and Zn — Comparison of results from experiments conducted inside and outside a greenhouse. *Environmental Pollution* 16:213–240.

Epstein, E., R.L. Chaney, C. Henry, and T.J. Logan. 1992. Trace elements in municipal solid waste compost. *Biomass and Bioenergy* 3:227–338.

Ewers, U., I. Freier, M. Turfeld, A. Brockhaus, I. Hofstetter, W. König, J. Leisner-Saaber, and T. Delschen. 1993. Heavy metals in garden soil and vegetables from private gardens located in lead/zinc smelter area and exposure of gardeners to lead and cadmium (in German). *Gesundheitswesen* 55:318–325.

Fließbach, R. Martens, and H.H. Reber. 1994. Soil microbial biomass and microbial activity in soils treated with heavy metal contaminated sewage sludge. *Soil Biology and Biochemistry* 26:1201–1205.

Friberg, L., C.-G. Elinder, T. Kjellstrom, and G.F. Nordberg (eds.). 1985. *Cadmium and Health: A Toxicological and Epidemiological Appraisal*. Volume 1. Exposure, dose, and metabolism. CRC Press, Boca Raton, Florida.

Fricke, K., W. Pertl, and H. Vogtmann. 1989. Technology and undesirable components on compost of separately collected organic wastes. *Agriculture, Ecosystems and the Environment* 27:463–469.

Gilmour, J.T. 1998. Carbon and nitrogen mineralization during co-utilization of biosolids and composts, p. 89–112. In: S. Brown, J.S. Angle, and L. Jacobs (eds.). *Beneficial Co-utilization of Agricultural, Municipal and Industrial Byproducts*. Kluwer Academic Publishers, Dordrecht, The Netherlands.

Gilmour, J.T. and V. Skinner. 1999. Predicting plant available N in land-applied biosolids. *Journal of Environmental Quality* 28:1122–1126.

Harrison, E.Z., M.B. McBride, and D.R. Bouldin. 1999. *The Case for Caution: Recommendations for Land Application of Sewage Sludges and an Appraisal of the U.S. EPA's Part 503 Sludge Rules*. Cornell Waste Management Institute Working Paper, Ithaca, New York, February 1999. 40 pp.

Hinesly, T.D., L.G. Hansen, and D.J. Bray. 1984. *Use of Sewage Sludge on Agricultural and Disturbed Lands*. U.S. Environmental Protection Agency. 600/S2-84–127. PB84-224419, National Technical Information Service, Springfield, Virginia.

Hornick, S.B., L.J. Sikora, S.B. Sterrett, J.J. Murray, P.D. Millner, W.D. Burge, D. Colacicco, J.F. Parr, R.L. Chaney, and G.B. Willson. 1983. Application of sewage sludge compost as a soil conditioner and fertilizer for plant growth. *USDA Agricultural Information Bulletin* 464:1–32.

Ibekwe, A.M., J.S. Angle, R.L. Chaney, and P. van Berkum. 1997a. Enumeration and N_2 fixing potential of *Rhizobium leguminosarum* biovar *trifolii* grown in soil with varying pH's and heavy metal concentrations. *Agriculture, Ecosystems and the Environment* 61:103–111.

Ibekwe, A.M., J.S. Angle, R.L. Chaney, and P. van Berkum. 1997b. Differentiation of clover *Rhizobium* isolated from biosolids-amended soils with varying pH. *Soil Science Society of America Journal* 61:1679–1685.

Jing, J. and T.J. Logan. 1992. Effects of sewage sludge cadmium concentration on chemical extractability and plant uptake. *Journal of Environmental Quality* 21:73–81.

Johnson, N.B., P.H.T. Beckett, and C.J. Waters. 1983. Limits of zinc and copper toxicity from digested sludge applied to agricultural land, p. 75–81. In: R.D. Davis, G. Hucker, and P. L'Hermite (eds.). *Environmental Effects of Organic and Inorganic Contaminants in Sewage Sludge.* D. Reidel Publishers, Dordrecht, The Netherlands.

King, L.D. and H.D. Morris. 1972. Land disposal of liquid sewage sludge: II. The effect on soil pH, manganese, zinc, and growth and chemical composition of rye (*Secale cereale* L.). *Journal of Environmental Quality* 1:425–429.

L'Huillier, L., J. d'Auzac, M. Durand, and N. Michaud-Ferriere. 1996. Nickel effects on two maize (*Zea mays*) cultivars: growth, structure, Ni concentration, and localization. *Canadian Journal of Botany* 74:1547–1554.

Li, Y.-M. and R.L. Chaney. 1998. Phytostabilization of zinc smelter contaminated sites — The Palmerton case, p. 197–210. In: J. Vangronsveld and S.D. Cunningham (eds.). *Metal-Contaminated Soils: In-situ Inactivation and Phytorestoration.* Landes Bioscience, Austin, Texas.

Li, Y.-M., R.L. Chaney, G. Siebielec, and B.A. Kershner. 2000. Response of four turfgrass cultivars to limestone and biosolids compost amendment of a zinc and cadmium contaminated soil at Palmerton. *Journal of Environmental Quality* 29:1440–1447.

Logan, T.J. and R.L. Chaney. 1983. Utilization of municipal wastewater and sludges on land — Metals, p. 235–323. In: A.L. Page, T.L. Gleason III, J.E. Smith, Jr., I.K. Iskander, and L.E. Sommers (eds.). *Proceedings of the 1983 Workshop on Utilization of Municipal Wastewater and Sludge on Land.* University of California, Riverside.

Lutrick, M.C., W.K. Robertson, and J.A. Cornell. 1982. Heavy applications of liquid-digested sludge on three Ultisols: II. Effects on mineral uptake and crop yield. *Journal of Environmental Quality* 11:283–287.

Mahler, R.J. and J.A. Ryan. 1988a. Cadmium sulfate application to sludge-amended soils: II. Extraction of Cd, Zn, and Mn from solid phases. *Communications in Soil Science and Plant Analysis* 19:1747–1770.

Mahler, R.J. and J.A. Ryan. 1988b. Cadmium sulfate application to sludge-amended soils: III. Relationship between treatment and plant available cadmium, zinc, and manganese. *Communications in Soil Science and Plant Analysis* 19:1771–1794.

Mahler, R.J., J.A. Ryan, and T. Reed. 1987. Cadmium sulfate application to sludge-amended soils. I. Effect on yield and cadmium availability to plants. *Science of the Total Environment* 67:117–131.

Marks, M.J., J.H. Williams, and C.G. Chumbley. 1980. Field experiments testing the effects of metal-contaminated sewage sludges on some vegetable crops, p. 235–251. In: *Inorganic Pollution and Agriculture.* Ministry of Agriculture, Fisheries and Food Reference Book 326. Her Majesty's Stationery Office, London, United Kingdom.

Mays, D.A. and P.M. Giordano. 1989. Landspreading municipal waste compost. *BioCycle* 30(3):37–39.

McBride, M.B. 1995. Toxic metal accumulation from agricultural use of sludge: are U.S. EPA regulations protective. *Journal of Environmental Quality* 24:5–18.

McGrath, S.P., P.C. Brookes, and K.E. Giller. 1988. Effects of potentially toxic metals in soil derived from past applications of sewage sludge on nitrogen fixation by *Trifolium repens* L. *Soil Biology and Biochemistry* 20:415–424.

McGrath, S.P., F.J. Zhao, S.F. Dunham, A.R. Crosland, and K. Coleman. 2000. Long-term changes in extractability and bioavailability of zinc and cadmium after sludge application. *Journal of Environmental Quality* 29:875–883.

National Research Council. 1996. *Use of Reclaimed Water and Sludge in Food Crop Production.* National Research Council, National Academy Press, Washington, D.C.

Nogawa, K., R. Honda, T. Kido, I. Tsuritani, and Y. Yamada. 1987. Limits to protect people eating cadmium in rice, based on epidemiological studies. *Trace Substances in Environmental Health* 21:431–439.

Page, A.L. 1974. *Fate and Effects of Trace Elements in Sewage Sludge when Applied to Agricultural Lands: A Literature Review Study.* U.S. Environmental Protection Agency (U.S. EPA) Report No. EPA-670/2-74-005. U.S. EPA, Washington, D.C.

Page, A.L., T.J. Logan, and J.A. Ryan (eds.). 1989. *W-170 Peer Review Committee Analysis of the Proposed 503 Rule on Sewage Sludge.* CSRS Technical Committee W-170, University of California-Riverside.

Parker, D.R., R.L. Chaney, and W.A. Norvell. 1995. Equilibrium computer models: applications to plant nutrition research, p. 163–200. In: R.H. Loeppert, A.P. Schwab, and S. Goldberg (eds.). *Chemical Equilibrium and Reaction Models.* Soil Science Society of America (SSSA), Madison, Wisconsin. SSSA Special Publication Number 42.

Richards, T.L. and P.B. Woodbury. 1992. The impact of separation on heavy metal contaminants in municipal solid waste compost. *Biomass and Bioenergy* 3:195–212.

Ryan, J.A. and R.L. Chaney. 1993. Regulation of municipal sewage sludge under the Clean Water Act Section 503: A model for exposure and risk assessment for MSW-compost, p. 422–450. In: H.A.J. Hoitink and H.M. Keener (eds.). *Science and Engineering of Composting: Design, Environmental, Microbiological and Utilization Aspects.* The Ohio State University, Columbus, Ohio.

Ryan, J.A. and R.L. Chaney. 1997. Issues of risk assessment and its utility in development of soil standards: the 503 methodology as an example, p. 393–414. In: R. Prost (ed.). *Contaminated Soils: Proceedings of the Third International Symposium on Biogeochemistry of Trace Elements*, Paris, France, 15–19 May 1995. Colloque No. 85, INRA Editions, Paris, France.

Sanderson, K.C. 1980. Use of sewage-refuse compost in the production of ornamental plants. *HortScience* 15:173–178.

Sarasua, S.M., M.A. McGeehin, F.L. Stallings, G.J. Terracciano, R.W. Amler, J.N. Logue, and J.M. Fox. 1995. *Final Report. Technical Assistance to the Pennsylvania Department of Health. Biologic indicators of exposure to cadmium and lead. Palmerton, Pennsylvania. Part II.* (May 1995). 57 pp. Agency for Toxic Substances and Disease Registry, U.S. Department of Health and Human Services, Atlanta, Georgia.

Siebielec, G. and R.L. Chaney.1999. Metals solubility in two contaminated soils treated with lime or compost, p. 784-785. In: W. Wentzel et al. (eds.). *Proceedings and Extended Abstracts from the 5th International Conference on the Biogeochemistry of Trace Elements*, Vienna, Austria, 11–15 July 1999. International Society for Trace Element Research, Vienna, Austria.

Sloan, J.J., R.H. Dowdt, M. S. Dolan, and D.R. Linden. 1997. Long-term effects of biosolids applications on heavy metal bioavailability in agricultural soils. *Journal of Environmental Quality* 26:966–974.

Smith, S. E. 1996. *Agricultural Recycling of Sewage Sludge and the Environment*. CAB International, Wallingford, United Kingdom, 382 pp.

Smith, G.S., D.M. Hallford, and J.B. Watkins III. 1985. Toxicological effects of gamma-irradiated sewage solids fed as seven percent of diet to sheep for four years. *Journal of Animal Science* 61:931–941.

Stern, A.H. 1993. Monte Carlo analysis of the U.S. EPA model of human exposure to cadmium in sewage sludge through consumption of garden crops. *Journal of Exposure Analysis and Environmental Epidemiology* 3:449–469.

Sterrett, S.B., R.L. Chaney, C.W. Reynolds, F.D. Schales, and L.W. Douglass. 1982. Transplant quality and metal concentrations in vegetable transplants grown in media containing sewage sludge compost. *HortScience* 17:920–922.

Sterrett, S.B., C.W. Reynolds, F.D. Schales, R.L. Chaney, and L.W. Douglass. 1983. Transplant quality, yield, and heavy metal accumulation of tomato, muskmelon, and cabbage grown in media containing sewage sludge compost. *Journal of the American Society for Horticultural Science* 108:36–41.

Sterrett, S.B., R.L. Chaney, C.E. Hirsch, and H.W. Mielke. 1996. Influence of amendments on yield and heavy metal accumulation of lettuce grown in urban garden soils. *Environmental Geochemistry and Health* 18:135–142.

Strehlow, C.D. and D. Barltrop. 1988. The Shipham Report — An investigation into cadmium concentrations and its implications for human health: 6. Health studies. *Science of the Total Environment* 75:101–133.

Stuczynski, T.I., W.L. Daniels, F. Pistelok, K. Pantuck, R.L. Chaney, and G. Siebielec. 2000. Application of sludges for remediation of contaminated soil environment, p. xxx. In: *Proceedings of the NATO Conference: Soil Quality in Relation to Sustainable Development of Agriculture and Environmental Security in Central and Eastern Europe*, Pulawy, Poland, 13–17 Oct. 1997. In press.

United States Department of Agriculture (USDA). 1993. Letter from Administrator Finney of the USDA-Agricultural Research Service to Administrator Browner of the U.S. EPA, May 19, 1993, reporting USDA comments and recommendations on the 40 CFR 503 Rule.

United States Environmental Protection Agency (U.S. EPA). 1989a. *Development of Risk Assessment Methodology for Land Application and Distribution and Marketing of Municipal Sludge*. EPA/600/6-89/001.

United States Environmental Protection Agency (U.S. EPA). 1989b. *Standards for the Disposal of Sewage Sludge*. Federal Register 54(23):5746–5902.

United States Environmental Protection Agency (U.S. EPA). 1990. *National Sewage Sludge Survey. Availability of Information and Data, and Expected Impacts on Proposed Regulations: Proposed Rule*. Federal Register 55(218):472448–47283.

United States Environmental Protection Agency (U.S. EPA). 1992. *Technical Support Document for the 40 CFR 503 Standards for the Use or Disposal of Sewage Sludge*. U.S. EPA, Washington, D.C.

United States Environmental Protection Agency (U.S. EPA). 1993. 40 CFR Part 257 et al. *Standards for the Use or Disposal of Sewage Sludge; Final Rules*. Federal Register 58(32):9248–9415.

White, M.C. and R.L. Chaney. 1980. Zinc, cadmium, and manganese uptake by soybean from two zinc- and cadmium-amended Coastal Plain soils. *Soil Science Society of America Journal* 44:308–313.

Woodbury, P.B. 1992. Trace elements in municipal solid waste composts: A review of potential detrimental effects on plants, soil biota, and water quality. *Biomass and Bioenergy* 3:239–259.

Wright, R.J. (ed.). 1999. *Agricultural Uses of Municipal Animal, and Industrial Byproducts.* Agricultural Research Service, Beltsville, Maryland. 127 pp. Available from USDA-ARS website: http://www.ars.usda.gov/is/np/agbyproducts/agbyintro.htm. Verified 2 June 2000.

CHAPTER 17

Human Pathogens: Hazards, Controls, and Precautions in Compost

Eliot Epstein

CONTENTS

I. INTRODUCTION

Waste materials used in the production of compost can contain primary pathogens and secondary pathogens. Biosolids, municipal solid waste (MSW), manures, yard waste, and food waste can contain pathogens from humans and domestic and farm animals.

Human pathogens include organisms, viruses, or substances that can infect humans. Most of us are familiar with pathogenic bacteria, such as *Salmonella*, or viruses. However, there are substances, such as endotoxins or prions, that are known

1-56670-460-X/01/$0.00+$.50

to cause diseases. Endotoxins are heat-stable lipopolysaccharide-protein complexes that are part of the outer cell wall of Gram-negative bacteria. A prion (proteinaceous infectious particle) is believed to be a natural protein that is converted to an abnormal (PrPsc) form. Bovine Spongioform Encephalopathy (BSE), or mad cow disease, is believed to be caused by a prion.

Heightened awareness of diseases has resulted in public concern about pathogens in foods and about the potential use of waste-derived products in crop production. Biosolids, a product of wastewater treatment, and biosolids compost are regulated by the U.S. Environmental Protection Agency (U.S. EPA) and state agencies. Compost derived from MSW, yard materials, and food waste is not regulated by the federal government, but many states use the U.S. EPA regulations to control the quality of these materials. Animal wastes and their products are generally not regulated. Pell (1997) indicated that there are numerous pathogens in livestock manure that can infect humans. The principal ones are *Cryptosporidium parvum*, *Giardia* spp., *Listeria monocytogenes*, *Escherichia coli* (*E. coli*) O157:H7, *Salmonella* spp., and *Mycobacterium paratuberculosis*.

Pathogens are generally classified as primary or secondary. Often the latter are also termed "opportunistic pathogens." A primary pathogen can invade and infect healthy persons, whereas a secondary, or opportunistic, pathogen invades and infects debilitated individuals or persons with highly suppressed immune systems or those on immunosuppressant medication.

The principal primary pathogens are classified as bacteria, viruses, protozoa, and helminths. The principal secondary pathogens of concern in composting are fungi and bacterial endotoxins. Infection by primary pathogens is usually through ingestion, whereas infection by secondary pathogens is through the respiratory system.

Pathogens are ubiquitous. They are present in the air we breathe, the water we drink, and the food we eat. We are continuously exposed to pathogens. However, to be infected, three conditions must occur. First, the individual must be exposed to the organism or substance; second, the individual must be susceptible; and third, there needs to be a sufficient quantity of the organism or substance to result in an infection. This latter condition is termed "infective dose." The infective dose is not only specific for different pathogens but is also a function of individual susceptibility. The ability of a pathogen to cause infection and disease in an exposed individual depends on the number of viable organisms and their virulence as well as susceptibility of the host. Table 17.1 shows some infective doses for several pathogens. For example, for *Salmonella*, it took 10,000 organisms to infect an individual, whereas for *Shigella*, it took only 10 organisms. Very low doses of enteric viruses have been shown to produce an infection (Smith and Farrell, 1996).

There are different pathways by which an individual can be infected by pathogens. These include ingestion, inhalation, and dermal contact. The primary potential mode for infection by individuals handling wastes or waste products is through ingestion. Thus, a key to prevention is good hygienic practices.

Table 17.1 Infective Dose for Several Pathogens

Pathogen	Infective Dose (No. of Organisms)	Range (No. of Organisms)	Reference
	Bacteria		
Shigella flexneri	10^2	10^2–10^9	Kowal, 1985
Shigella dysenteria	10–10^2	10–10^9	Kowal, 1985 Keswick, 1984 Levine et al., 1973
Salmonella sp.	10^2	10^2–10^9	Kowal, 1985 Keswick, 1984
E. coli	10^4	10^4–10^{10}	Kowal, 1985
Vibrio cholerae	10^3	10^3–10^{11}	Kowal, 1985 Keswick, 1984
Streptococcus faecalis	10^9	10^9–10^{10}	Kowal, 1985
Clostridium perfrigens	10^6	10^6–10^{10}	Kowal, 1985
	Parasites		
Entamoeba coli	1–10 cysts	1–10 cysts	Kowal, 1985
Giardia lamblia	1 cyst	Not reported	Kowal, 1985
Cryptosporidium	10 cysts	10–100 cysts	Casemore, 1991
Helminths	1 egg	Not reported	Kowal, 1985
	Viruses		
Echovirus 12	HID50 919 PFU[z]	17–919 PFU	Kowal, 1985
Poliovirus	1TCID50, <1 PFU	4×10^7 TCID50[y]	Kowal, 1985
Rotavirus	HID50 10 ffu[x] HID25 1 ffu est.	0.9–9×10^6 ffu	Ward et al., 1986

[z] PFU = plaque forming units.
[y] TCID50 = tissue culture infection dose for 50% response.
[x] ffu = focus forming units.

II. PATHOGENS IN COMPOST FEEDSTOCKS

A. Primary Pathogens

There are many different feedstocks used for composting, including biosolids, MSW, food waste, yard waste, and animal waste. Primary pathogens in biosolids compost are regulated by the U.S. EPA under the regulations titled 40 CFR Part 503, promulgated February 19, 1993 (U.S. EPA, 1994). States have adopted these regulations and, in several cases, applied these regulations to MSW, food waste, and yard waste composts.

Tables 17.2a, 17.2b, 17.2c, 17.3, and 17.4 provide information on some of the more common pathogens found in the different feedstocks that are used for composting. In biosolids and MSW, the primary source of pathogens is human wastes, whereas in yard waste, domestic animal waste is the major contributor.

Table 17.2a Some Bacteria Found in Biosolids and the Diseases They Transmit

Bacteria	Disease	Reference
Salmonella (approximately 1,700 types)	Salmonellosis, gastroenteritis	Akin et al., 1978
Salmonella typhi	Typhoid fever	Akin et al., 1978
Mycobacterium tuberculosis	Tuberculosis	Akin et al., 1978
Shigellae spp.(4 species)	Shigellosis, bacterial dysentery, gastroenteritis	Akin et al., 1978
Campylobacter jejuni	Gastroenteritis	Smith and Farrell, 1996
E. coli (pathogenic strains)	Gastroenteritis	Akin et al., 1978; Smith and Farrell, 1996
Yarsinia spp.	Yersinosis	Smith and Farrell, 1996
Vibrio cholerae	Cholera	Smith and Farrell, 1996

Table 17.2b Some Viruses Found in Biosolids and the Diseases They Transmit

Virus	Disease	Reference
Adenovirus (31 types)	Conjectivitis, respiratory infections, gastroenteritis	Akin et al., 1978
Poliovirus	Poliomyelitis	Smith and Farrell, 1996
Coaxsachievirus	Aseptic meningitis, gastroenteritis	Smith and Farrell, 1996
Echovirus	Aseptic meningitis	Smith and Farrell, 1996
Reovirus	Respiratory infections, gastroenteritis	Smith and Farrell, 1996
Norwalk Agents	Epidemic gastroenteritis	Smith and Farrell, 1996
Hepatitis A virus	Infectious hepatitis	Smith and Farrell, 1996
Rotavirus	Gastroenteritis, infant diarrhea	Smith and Farrell, 1996

In the past few years, there has been concern that newer pathogens, resistant forms of existing pathogens, and pathogenic substances or toxins may be transmitted to humans through wastes. Much of this apprehension has been the result of outbreaks of diseases due to water and food contamination. In Milwaukee, WI, an outbreak of diarrhea resulting from *Cryptosporidium parvum* in drinking water infected 403,000 persons (MacKenzie et al., 1994). There have been cases of *E. coli* O157:H7 in hamburgers and *Salmonella* spp. in eggs and ice cream (Hennessy et al., 1996). Public concern has been focused on human immunodeficiency virus (HIV), which causes acquired immunodeficiency syndrome (AIDS); BSE, amyotrophic lateral sclerosis (ALS, or Lou Gehrig's disease); *Legionella* spp.; *E. coli* O157:H7; *Cryptosporidium parvum*; and *Campylobacter jejuni*.

The HIV virus is contracted through contact with blood or other body fluids of an infected individual. The conditions in the wastewater system are not favorable for survival of the virus. Also, the additional treatment that wastewater goes through to become Class A or B biosolids makes it virtually impossible that biosolids would contain the HIV virus (WEF/U.S. EPA Fact Sheet, 1997).

Table 17.2c Some Protozoa and Helminth Parasites Found in Biosolids and the Diseases They Transmit

Organism	Disease	Reference
	Protozoa	
Entamoeba histolytica	Amoebic dysentery, amebiasis, acute enteritis	Smith and Farrell, 1996
Giardia lamblia	Giardiasis, diarrhea, weight loss	Smith and Farrell, 1996
Balantidium coli	Balantidiasis, diarrhea, dysentery	Reimers et al., 1998
Cryptosporidium	Gastroenteritis	Reimers et al., 1998
Toxoplasma gondii	Toxoplasmosis	Smith and Farrell, 1996
	Helminths – Nematodes	
Ascaris suum	Fever, respiratory effects	Smith and Farrell, 1996
Ascaris lumbricoides	Ascariasis, digestive and nutritional disturbances, abdominal pain, vomiting	Smith and Farrell, 1996
Ancylostoma duodenale	Hook worm disease, Ancylostomiasis	Akin et al., 1978
Necator americanus	Hookworm disease	Smith and Farrell, 1996
Enterobius vermicularis (pinworm)	Enterobiasis, intestinal inflammation, mucosal necrosis	Akin et al., 1978
Strongyloides stercoarlis (threadworm)	Strongyloidiasis, abdominal pain, diarrhea	Akin et al., 1978
Toxocara canis (dog roundworm)	Fever, abdominal pain, neurological symptoms	Smith and Farrell, 1996
Trichuris trichiura (whip worm)	Trichuriasis, abdominal pain, diarrhea, anemia	Akin et al., 1978
	Helminths – Cestodes	
Taenia saginata (Beef tapeworm)	Taeniasis, nervousness, insomnia, anorexia, abdominal pain, digestive disturbances	Smith and Farrell, 1996
Taenia solium (pork tapeworm)	Taeniasis, nervousness, insomnia, anorexia, abdominal pain, digestive disturbances	Smith and Farrell, 1996
Hymenolepis nana (dwarf tapeworm)	Taeniasis	Smith and Farrell, 1996

The pathway of BSE transmission is through the ingestion of meat or bone meal from infected animals. There is no evidence that the BSE prion protein is shed in feces or urine and can enter the waste stream (Epstein, 1997).

ALS was once thought to be caused by heavy metals, and hence the concern that heavy metals in wastes could result in the disease (Epstein, 1997). Kurland (1987) indicated that the most promising leads in the etiology of ALS relate to a genetic factor or predisposition and an increasing likelihood that a specific neurotoxic amino acid is present in certain foods. There is no evidence that this disease can be caused through the handling of wastes or compost or that it is related to heavy metals.

Legionella longbeachae has been identified in potting soils, soils, composted bark, fresh pine (*Pinus* sp.) sawdust, and composted pine and eucalyptus (*Eucalyptus* sp.) sawdust in Australia (Hughes and Steele, 1994; Steele et al., 1990a, 1990b). The organism was found in 73% of 45 potting soils produced by 13 manufacturers, but was not detected in 19 potting soils from Greece, Switzerland, and the U.K. No

Table 17.3 Some Pathogens in Municipal Solid Waste (MSW) and Diseases They Can Cause

Organism	Disease
Hepatitis virus	Hepatitis A&E
Staphylococcus spp.	Skin infections, osteomyelitis, pneumonia, impetigo
Diplococcus pneumoniae	Loblar type pneumonia, osteomyelitis, arthritis, endocarditis
Arenavirus	Hemorrhagic fevers
Neisseria meningitides	Meningococcal
Arboviruses	HIV, hepatitis, encephalitis
Clostridium tetani	Tetanus
Mycobacterium tuberculosis	Pulmonary tuberculosis

Adapted from Geyer, 1994.

Table 17.4 Number of Pathogenic Organisms in Food Waste, Yard Waste, and Wood Wastes

Organism	Food Waste	Yard Waste	Yard Waste and Waste Paper	Wood Waste
		Indicator Organism		
Total coliform	5.00×10^6	8.0×10^5	5.00×10^5	1.30×10^6
Fecal coliform	2.00×10^4	8.00×10^5	5.00×10^5	1.30×10^6
E. coli	3.50×10^3	8.00×10^5	3.00×10^5	1.30×10^6
Fecal *Streptococcus*	8.00×10^6	1.60×10^6	1.60×10^6	1.60×10^6
Enterococcus	1.30×10^5	2.30×10^5	1.30×10^5	3.00×10^5
		Pathogens		
Salmonella spp.	<0.002	<0.002	0.36	<0.002
Staphylococcus	32.2	0.8	4.4	3.8
Listeria spp.	<0.02	0.02	<0.02	<0.02
Parasites	Protozoa	negative	negative	negative

After E&A Environmental Consultants, Inc. 1994.

data on this organism have been reported in the U.S. Composting temperatures exceeding 43°C would destroy this organism.

E. coli is a species that occurs normally in the intestines of man and other vertebrates and is ubiquitous. Some species can be pathogenic. *E. coli* O157:H7 is a verotoxin-producing strain and was recognized as a pathogen in 1982. As many as 20,000 cases and 250 deaths per year have been reported in the U.S. (Boyce et al., 1995; Council for Agricultural Science and Technology, 1994). Consumption of ground beef, lettuce (*Lactuca sativa* L.), raw apple (*Malus* x*domestica* Borkh.) cider, raw milk, and untreated water have been implicated in outbreaks (Pell, 1997). Composting at temperatures that meet U.S. EPA 40 CFR Part 503 regulations would eliminate this organism.

Cryptosporidium pavum is an enteric coccidian protozoan parasite and has been recognized as a human pathogen since 1976. It can cause diarrhea, nausea, abdominal cramps, and low-grade fever. Immunocompromised individuals are more susceptible. For humans, the infective dose can be as few as 30 oocysts (DuPont et al., 1995). As with other parasites, thermophilic composting temperatures would destroy this organism.

Disorders caused by *Campylobacter* sp. are among the most frequently occurring acute gastroenteritis diseases in humans (Koenraad et al., 1997). *Campylobacter jejuni* is now considered the leading cause of food-borne bacterial infection (Tauxe, 1992). Undercooked poultry products, raw milk, untreated surface water, and pets are considered the dominant vectors in transmission of this *Campylobacter* (Koenraad et al., 1997; Tauxe, 1992). *Campylobacter* survives in anaerobically digested biosolids but not in aerobically digested biosolids. It appears to be sensitive to oxygen. Koenraad et al. (1997) indicated that the land application of aerobically digested biosolids represents a low risk, but the application of anaerobically digested biosolids may present a possible hazard.

B. Secondary Pathogens

The secondary pathogens in compost are generally released as bioaerosols. As secondary pathogens, they principally infect highly debilitated or immunosuppressed individuals. Bioaerosols are airborne materials comprised of organisms or biological agents that may affect humans through infectivity, allergenicity, toxicity, or other means. Compost bioaerosols are organisms or biological agents that can be dispersed through the air and affect human health. These bioaerosols can contain living organisms, including bacteria, fungi, actinomycetes, arthropods, and protozoa, as well as microbial products, such as endotoxins, microbial enzymes, β-1,3 glucans, and mycotoxins (Millner et al., 1994). Organic dust may contain bioaerosols, such as endotoxins, fungi, and other materials, which can result in a condition termed "organic dust toxic syndrome" (ODTS). These bioaerosols are present in our daily environment and are not exclusive to compost. They are present in our homes, gardens, and other outdoor environments and in our work places (Epstein, 1997; Millner et al., 1994). As bioaerosols, their primary mode of infection is through the respiratory system. The primary bioaerosols of concern are the fungus *Aspergillus fumigatus* and endotoxins. Others of much less importance are glucans and mycotoxins (Millner et al., 1994).

Aspergillus fumigatus is a very common fungus that is associated with decaying organic matter and soil throughout the world. It is ubiquitous, found in the air, dust, furniture, air filters, potted plants, soil, food, homes (Hirsch and Sosman, 1976; Millner et al., 1994; Solomon, 1975), lawn clippings (Domsch et al., 1980; Slavin and Winzenburger, 1977), wood chips (Passman, 1980), compost (Kothary et al., 1984), and agricultural environments (Domsch et al., 1980; Millner et al., 1994).

Aspergillus fumigatus is considered an opportunistic, or secondary, pathogen and principally affects highly debilitated or immunosuppressed individuals. The majority of aspergillosis cases cited in the literature have occurred in hospitals (Patterson, 1995; Wingard, 1995). It is rare that nondebilitated individuals are infected. Exposure to *Aspergillus fumigatus* may result in allergenic reactions, including respiratory inflammation and irritation.

Endotoxins are lipopolysaccharides, nonliving components of the cell wall of Gram-negative bacteria. Endotoxins are found in water, soil, and living organisms throughout the world, since they are components of cell walls of ubiquitous Gram-negative bacteria. Endotoxins can trigger a response that may result in various levels of respiratory distress. Reactions to high concentrations of endotoxins may include acute fever, chest tightness, coughing, shortness of breath, and wheezing. Long-term exposure may lead to decreased pulmonary function and chronic bronchitis.

The primary concern with these two secondary pathogens is for workers exposed to high concentrations on a regular basis. Cases of ODTS have been reported among workers in recycling and composting operations in Europe (Sigsgaard et al., 1994). Aspergillosis, irritation due to *Aspergillus fumigatus*, and reactions to endotoxins have not been reported among workers in over 4000 composting facilities in the U.S. and Canada. During composting operations where large quantities of dust are generated, as with yard waste grinding and screening of compost, workers are advised to use dust masks or other respiratory protection devices.

III. PATHOGEN DESTRUCTION DURING COMPOSTING

Composting, if carried out properly, is very effective in the destruction of pathogens. Composting is a method of disinfection and not sterilization. The main means of pathogen destruction in composting are:

- High temperatures for established periods of time
- Ammonia disinfection
- Presence of competing organisms or antagonistic substances

A. The Time–Temperature Relationship

High temperatures for established periods of time are extremely effective in the destruction of pathogens. At higher temperatures, pathogens are destroyed more quickly; at lower temperatures, a longer time is needed. The time–temperature relationship is shown in Tables 17.5 and 17.6. The D value is the amount of time required to cause a tenfold (one log) reduction in the number of organisms (Table 17.5). Poliovirus was reduced by a factor of ten within 32 min at 55°C, whereas it took 89 min for the same level of reduction for *Salmonella senftenberg*. Relatively hardy organisms, such as *Salmonella* and bacteriophages, are therefore used to indicate the potential destruction of other pathogens.

Temperature is very effective in destroying bacteria, viruses, and parasites (Table 17.6). Figure 17.1 shows the effect of temperature on the inactivation of *Salmonella enteritidis* serotype Montevideo. Figure 17.2 shows the die-off rates for fecal *Streptococcus*. This temperature–time relationship is the basis of the U.S. EPA biosolids "Processes to Further Reduce Pathogens" (PFRP), as stated in 40 CFR Part 503 (U.S. EPA, 1994). During composting, temperature variation can occur throughout the mass. Therefore, the U.S. EPA, as well as many states in the U.S., specify different temperature–time criteria for different composting systems.

Table 17.5 Time-Temperature as Indicated by D Values for Destruction of Various Microorganisms by Heat

Organism	D values (min at Given Temperatures) 55°C	60°C
Adenovirus, 12 NIAID	11	0.17
Poliovirus, type 1	32	19
Ascaris ova	—	1.3
Histolytica cysts	44	25
Salmonella senftenberg 775W	89	7.5
Bacteriophage, f_2	267	47

Note: The D value (min) is the amount of time required to cause a 10-fold (one log) reduction in the number of organisms.

After Burge, 1983.

Table 17.6 Temperature–Time Relationship Required for the Destruction of Several Pathogens

Organism	Time (in min) for the Destruction of Organisms at Several Temperatures 50°C	55°C	60°C	65°C	70°C
	Bacteria				
Salmonella typhi	—	—	30	—	4
E. coli	—	—	60	—	5
Mycobacterium tuberculosis	—	—	—	—	20
Shigella spp.	60	—	—	—	—
Mycobacterium diphtheria	45	—	—	—	—
Brucella abortus	—	60	—	3	—
Corynebacterium diphtheriae	—	45	—	—	4
	Viruses				
Viruses	—	—	—	—	25
	Protozoa				
Entamoeba histolytica cysts	5	—	—	—	—
	Helminths				
Ascaris lumbricoides eggs	60	7	—	—	—
Necator americanus	50	—	—	—	—
Taenia saginata	—	—	—	—	5

After Stern, 1974.

This time–temperature relationship is the basis of pasteurization (70°C for 30 min or longer). The heat generated during composting as a result of the oxidative activities of thermophilic microorganisms is capable of destroying all four groups of primary pathogens.

Several studies using different feedstocks have illustrated the time–temperature relationship for pathogen destruction by composting. Knoll (1961) described several experiments where he subjected different *Salmonella* strains to composting temperatures at the Baden-Baden biosolids-refuse composting plant. After 14 days of

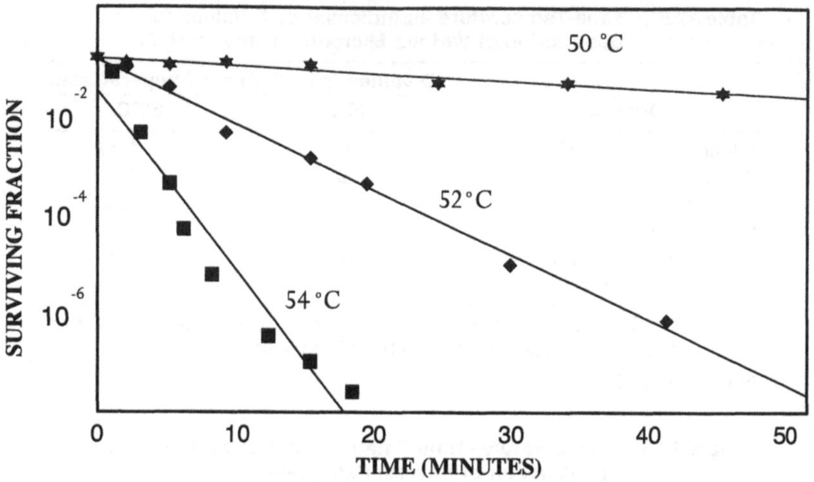

Figure 17.1 Heat inactivation of *Salmonella enteritidis* serotype Montevideo in composted sludge. (From Ward, R.L. and J.R. Brandon, 1978. Effect of heat on pathogenic organisms found in wastewater sludge, p. 122–134. In: *1977 National Conference on Composting Municipal Residues and Sludges.* Information Transfer, Inc., Rockville, Maryland. With permission.)

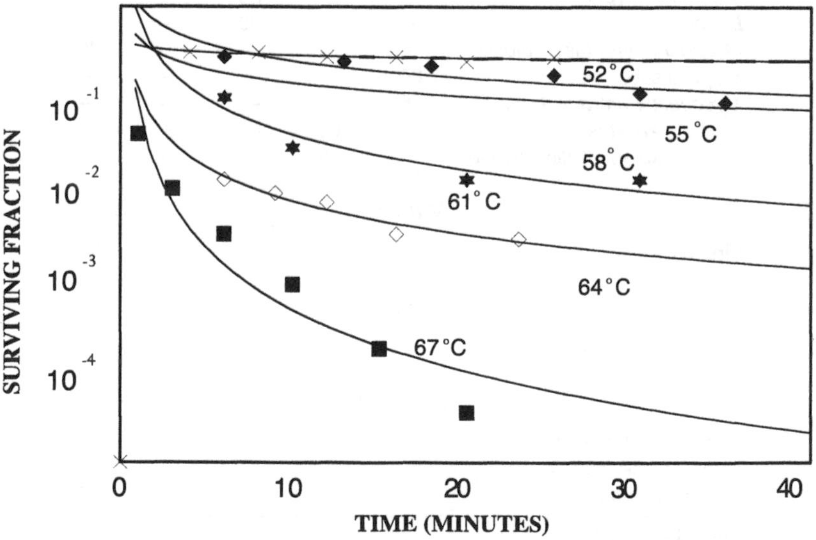

Figure 17.2 Heat inactivation of fecal *Streptococcus* bacteria in composted sludge. (From Ward, R.L. and J.R. Brandon, 1978. Effect of heat on pathogenic organisms found in wastewater sludge, p. 122–134. In: *1977 National Conference on Composting Municipal Residues and Sludges.* Information Transfer, Inc., Rockville, Maryland. With permission.)

reactor time, with temperatures of 55 to 60°C and a moisture content of 40 to 60%, the product did not contain *Salmonella cairo*, *Salmonella typhi*, and *Salmonella paratyphi B*.

Wiley and Westerberg (1969) reported that forced aeration composting in a bin destroyed Poliovirus type 1, *Salmonella newport, Candida albicans,* and *Ascaris lumbricoides* ova. They indicated that all pathogens would be destroyed if temperatures of 60 to 70°C were achieved for 3 days. Morgan and MacDonald (1969) also found that with windrow composting of MSW and biosolids, *Mycobacterium tuberculosis* was destroyed in 14 days when the temperatures exceeded 65°C.

In a study of windrow composting of MSW and biosolids, Gaby (1975) found that *Salmonella* and *Shigellae* disappeared in 7 to 21 days. *Leptospira philadephia* was destroyed in 2 days. Type 2 Poliovirus was not detected after 3 to 7 days. Human parasitic cysts, *Endolimax nana, Entamoeba histolytica,* as well as hookworm ova, *Necator americus* and *Ancylostoma duodenale*, disintegrated after 7 days.

Krogstad and Gudding (1975) inoculated *Salmonella typhimurium, Serratia marcescens,* and *Bacillus ceres* into a composting drum containing septage, biosolids, and MSW. After 3 days of composting with temperatures of 60 to 65°C, *Salmonella typhimurium* and *Serratia marcescens* were not detected. *Bacillus ceres* was found after 7 days at these temperatures, but after the temperature in the drum reached 70°C, it was not detected.

Ward and Brandon (1978) provided data on the destruction of Poliovirus in 40% solids compost as related to temperature–time (Table 17.7). At 35°C, Poliovirus survived a much greater period of time than at 47°C. Data on heat inactivation of total coliforms, fecal coliforms, fecal streptococcus, and *Salmonella enteritidis* serotype Montevideo also showed great reduction of organisms when the temperature exceeded 55 to 65°C (Ward and Brandon, 1978).

Table 17.7 Inactivation of Poliovirus in Composted Biosolids at 60% Moisture

Temperature (°C)	Time (Min.)	Recovery of Plaque-Forming Units (%)
35	20	30
39	20	7.2
43	20	0.087
47	5	0.003

After Ward and Brandon, 1978.

Burge et al. (1978) showed that *Salmonella* spp. were destroyed in 10 days by aerated static pile composting of raw sludge and in 15 days in turned windrows. The f_2 bacteriophage, a hardy indicator organism, survived for 13 to 20 days in an aerated static pile and up to 70 days in windrows when the weather was rainy.

Pereira-Neto et al. (1986) evaluated both windrow and aerated static pile composting and determined that the aerated static pile was more effective in pathogen destruction. *E. coli* and *Salmonella* were not detected after 16 days in the static pile but were still present after 60 days in windrows. After 32 days in an aerated static pile, fecal streptococci were below 10^2 colony forming units/gram wet weight (cfu/gww), and after 60 days in windrows were 10^2 to 10^3 cfu/gww.

A comparison was made between windrow, negative aerated static pile, and positive aerated static pile composting with regard to survival of several pathogens and indicator organisms (de Bertoldi et al., 1988). They indicated that negative aerated static pile composting was the most effective in destroying *Salmonella* spp., fecal streptococci, and fecal coliform organisms. Windrowing was the least effective of the three systems.

Yanko (1988) measured pathogen levels in several composting facilities and concluded that well-run facilities controlled pathogens in compost, whereas poorly run facilities had contaminated compost. Hays (1996) showed that in a well-managed biosolids windrow composting operation, indicator organisms and pathogens are destroyed (Table 17.8).

Table 17.8 Destruction of Several Indicator Organisms and Pathogens in a Well-Managed Windrow Biosolids Composting Operation

Organism/Unit	Initial Concentration	Final Concentration
Total coliforms, MPN[z] per dry g	$\geq 3.0 \times 10^6$	5
Fecal coliforms, MPN per dry g	$\geq 1.7\times 10^6$	< 3
Salmonella sp., MPN per dry g	7.6×10^3	0.2
Viable *Ascaris* ova, ova per dry g	1.3	< 0.3
Virus, PFU[y] per dry g	64	< 0.05

[z] MPN = most probable number.
[y] PFU = plaque-forming units.
After Hays, 1996.

Gibbs et al. (1998) investigated the effect of two windrow composting operations on the survival of *Salmonella* and *Giardia*. At Facility 1, *Salmonella* spp. were not detected in 13 samples, whereas at Facility 2, *Salmonella* spp. were detected in 7 of 11 samples. *Giardia* cysts were detected in compost from the two facilities. Since the viability of the *Giardia* cysts was not determined, these results do not indicate whether the cysts would create a public health risk. At Facility 1, temperatures near the edge of the pile achieved 52°C for 15 days, whereas in the deeper zones, 53°C was achieved for only a few days. It is surprising that the upper reaches of the windrow were hotter than the interior; this is not typical. At Facility 2, temperatures over a period of 3 to 15 weeks ranged from 30 to 40°C. There were no data given on the frequency of windrow turning, which would affect temperature. The U.S. EPA regulations state that for windrow composting, the temperature of the biosolids is maintained at 55°C (131°F) or higher for 15 days or longer with a minimum of five turnings during that time.

Since curing is a necessary extension of the composting process in order to produce a stable, marketable product, there is further potential for the destruction of pathogens. Burge et al. (1978) reported that considerable die-off of fecal coliforms and f_2 bacteriophage indicator organisms can occur during curing and stockpiling. Furthermore, curing reduces the potential for the attraction of vectors that could transmit diseases. The U.S. EPA's 40 CFR Part 503 regulations (U.S. EPA, 1994) require that vector attraction reduction (VAR) be instituted during the treatment of biosolids. The regulations specify 12 options. However, only one option is applicable to composting. This option requires that during composting, the biosolids must be maintained at 40°C (104°F) for at least 14 days, with an average temperature of over 45°C (113°F).

Parasites and viruses do not reproduce outside their hosts; however, bacteria can. There have been numerous studies on the regrowth or repopulation of *Salmonella* spp. and fecal coliforms in compost (Brandon et al., 1977; Burge et al., 1987a; Burge et al., 1987b; Russ and Yanko, 1981; Soares et al., 1995).

The data on pathogen destruction during composting are derived from the use of biosolids and MSW. The author found no reports involving the use of other feedstocks.

B. Ammonia Disinfection

During composting, ammonia (NH_3) is volatilized. Ammonia has been shown to be toxic to numerous bacteria and viruses (Taylor et al., 1978; Ward and Ashley, 1976). Ward and Ashley (1976) reported that NH_3 was a major virucidal agent in either raw or digested biosolids that inactivates poliovirus at pH values between 4.5 and 9.5. Sikora et al. (1985) also reported that NH_3 was virucidal in biosolids and in ammonium chloride (NH_4Cl) solutions. Heat increases the rate of viral inactivation by NH_3 in sludge (Ward and Ashley, 1976). Reimers et al. (1998) stated that NH_3 was significantly effective in inactivating *Ascaris* eggs.

C. Presence of Competing Organisms or Antagonistic Substances

Knoll (1961) reported that *Salmonella cairo* and *Salmonella paratyphi B* bacteria were inhibited by pure compost extracts. This suggested the presence of an inhibitory substance, possibly an antibiotic. In sterilized compost, inoculated fecal *Streptococcus* and *Salmonella* grew to much higher numbers than in nonsterilized or normal compost (Ward and Brandon, 1978).

Millner et al. (1987) studied the growth suppression of *Salmonella* by compost microflora. They indicated that the types and amounts of different microorganisms affect the growth of *Salmonella* organisms. The presence of coliforms or metabolically active bacteria and actinomycetes resulted in the death of *Salmonella* in compost.

IV. PATHOGEN SURVIVAL IN SOILS AND ON PLANTS

In general, most pathogens do not survive in soils or on plants for very long periods of time (Akin et al., 1978; Golueke, 1983; Lance, 1977; Rudolfs et al., 1950, 1951; Sorber and Moore, 1986). There are several environmental factors that affect the potential for pathogen survival. Organisms on the surface of soils are destroyed by temperature, desiccation, and ultraviolet light. Other factors, including pH, organic matter, soil colloidal matter, soil temperature, and competitive or antagonistic organisms, will affect pathogen survival in soils.

Data on the survival of various pathogens in soils and on plant materials are presented in Table 17.9. Most bacteria survive in soils or on plants for relatively short periods of time. Bagdasaryan (1964) investigated the survival of Poliovirus 1, ECHO-7, ECHO-9 and Coxsackie B3 viruses in soils and on vegetables. He con-

Table 17.9 Survival of Some Pathogens in Soil and on Plants

Organism	Media	Survival (Days)	Reference
Salmonella typhimurium	Soil	< 28–70	Golueke, 1983
Salmonella typhi	Soil	30–120	Golueke, 1983
Salmonella typhi	Vegetables, fruit	< 1–68	Parsons et al., 1975
Fecal streptococci	Soil	23–67	Rudolfs et al., 1950
Shigella	Vegetables	2–10	Rudolfs et al., 1950 Parsons et al., 1975
Shigella alkalescens	Tomatoes	6	Rudolfs et al., 1950
Tubercle bacilli	Soil	> 180	Rudolfs et al., 1950
Poliovirus	Sand	< 77–91	Golueke, 1983
Poliovirus 1	Loamy find sand	84	Golueke, 1983
Poliovirus 1	Lettuce, radishes	36	Larkin et al., 1978
Coxsackie B3	Clay	< 161	Golueke, 1983
Enteroviruses	Sandy soil	90–170	Golueke, 1983
Enteroviruses	Loam	90–150	Golueke, 1983
Enteroviruses	Vegetables	4–6	Parsons et al., 1975
Enteric viruses	Rooted and leafy vegetables	< 60	Golueke, 1983
Entamoeba histolytica	Loam and sand	8–10	Golueke, 1983
Ascaris ova	Sandy soil	< 90	Golueke, 1983
Ascaris ova	Vegetables, fruit	27–35	Rudolfs et al., 1950 Parsons et al., 1975
Hookworm larvae	Sand	< 120	Golueke, 1983
Hookworm larvae	Soil	42–< 180	Golueke, 1983
Protozoan cysts	Rooted and leafy vegetables	< 2	Golueke, 1983
Helminth ova	Rooted and leafy vegetables	< 30	Golueke, 1983

cluded that these enteroviruses survive for long periods of time in sandy or loam soils. Soil temperature and moisture play a significant role in survival of poliovirus (Duboise et al., 1976a). Although Larkin et al. (1978) indicated that Poliovirus 1 survived on lettuce and radishes (*Raphanus sativus* L.) for 14 to 36 days, most viral particles (99%) were lost in the first 5 to 6 days. Wellings et al. (1975) found that some viruses (Polio-1, Polio-s, Echo 22/23) survive in the soil for at least 28 days. Viruses can be adsorbed by the soil, desorbed, and move with subsurface water. In a laboratory study, Duboise et al. (1976b) investigated the die-off rate of Poliovirus in a forest soil as related to temperature. At 4°C, the virus was reduced by less than 1 log unit in 84 days, whereas at 20°C, there was a 5 log reduction in the same time. Much of the literature is based on studies in which irrigated effluent was applied to soils. It is difficult to extrapolate those data to compost or biosolids applied as a semisolid, since movement through soils is a function of water movement. Data on virus survival in field soils are more limited. Most of the data in the literature are on movement of viruses through soils. The predominance of data are from column studies using liquids.

Parasites tend to survive longer in soil than most other pathogens (Table 17.9). Soil conditions, especially moisture, temperature, and sunlight, greatly affect the survival of parasites. Protozoa cysts are very sensitive to drying and survive for only a few days (Parsons et al., 1975). Kowal (1982) indicated that *Entamoeba histolytica* survived for 18 to 24 hours in dry soil and 42 to 72 hours in moist soil. Data on parasite survival in soils and on plants are meager.

There are limited data on the antagonistic effects of other organisms on pathogen survival in soils. The growth of *Salmonella* and dysentery bacilli was reported to be suppressed in soil by actinomycetes (Bryanskaya, 1966). In sterilized soil, inoculated fecal *Streptococcus* and *Salmonellae* grew to much higher numbers than in nonsterilized soil (Ward and Brandon, 1978).

V. PUBLIC AND WORKER HEALTH ASPECTS

Composting can be very effective in the destruction of pathogens. The soil environment is generally very hostile to pathogen survival. The potential for survival of pathogens in compost or soil used in horticulture is much less than in field soils. Horticultural soils generally undergo much greater exposure to edaphic and environmental conditions. Thus, the potential for public or worker health impacts from the handling of compost is extremely minimal. Public health exposure from domestic animal waste in gardens is much greater than from composted wastes since composted wastes have undergone disinfection, whereas animal wastes generally have not. There is no evidence in the medical literature that public exposure to composted wastes has resulted in infection or disease from primary pathogens.

Workers handling organic materials, including compost, can be exposed to bioaerosols and organic dusts. The predominant exposure by workers to bioaerosols is through the respiratory system (Centers for Disease Control, 1994; Epstein, 1993; Millner et al, 1994). In some cases, dermal irritation has been indicated. Most of the reports in the U.S. and Europe have been with workers in composting facilities. Several cases of respiratory complications were reported where workers shoveled composted wood chips, fresh wood chips, and leaves (Centers for Disease Control, 1994). Weber et al. (1993) reported that a 52-year-old male developed fever, myalgia, and marked dyspnea 12 h after shoveling composted wood chips and leaves. An investigation of conditions at the site indicated that fine dust particles contained from 636 to 16,300 endotoxin units per m^3. The staff at E&A Environmental Consultants, Inc. have been involved in over 150 composting operations in the U.S. and Canada and have become aware of two incidents involving workers. In one case at an MSW/biosolids composting facility, a front-end loader operator who was a heavy smoker reported that on Mondays he felt that his breathing habits were better than on Fridays. In a second case, a worker at a biosolids composting facility developed a facial skin rash.

There are several mitigation measures that can be implemented at composting facilities. Moisture control is most important. Dust is predominantly generated during screening and pile construction or windrow turning. Maintaining moisture in the compost in the 40 to 50% range will reduce dust emissions. In buildings, good ventilation or dust control systems will reduce worker exposure to dust and endotoxins. Personal hygiene devices such as dust masks or respirators could be used to reduce exposure to airborne contaminants. Workers should exercise good hygienic practices, such as washing hands before eating, refraining from smoking with dirty hands, and leaving work clothes at the work place.

VI. CONCLUSION

Numerous feedstocks, such as animal manures, yard waste, food waste, biosolids, and MSW, are being composted and used in horticulture. All of these feedstocks can contain different types and amounts of pathogenic organisms. The application of untreated feedstocks to soils can result in the introduction of pathogens to those soils, but survival time of the pathogens varies, depending on the organism and the soil and climatic conditions.

Composting is a very effective method for disinfection and destruction of pathogens. The principal means of pathogen destruction in composting is the exposure to high temperatures for a prolonged period of time. Therefore, it is essential for the composting process to be carried out effectively so as to achieve high temperatures uniformly throughout the mass. The temperature in the composting mass that has been shown to be effective is 55°C (131°F) for several days, depending on the composting system used. This is the basis for the U.S. EPA 40 CFR Part 503 regulations for composting biosolids, the only feedstock regulated by the U.S. Federal government. Several states use the U.S. EPA regulations and apply the same standards to other feedstocks.

There is no documented evidence in the medical or other scientific literature that humans or animals have been subjected to increased risk of infection through the use of properly composted materials. Many uses of compost in horticulture involve soil surface applications, where desiccation and exposure to sunlight and ultraviolet rays can largely eliminate pathogens. Other uses involve landscape and turfgrass industries, where non-food-chain plants are generally used, further reducing the potential for human infection. With proper precautions, the risks from human pathogens in compost utilized in horticultural cropping and management systems should be minimal.

REFERENCES

Akin, E.W., W. Jakubowski, J.B. Lucas, and H.R. Pahren. 1978. Health hazards associated with wastewater effluent and sludge: microbiological considerations, p. 9–26. In: B.P. Sagik and C.A. Sorber (eds.). *Risk Assessment and Health Effects of Land Application of Municipal Wastewater and Sludge*. University of Texas, San Antonio.

Bagdasaryan, G.A. 1964. Survival of viruses of the enterovirus group (Poliomyelitis, ECHO, Coxsackie) in soil and on vegetables. *Journal of Hygiene, Epidemiology, Microbiology, and Immunology* 7:497–505.

Boyce, T.G., D.L. Swerdlow, and P.M. Griffin. 1995. *Escherichia coli* O157:H7 and the hemolytic-uremic syndrome. *New England Journal of Medicine* 333:364–368.

Brandon, J.R., W.D. Burge, and N.E. Enkiri. 1977. Inactivation by ionizing radiation of *Salmonella enteritidis* serotype Montevideo growth in composted sewage sludge. *Applied and Environmental Microbiology* 33:1011–1012.

Bryanskaya, A.M. 1966. Antagonistic effect of actinomyces on pathogenic bacteria in soil. *Hygiene and Sanitation* 31:123–125.

Burge, W.D. 1983. Monitoring pathogen destruction. *BioCycle* 24(2):48–50.

Burge, W.D., W.N. Cramer, and E. Epstein. 1978. Destruction of pathogens in sewage sludge by composting. *Transactions of the American Society of Agricultural Engineers* 21(3):510–514.

Burge, W.D., N.K. Enkiri, and D. Hussong. 1987a. *Salmonellae* regrowth in compost as influenced by substrate. *Microbiological Ecology* 14:243–253.

Burge, W.D., P.D. Millner, N.K. Enkiri, and D. Hussong. 1987b. *Regrowth of* Salmonellae *in composted sewage sludge*. EPA/600/S2-86/106. U.S. Environmental Protection Agency, Water Engineering Research Laboratory, Cincinnati, Ohio.

Casemore, D.P. 1991. The epidemiology of human cryptosporidiosis and the water route of infection. *Water Science and Technology* 24(2):157–164.

Centers for Disease Control (CDC). 1994. *Preventing Organic Dust Toxic Syndrome*. DHHS (NIOSH) Publication 94–102. Centers for Disease Control, National Institute for Occupational Safety and Health (NIOSH), Public Health Service, U.S. Department of Health and Human Services (DHHS), Cincinnati, Ohio.

Council for Agricultural Science and Technology (CAST). 1994. *Foodborne Pathogens: Risks and Consequences*. Task Force Report No. 122. Council for Agricultural Science and Technology, Ames, Iowa.

de Bertoldi, M., F. Zucconi, and M. Civilini. 1988. Temperature, pathogen control and product quality. *BioCycle* 29(2):43–50.

Domsch, K.H., W. Gams, and T.H. Anderson. 1980. *Compendium of Soil Fungi*. Two volumes. Academic Press, London, England.

Duboise, S.M., B.E. Moore, B.P. Sagik, and C.A. Sorber. 1976a. The effects of temperature and specific conductance on poliovirus survival and transport in soil. In: *National Conference on Environmental Research, Development and Design*, University of Washington, Seattle.

Duboise, S.M., B.E. Moore, and B.P. Sagik. 1976b. Poliovirus survival and movement in a sandy forest soil. *Applied and Environmental Microbiology* 31(4):536–543.

DuPont, H.L., C.L. Chappell, C.R. Sterling, P.C. Okhuysen, J.B. Rose, and W. Jakubowski. 1995. The infectivity of *Cryptosporidium parvum* in health volunteers. *New England Journal of Medicine* 332:855–859.

Epstein, E. 1993. Neighborhood and worker protection for composting facilities: issues and actions, p. 319–338. In: H.A.J. Hoitink and H.M. Keener (eds.). *Science and Engineering of Composting: Design, Environmental, Microbiological and Utilization Aspects*. Renaissance Publications, Worthington, Ohio.

Epstein, E. 1997. *The Science of Composting*. Technomic Publishing Company, Lancaster, Pennsylvania.

E&A Environmental Consultants, Inc. 1994. *Food Waste Collection and Composting Demonstration Project for City of Seattle Solid Waste Utility*. City of Seattle Solid Waste Utility, Seattle, Washington.

Gaby, W.L. 1975. *Evaluation of Health Hazards Associated with Solid Waste/Sewage Sludge Mixtures*. EPA-670/2-75-023. U.S. Environmental Protection Agency, National Environmental Research Center, Office of Research and Development, Cincinnati, Ohio.

Geyer, M.D. 1994. *Bloodborne Pathogen Rule Compliance*. Presentation at SWANA 32nd Annual International Solid Waste Exposition, San Antonio, Texas, 1–4 Aug. 1994. Solid Waste Association of North America (SWANA), Silver Spring, Maryland.

Gibbs, R.A., C.J. Hu, J. Sidhu, and G.E. Ho. 1998. Risks associated with human pathogens in composted biosolids, p. 1–12. In: *Water TECH*, Australian Water Wastewater Association, Brisbane, Queensland, Australia.

Golueke, C.G. 1983. Epidemiological aspects of sludge handling and management, Part II. *BioCycle* 24(4):52–58.

Hays, J.C. 1996. Pathogen destruction and biosolids composting. *BioCycle* 37(6):67–76.

Hennessy, T.W., C.W. Hedberg, L. Slutsker, K.E. White, J.M. Besser-Wiek, M.E. Moen, J. Feldman, W.W. Coleman, L.M. Edmonson, K.L. MacDonald, and M.T. Osterholm. 1996. A national outbreak of *Salmonella enteriditis* infections from ice cream. *New England Journal of Medicine* 334:1281.

Hirsch, S.R. and J.A. Sosman. 1976. A one-year survey of mold growth inside twelve homes. *Annals of Allergy* 36:30–38.

Hughes, M.S. and T.W. Steele. 1994. Occurrence and distribution of *Legionella* species in composted plant materials. *Applied and Environmental Microbiology* 60:2003–2005.

Keswick, B.H. 1984. Sources of ground water pollution, p. 39–64. In: *Ground Water Pollution Microbiology*. John Wiley & Sons, New York.

Knoll, K.H. 1961. Public health and refuse disposal. *Compost Science* 2(1):35–40.

Koenraad, P.M.F.J., F.M. Rombouts, and S.H.W. Notermans. 1997. Epidemiological aspects of thermophilic *Campylobacter* in water-related environments: a review. *Water Environment Research* 69:52–63.

Kothary, M.M., T. Chase Jr., and J.D. Macmillan. 1984. Levels of *Aspergillus fumigatus* in air and in compost at a sewage sludge composting site. *Environmental Pollution*, Series A 34(1):1–14.

Kowal, N.E. 1982. *Health Effects of Land Treatment: Microbiological*. EPA/600/1-82/007. U.S. Environmental Protection Agency, Cincinnati, Ohio.

Kowal, N.E. 1985. *Health Effects of Land Application of Municipal Sludge*. EPA/600/1-85/015. U.S. Environmental Protection Agency, Cincinnati, Ohio.

Krogstad, O. and R. Gudding. 1975. The survival of some pathogenic micro-organisms during reactor composting. *Acta Agriculture Scandinavica* 25:281–284.

Kurland, L.T. 1987. Letter of 14 March 1987 to Dr. H.M. Goldberg, Medical Director, Milwaukee Industrial Clinics, Milwaukee, Wisconsin.

Lance, J.C. 1977. Fate of pathogens in saturated and unsaturated soils, p. 135–141. In: *National Conference on Composting of Municipal Residues and Sludges*, Silver Spring, Maryland. Information Transfer, Inc., Rockville, Maryland, and Hazardous Materials Control Research Institute.

Larkin, E.P., J.T. Tierney, J. Lovett, D. Van Dorsal, and D.W. Francis. 1978. Land application of sewage wastes: potential for contamination of foodstuffs and agricultural soil by viruses, p. 102–115. In: B.P. Sagic and C.A. Sorber (eds.). *Risk Assessment and Health Effects of Land Application of Municipal Wastewater and Sludges*. University of Texas at San Antonio.

Levine, M.M., H.L. DuPont, and S.B. Formal. 1973. Pathogenesis of *Shigella Dysenteria* (Shinga) dysentery. *Journal of Infectious Diseases* 154(5):871–880.

MacKenzie, W.R., N.J. Hoxie, M.E. Proctor, M.S. Gradus, K.R. Blair, D.E. Peterson, J.J. Kazmierczak, D.G. Addiss, K.R. Fox, J.B. Rose, and J.P. Davis. 1994. A massive outbreak in Milwaukee of *Cryptosporidium* infection transmitted through the public water supply. *New England Journal of Medicine* 331:161–167.

Millner, P.D., K.E. Powers, N.K. Enkiri, and W.D. Burge. 1987. Microbial mediated growth suppression and death of *Salmonella* in composted sewage sludge. *Microbiological Ecology* 14:255–265.

Millner, P.D., S.A. Olenchock, E. Epstein, R. Rylander, J. Haines, J. Walker, B.L. Ooi, E. Horne, and M. Maritato. 1994. Bioaerosols associated with composting facilities. *Compost Science & Utilization* 2(4):6–57.

Morgan, M.T. and F.W. MacDonald. 1969. Test show MB tuberculosis doesn't survive composting. *Journal of Environmental Health* 32(1):101–108.

Parsons, H.R., C. Brownlee, D. Wetter, A. Maurer, E. Haughton, L. Kordner, and M. Slezak. 1975. *Health Aspects of Sewage Effluent Irrigation.* Pollution Control Branch, British Columbia Water Resources Service, Victoria, British Columbia, Canada.

Passman, F.J. 1980. *Monitoring of* Aspergillus fumigatus *Associated with Municipal Sewage Sludge Composting Operations in the State of Maine.* Report prepared for Portland Water District, Portland, Maine.

Patterson, T.F. 1995. Invasive aspergillosis in organ transplantation, p. 11–14. In: Cunha (ed.). Aspergillosis: Special Report. *Internal Medicine*, Medical Economics, Montvale, New Jersey.

Pell, A.N. 1997. Manure and microbes; public and animal health problem. *Journal of Dairy Science* 89:2673–2681.

Pereira-Neto, J.Y., E.I. Stenfiord, and D.V. Smith. 1986. Survival of fecal indicator microorganisms in refuse/sludge composting using the aerated static pile system. *Waste Management & Research* 4:397–406.

Reimers, R.S., E.R. DeSocio, W.S. Bankston, and J.A. Oleszkiewicz. 1998. Current/future advances in biosolids disinfection processing, p. 445–459. In: *WEFTEC '98 Residuals & Biosolids Management*, Volume 2. Water Environment Federation, Orlando, Florida.

Rudolfs, W., L.L. Falk, and R.A. Ragotzkie. 1950. Literature review of the occurrence and survival of enteric pathogenic, and relative organisms in soil, water, sewage, and sludge and on vegetation. *Sewage and Industrial Wastes* 22:1261–1281.

Rudolfs, W., L.L. Falk, and R.A. Ragotzkie. 1951. Contamination of vegetables grown in polluted soil. *Sewage and Industrial Wastes* 23:253–268.

Russ, C.F. and W.A. Yanko. 1981. Factors affecting *Salmonella*e repopulation in composted sludges. *Applied and Environmental Microbiology* 41:597–602.

Sikora, L.J., P.D. Millner, and W.D. Burge. 1985. *Chemical and Microbial Aspects of Sludge Composting and Land Application.* Report to U.S. EPA Office of Research and Development, Cincinnati, Ohio, Interagency Agreement Number AD-12-F-2-534. U.S. Department of Agriculture, Agricultural Research Service, Beltsville, Maryland.

Sigsgaard, T., A. Abel, L. Donbaek, and P. Malmros. 1994. Lung function changes among recycling workers exposed to organic dust. *American Journal of Industrial Medicine* 25:69–72.

Slavin, R.G. and P. Winzenburger. 1977. Epidemiologic aspects of allergic aspergillosis. *Annals of Allergy* 38: 215–218.

Smith, J.E. Jr., and J.B. Farrell. 1996. Current and future disinfection – Federal perspectives, p. 1–16. In: *Proceedings of the WEF 69th Annual Conference and Exposition*, Charlotte, North Carolina, 5 Oct. 1996. Water Environment Federation (WEF), Alexandria, Virgina.

Soares, H.M., B. Cardenas, D. Weir, and M.S. Switzenbaum. 1995. Evaluating pathogen regrowth in biosolids compost. *BioCycle* 36(6):70–76.

Solomon, W.R. 1975. Assessing fungus prevalence in domestic interiors. *Journal of Allergy Clinical Immunology* 56:235–242.

Sorber, C.A. and B.E. Moore. 1986. Survival and transport of pathogens in sludge-amended soil, p. 25–32. In: *Proceedings of the National Conference on Municipal Treatment Plant Sludge Management*, Orlando, Florida. Information Transfer, Inc., Rockville, Maryland.

Steele, T.W., J. Lanser, and N. Sangster. 1990a. Isolation of *Legionella longbeachae* serogroup 1 from potting mixes. *Applied and Environmental Microbiology* 56(10):49–53.

Steele, T.W., C.V. Moore, and N. Sangster. 1990b. Distribution of *Legionella longbeachae* serogroup 1 and other *Legionellae* in potting soils in Australia. *Applied and Environmental Microbiology* 56:2984–2988.

Stern, G. 1974. Pasteurization of liquid digested sludge. In: *Proceedings of the National Conference on Composting Municipal Sludge Management.* Information Transfer, Inc., Silver Spring, Maryland.

Tauxe, R.V. 1992. Epidemiology of *Campylobacter jejuni* infections in the U.S. and other industrialized nations. In: I. Nachamkin, S. Tompkins, and M. Blase (eds.). *Campylobacter jejuni: Current Status and Future Trends.* American Society of Microbiology, Washington, D.C.

Taylor, J.M., E. Epstein, W.D. Burge, R.L. Chaney, J.D. Menzies, and L.J. Sikora. 1978. Chemical and biological phenomena observed with sewage sludges in simulated trenches. *Journal of Environmental Quality* 7:477–482.

United States Environmental Protection Agency (U.S. EPA). 1994. *A Plain English Guide to the EPA Part 503 Biosolids Rule.* EPA/832/R-93/003. Office of Wastewater Management, Washington, D.C.

Ward, R.L. and C.S. Ashley. 1976. Inactivation of poliovirus in digested sludge. *Applied and Environmental Microbiology* 33:921–930.

Ward, R.L. and J.R. Brandon. 1978. Effect of heat on pathogenic organisms found in wastewater sludge, p. 122–134. In: *1977 National Conference on Composting Municipal Residues and Sludges.* Information Transfer, Inc., Rockville, Maryland.

Ward, R.L., D.I. Berstein, E.C. Young, J.R. Sherwood, D.R. Knowlton, and G.M. Schiff. 1986. Human rotavirus studies in volunteers: determination of infective dose and serological response to infection. *Journal of Infectious Diseases* 154(5):871–880.

Weber, S., G. Kullman, E. Petsonk, W.G. Jones, S. Olenchock, W. Sorensen, J. Parker, R. Marcelo-Baciu, D. Fraser, and V. Castranova. 1993. Organic dust exposures from compost handling: Case presentation and respiratory exposure assessment. *American Journal of Industrial Medicine* 24(4):365–374.

WEF/U.S. EPA Biosolids Fact Sheet Project. 1997. *Can AIDS be Transmitted by Biosolids?: Biosolids Fact Sheet.* Water Environment WEB http://www.wef.org/docs/biofact/aids.fct.html.

Wellings, F.M., A.L. Lewis, C.W. Mountain, and L.M. Stark. 1975. Virus consideration in land disposal of sewage effluent and sludge. *Florida Scientist* 38:202–207.

Wiley, J.S. and S.C. Westerberg. 1969. Survival of human pathogens in composted sewage. *Applied and Environmental Microbiology* 18:994–1001.

Wingard, J.R. 1995. Aspergillus infections in immunocompromised cancer patients, p. 4–10. In: B. Cunha (ed.). *Aspergillosis: Special Report. Internal Medicine*, Medical Economics, Montvale, New Jersey.

Yanko, W.A. 1988. *Occurrence of Pathogens in Distribution and Marketing.* Report Number EPA/1-87/014 (NTIS PB #88-154273/AS). Los Angeles County Sanitation Districts, Los Angeles, California.

CHAPTER 18

U.S. Environmental Protection Agency Regulations Governing Compost Production and Use

John M. Walker

CONTENTS

1-56670-460-X/01/$0.00+$.50

I. INTRODUCTION

Composting is an outstanding method for stabilizing and disinfecting organic byproducts that often contain pathogens. Byproducts that can be composted include but are not limited to biosolids, animal manures, grass and yard trimmings, food processing residues, and municipal solid wastes (MSW). There are many beneficial uses for these composted byproducts. They can be used to fertilize and condition soil for purposes such as landscaping and the production of high-quality foods and ornamentals (U.S. EPA, 1993, 1994b). In addition, composted materials like biosolids can be used by themselves or in combination with other byproducts to sustainably support the revegetation of drastically disturbed soils that have virtually no organic matter and that may also contain elevated levels of toxic metals (Henry and Brown, 1997). Composted biosolids and other specially formed composts can be used to suppress plant pathogens (Hoitink et al., 1992). Composts can even reduce the bioavailability of metals already in soils like lead (Pb) such that children playing on and potentially ingesting such amended soils would be at reduced risk from potential Pb poisoning (Heneghan et al., 1994).

The content of pathogens, heavy metals, refractory organic compounds, and nutrients can be quite varied in these materials that are being composted depending upon the nature and source of the byproduct (Bastian, 1997; Logan and Chaney, 1983; U.S. EPA, 1999a; Walker and O'Donnell, 1991; Wright et al., 1998). Although many composts are beneficial, they can also be less useful and possibly detrimental if improperly composted and used (Millner et al., 1994; Ryan and Chaney, 1993). Hence, the composting process and recycling of the compost into the environment should be properly controlled so that the degree of disinfection and stabilization is appropriate for the intended end use, the release of malodorous volatile compounds (which can be offensive to neighbors) is minimized, and the levels of potentially toxic pollutants in the composts being recycled into the environment are not harmful.

This chapter discusses the current status of U.S. Environmental Protection Agency (EPA) regulations, guidance, and controls that apply directly or indirectly to the production and use of compost and uncomposted byproducts. This chapter will also broadly discuss other regulations, guidance and control mechanisms that apply, which are outside of EPA's regulatory authority. *The only rules discussed in this chapter are federal rules. The many additional state and local rules that pertain to the management of byproducts and wastes have not been included in these discussions.*

II. REGULATIONS

There are few EPA rules that generically apply to the production and use of all types of composts. On the other hand, several regulations have been developed to govern the production and use of specific composts, especially composts that contain biosolids. Lacking other standards, the biosolids regulations are unofficially applied to many other composted products.

Federal EPA regulations that can impact the composting and use of composted byproducts are listed in Table 18.1. The first regulation that was developed with standards applicable to the production and use of some composts was 40 CFR part 257.

Table 18.1 EPA Regulations that Can Impact Compost Production and Use

Source	Specific Purpose	Contents	Impacts On Composting
40 CFR Part 257, RCRA	Rules for nonhazardous solid wastes	Specific requirements for preparation and use of composted solid wastes other than biosolids that are land-applied	Indicates required practices
40 CFR Parts 261-268, RCRA	Rules for hazardous wastes	Cradle to grave care of hazardous wastes including toxicity and other tests	Because of the generally low pollutant content in composted "solid wastes," most generators do not have to be concerned with the hazardous waste rule
40 CFR Part 761, TSCA	Rules for toxic substances	PCBs are the primary pollutant of concern	Composts subjected to a minimum concentration of PCBs
CERCLA	An act dealing with superfund liability	Information about superfund liability	Indicates that proper land use of composts can exclude them from CERCLA site clean up liability
40 CFR Part 503, CWA	Technical rules governing the use and disposal of biosolids	Pollutant limits Operational standards Management practices Monitoring, record keeping, & reporting requirements	Each of these parts of the regulation apply specifically to composting and the use and disposal of composted byproducts that contain any amount of biosolids The Part 503 rule has been applied generically to composting and use and disposal of all composts.

Note: CERCLA = Comprehensive Environmental Response, Compensation and Liability Act; CFR = Code of Federal Regulations; CWA = Clean Water Act; PCBs = polychlorinated biphenyls; RCRA = Resource Conservation and Recovery Act; TSCA = Toxic Substances Control Act.

A. 40 CFR Part 257

Part 257 contains the requirements that have to be met when solid waste is disposed on the land. Under 40 Part 257, solid waste includes, but is not limited to, garbage, refuse, sludge from water treatment plants, air pollution control facility solids, and other discarded material. Prior to publication of the Standards for the Use or Disposal of Sewage Sludge (Biosolids) (40 CFR Part 503), the requirements in Part 257 had to be met when biosolids were applied to the land or placed on a surface disposal site, including standards for cadmium (Cd). Except for one limited circumstance, the Part 257 rule no longer applies to biosolids unless they have been generated during the treatment of a combination of domestic sewage (i.e., waste from humans and household operations) and industrial wastewater, and the treatment works is located at an industrial facility. Also contained in the Part 257 rule are requirements for treatment to control pathogens in regulated wastes, including time and temperature requirements for composting. Part 257 was published in 1979 under the authority of both the Resource Conservation and Recovery Act (RCRA) and the Clean Water Act (CWA).

Specifically excluded from coverage are animal manures, dead animals, and crop residues that are returned to soil as a fertilizer or soil conditioner. Also excluded are byproducts recycled in the back yard. For example, if leaves, grass, and other yard trimmings are taken away from the home site and recycled, they would be covered by part 257 if applied anywhere other than in the home or home yard. Even if not required, it would be prudent to apply the Part 257 time and temperature requirements for composting to control those pathogens that may be present in such materials and to assure the adequate reduction of attractiveness of the compost product to vectors. A summary of the Part 257 time and temperature requirements for composting and the Cd and polychlorinated biphenyls (PCBs) requirements for land application of those composts, which are not excluded from coverage by Part 257, is presented in Table 18.2.

B. 40 CFR Parts 261–268

The criteria for identifying hazardous wastes are in 40 Part 261. A waste can either be listed as a hazardous waste or be determined to be hazardous because it exhibits a defined characteristic, e.g., corrosivity, ignitability, reactivity, or toxicity. Parts 262 through 268 contain standards for generators of hazardous waste; transporters of hazardous waste; treatment, storage, and disposal of hazardous waste; special hazardous waste; and land disposal of hazardous waste.

Toxicity is the most likely characteristic causing byproducts destined for composting to be considered hazardous. When it is believed that a byproduct may exhibit the toxicity characteristic, it can be tested using the Toxicity Characteristic Leaching Procedure (TCLP) (U.S. EPA, 1991a). Most organic byproducts would probably not fail the TCLP unless they contain toxic constituents at levels that would cause them to exceed the limiting parameters. As an example, almost no biosolids have failed the TCLP (U.S. EPA, 1991a).

Table 18.2 Selected Requirements under 40 CFR Part 257

Time/Temperature Pathogen Reduction Requirements

Configuration	Time/Temperature Requirement
To Significantly Reduce Pathogens	
Within-vessel, static aerated pile or windrow	The temperature of biosolids is raised to 40°C or greater for 5 days with the compost being at 55°C or greater for 4 hours during that period
To Further Reduce Pathogens	
Within-vessel or static aerated pile	Maintained at 55°C or higher for 3 days
Windrow	≥ 55°C for 15 days or longer and the windrow must be turned a minimum of 5 times

Requirements for Solid Waste Applied for the Production of Food Chain Crops

Pollutant	Requirement
Cadmium for all food-chain crops	(1) The pH of the solid waste is at 6.5 or greater at time of application unless cadmium content is 2 $mg \cdot kg^{-1}$ or less
	(2) The annual application of cadmium does not exceed 0.5 $kg \cdot ha^{-1}$
	(3) The cumulative application of cadmium from solid waste does not exceed 5 $kg \cdot ha^{-1}$ if the cation exchange capacity (CEC) = < 5 meq per 100 g; 10 $kg \cdot ha^{-1}$ if the CEC = 5–15 meq per 100 g; and 20 $kg \cdot ha^{-1}$ if the CEC = > 15 meq per 100 g
Cadmium with the only food-chain crop being animal feed	(1) pH of soil at 6.5 or greater when the crop is planted and grown
	(2) A plan is prepared that demonstrates how the animal feed will be distributed to preclude ingestion by humans and how to safeguard against cadmium entering the food chain from alternate uses of the soil
	(3) A deed stipulation indicating potentially high additions of cadmium
Polychlorinated biphenyls (PCBs)	(1) Solid waste with ≥ 10 $mg \cdot kg^{-1}$ PCBs is incorporated into the soil
	(2) Soil incorporation not needed if the PCB content in meat is < 0.2 $mg \cdot kg^{-1}$ or < 1.5 $mg \cdot kg^{-1}$ in milk

Some byproducts may contain listed substances that could cause them to be considered hazardous no matter what the concentration of the substances. However, there are special provisions for delisting such byproducts for cause, particularly when byproducts are being used beneficially. Biosolids are exempt from being listed, even though they may contain listed hazardous constituents. This is because of the comprehensive Part 503 rule that regulates biosolids use and disposal.

As with the Part 257 rule, farm residues returned to soils are not hazardous wastes under the part 261 rule or solid wastes by definition. In contrast, household wastes including refuse and garbage and solid wastes from universities and government buildings would be covered under part 257 and, if hazardous, under Part 261–268.

C. 40 CFR Part 761

Part 761 establishes the requirements for the manufacture, processing, distribution in commerce, storage, use, and disposal of PCBs and PCB items. Part 761 is being revised to establish requirements for the use of PCB remediation waste, which includes any environmental media with any concentration of PCBs. However, biosolids are not subject to the Part 761 disposal requirements for PCB remediation waste if the concentration of PCBs is less than 50 $mg \cdot kg^{-1}$ and if the requirements in Parts 257, 258, or 503 are met when the biosolids are used or disposed. Other byproducts may potentially be impacted by the Part 761 rule if they contain PCBs in any concentration.

D. Comprehensive Environmental Response, Compensation and Liability Act (CERCLA)

The following information was derived from a letter to Congressman Gary Condit from Acting EPA Assistant Administrator, Martha G. Prothro, dated 27 September 1993:

> Section 107 of CERCLA generally imposes liability for cleanup costs on, among others, persons who own or operate facilities at which hazardous substances are disposed. Section 107 liability extends to the costs of cleanup necessitated by a release or threat of release of a hazardous substance. Virtually all soil amendments contain hazardous substances, albeit in small amounts, and CERCLA does not distinguish among amounts. However, Section 101[22] of CERCLA defines "release" to exclude the "normal application of fertilizer under the Part 503 rule," If the placement of biosolids on land were considered to be "the normal application of fertilizer," the placement could not give rise to liability under CERCLA.
>
> Under CERCLA, protection from liability is also provided when there is a release of a CERCLA hazardous substance and the release occurs pursuant to Federal authorization. Thus, under CERCLA, in defined circumstances the application of biosolids to land in compliance with a permit required by section 405 of the CWA is a Federally permitted release. Recovery for response costs for damages under Section 107 of CERCLA is not authorized for Federally permitted releases. The Act defines Federally permitted releases as, among others, discharges in compliance with a National Pollutant Discharge Elimination System (NPDES) permit under section 402 of the CWA.
>
> Thus, in circumstances described herein, application of biosolids to land for a beneficial purpose would not give rise to CERCLA liability for the municipality generating the biosolids, the land applier, the land user, or the land owner. It is also EPA's long standing position that the beneficial application of biosolids to land to provide crop nutrients or to condition the soils is not only safe, but good public policy so long as appliers comply with all applicable requirements of Federal, state and local law.

Although not addressed in the Prothro to Condit letter, it could be argued that other composted byproducts similar to composted biosolids could be land applied without CERCLA liability. Final determinations of the interpretations in this letter of applicable guidance and rules to composted biosolids and other similar composted byproducts would ultimately have to be given by the responsible permitting party within EPA.

E. 40 CFR Part 503

Perhaps the most universally adopted parts of EPA rules regarding composting and the use of composts are from 40 CFR Part 503 (U.S. EPA, 1994a). This regulation contains time and temperature requirements that were developed for biosolids composting to ensure pathogen destruction and the reduction of vector attractiveness. This regulation also contains tables that list metal pollutant standards for biosolids.

1. Standards for Pathogen and Vector Attraction Reduction

Table 18.3 presents the time and temperature composting requirements for pathogen reduction. Class A pathogen reduction allows the use of biosolids with few restrictions because pathogen densities are below detectable levels. The table shows procedures that are suitable for producing Class A compost from materials containing biosolids via different composting configurations, as well as associated pathogen testing requirements. These time and temperature requirements are directly applicable to the elimination of pathogens from any byproduct that may be composted; however, only byproducts that contain any amount of biosolids are regulated by Part 503.

2. Standards for Pollutant Limits

Some requirements from the Part 503 pollutant limits have been summarized in Table 18.4. Biosolids containing pollutants in concentrations at or below the lowest concentrations in Table 18.4 (column on the right) can be used without having to track the cumulative level of pollutants added to soil. If the biosolids have also been composted in accordance with provisions of the part 503 rule to produce Class A biosolids with respect to pathogens, they have been called "exceptional quality" and can be used with minimal regulatory control.

Although not included in the Part 503 rule, pollutant limits were also calculated for toxic organic pollutants in the biosolids (Table 18.5). These limits were not included in the Part 503 rule because (1) the pollutants have been banned or restricted for use in the U.S., or are no longer manufactured in the U.S., (2) the pollutants are not present in biosolids at significant frequencies of detection (i.e., 5%) based on data gathered in the National Sewage Sludge Survey (NSSS) (U.S. EPA, 1990), or (3) the limits for the pollutants identified in the biosolids risk assessments are not expected to be exceeded in biosolids that are used or disposed, based on data in the NSSS.

Table 18.3 40 CFR Part 503 Time/Temperature Pathogen and Vector Attraction Reduction Requirements

Time/Temperature Pathogen Reduction Requirements	
Configuration	**Time/Temperature Requirement**
Within vessel or static aerated pile	≥ 55°C for 3 days plus analytical requirements below
Windrow	≥ 55°C for 15 days or longer and the windrow must be turned a minimum of 5 times plus an analytical requirement below
	Analytical Requirement
For all configurations	Density of: Fecal coliform less than 1000 most probable numbers (MPN) per gram total solids (dry weight basis [DW]) *or* *Salmonella* sp. less than 3 MPN per 4 grams of total solids (DW) Either of the analytical requirements must be met when Biosolids are used or disposed, Biosolids are prepared for sale or donation, or Biosolids or derived materials are prepared to meet the requirements for *Exceptional Quality* biosolids
Other Pathogen Reduction Requirements	
Configuration	**Testing**
Unknown	*Salmonella* sp. or fecal coliform bacteria, enteric viruses, and viable helminth ova at the time biosolids are used or disposed
Other	Other processes might be used, but seldom if ever used to establish pathogen reduction requirements
Vector Attraction Reduction Requirements (VAR)	
Use aerobic processes at >40°C for 14 days	

Note: For all configurations, pathogen reduction must take place before or at the same time as vector attraction reduction, except when the pH adjustment, percent solids vector attraction, injection, or incorporation alternatives are met.

3. *Applicability of the Part 503 Rule to Other Processing and Composting of Byproducts*

Although not specifically developed for wastes other than biosolids, the pollutant concentration limits in Part 503 (Table 18.4, column on the right) have been used as a reference for discussion of the potential safety and usefulness of a large spectrum of other byproducts and wastes that are being considered for recycling. In many instances, the pollutant limits should serve as a first approximation when determining the potential safety of using the byproduct under consideration (Logan et al., 1997; Walker, 1994, 1998; Walker and O'Donnell, 1991; Walker et al., 1995, 1997).

The ultimate applicability of the Part 503 limits to other byproducts and wastes will depend on their characteristics, especially the presence of oxides of iron (Fe),

Table 18.4 Limits for Metals Regulated Under the Part 503 Rule

Pollutant	Ceiling Concentration Limit[z] (mg·kg^{-1})	Pollutant Concentration Limit[y] (mg·kg^{-1})
Arsenic (As)	75	41
Cadmium (Cd)	85	39
Copper (Cu)	4300	1500
Lead (Pb)	840	300
Mercury (Hg)	57	17
Molybdenum (Mo)	75	to be added
Nickel (Ni)	420	420
Selenium (Se)	100	100
Zinc (Zn)	7500	2800

[z] From Table 1, Part 503, Section 503.13.
[y] From Table 3, Part 503, Section 503.13.

Table 18.5 Limits for Toxic Organic Compounds for Land Applied Biosolids (Calculated But Not Incorporated into the Part 503 Rule)[z]

Pollutant	Pollutant Concentration Limit (µg·kg^{-1})
Aldrin/Dieldrin	2.7
Benzo(a)Pyrene	15
Chlordane	86
DDT/DDD/DDE	120
N-Nitrosodimethylamine	2.1
Heptachlor	7.4
Hexachlorobenzene	29
Hexachlorobutadiene	600
Lindane	84
PCBs	4.6
Toxaphene	10
Trichlorethylene	10,000

[z] From the land application risk assessment for the Part 503 rule.

aluminum (Al), and manganese (Mn) and also of phosphates. The ability of these oxides and phosphates to bind and make pollutants less bioavailable in biosolids is strong (Chaney and Ryan, 1993; Corey et al., 1987; Ryan and Chaney, 1993, 1995; Walker, 1998). The binding effects persist after they have been applied to soils, even after decomposition of the added organic matter. The bioavailability of pollutants in the added biosolids remains much lower than if the pollutants had been added to soil in the form of inorganic salts. An important exception to this strong binding of the added biosolids' metals is when the soil pH is low. At low pHs, added biosolids' metals increase in solubility. Simultaneously, however, there is also an increased solubility of the Al universally present within the mineral fraction of soils. It is the solubilized Al which generally has a dominating toxic effect on most plants. This is why pH management is a normal farming practice.

Examination of field data, gathered as many as 60 to 100 years after the use of irrigation wastewater and/or biosolids on soils, supports the concept of the continuity of the binding effect by the oxides and phosphate fractions of the biosolids even after the biosolids' organic matter has had time to degrade. This binding phenomenon and the extent of applicability of the Part 503 rule to other wastes are discussed more fully by Chaney and Ryan (1993) and Ryan and Chaney (1993).

4. Comparison of EPA Part 503 Pollutant Limits with Limits in Other Countries

Some other countries have pollutant limits for biosolids (and in some cases for composts) that are more or less restrictive than the pollutant limits for biosolids in the U.S. A comparison of limits for a few countries are given in Table 18.6. The EPA Part 503 rule was developed realizing that the use or disposal of biosolids will result in environmental changes, as does the use of all other fertilizers, the construction of buildings, and many other aspects of human activity (Walker, 1998). The biosolids risk assessment process provides a scientific basis for determining acceptable environmental change when biosolids are used or disposed. Acceptable change means that even though changes have occurred as a result of the use or disposal of biosolids, (e.g., increases in nutrients, organic matter, and pollutants), public health and the environment are still protected from the reasonably anticipated adverse effects of pollutants in biosolids. The EPA approach is quite different than the no net degradation and precautionary principle policy-driven approaches used by some other countries (e.g., the Netherlands and Sweden) as a basis for setting limits for biosolids. Countries using this approach generally allow only small, incremental increases of pollutants from the use or disposal of biosolids over background levels of pollutants in the environment.

III. EPA GUIDANCE

In addition to the regulations already discussed, there have been a number of studies and guidances that have yielded information that can be beneficially applied to the production and use of all composts. A number of applicable EPA and United States Department of Agriculture (USDA) guidances are listed in Table 18.7.

IV. OTHER FACTORS IMPACTING THE PRODUCTION AND BENEFICIAL USE OF COMPOSTED BYPRODUCTS

"Nuisance factors" are the most important limitations to the processing and use of composted organic byproducts (Haug, 1990, 1993; Walker, 1991, 1992, 1998). Concerns due to odor, coupled with the public's perception that the use of many byproducts is potentially detrimental, have resulted in lack of acceptance for

Table 18.6 Comparison of U.S. EPA Part 503 Pollutant Limits for Land Applied Biosolids with Limits in Other Countries

Pollutant	U.S. EPA (Part 503)	European Union (EU)	Ontario, Canada	Germany	Netherlands	Sweden	U.S. Median[z]
Cadmium (Cd)	39	20–40	34	10	1.25	2	3.2
Zinc (Zn)	2800	2500–4000	4200	2500	300	800	754
Copper (Cu)	1500	1000–1750	1700	800	75	600	451
Nickel (Ni)	420	300–400	420	200	30	50	19
Lead (Pb)	300	750–1200	1100	900	100	100	78
Arsenic (As)	41	—	170	—	25	—	4.7
Mercury (Hg)	17	16–25	11	8	0.75	2.5	1.6
Chromium (Cr)	—	—	2800	900	75	100	44
PCBs[y]	—	—	—	0.2	—	0.4	—

Note: Values are $mg \cdot kg^{-1}$ dry weight.

[z] From the National Sewage Sludge Survey (NSSS) (U.S. EPA, 1990).

[y] PCBs = polychlorinated biphenyls.

From Harrison, E.A. et al., 1999. *The Case for Caution: Recommendations for Land Application of Sewage Sludge and an Appraisal of the U.S. EPA's Part 503 Sludge Rules*. Cornell Waste Management Institute, Cornell University, Ithaca, New York. With permission.

recycling to land and the demand for careful oversight of land applied byproducts. Nuisances are generally not part of federal regulations and, unfortunately, resources at the federal and state levels are often not sufficient to support comprehensive oversight of those rules that do exist. It is therefore important that a strategy be developed to help overcome these problems.

A. Development of a Holistic Management Strategy for Byproducts and Wastes

Some key components of a strategy for the holistic management of organic and inorganic byproducts and wastes in a manner that is publicly acceptable and agriculturally sustainable, including oversight and gaining public acceptance, are first summarized and then discussed in greater detail. This discussion uses as its primary example the activities that have been undertaken to foster the beneficial use of biosolids. These components of the strategy are (Walker, 1998):

- Becoming proactive
- Implementing a sound communication strategy (Powell Tate, 1993)
- Providing consistent technical information
- Forming partnerships and teams (NBP, 1999a; Thompson et al., 1996; Touart, 1997, 1998) with early and continued involvement of stakeholders (including environmentalists)
- Ensuring oversight and compliance by the active participation of stakeholders (NBP, 1999b)
- Promoting relevant research to provide a sound basis for utilization practices
- Promoting the beneficial use of byproducts and wastes and educating the public about the benefits of byproduct recycling
- Changing the paradigm: Holistic consideration of all byproducts and wastes (Galloway and Walker, 1997; Walker, 1998)

Table 18.7 EPA Guidance That Can Impact Compost Production and Use

Source	Specific Purpose	Contains	Impacts On Composting
EPA 832-R-93-003	A plain English guide to the EPA Part 503 biosolids rule (U.S. EPA, 1994a)	Guidance to the biosolids rule	Has been unofficially applied to all composting and composts
EPA 832-B-93-005	A guide to the biosolids risk assessments for the EPA Part 503 (U.S. EPA, 1995)	Explains scientific risk assessment process	Understanding the risk assessment process helps some parties accept the soundness of rules that impact composting
EPA 832-R-94-009	Biosolids recycling: beneficial technology for a better environment (U.S. EPA, 1994b)	Discusses beneficial use	Makes available in short summary form the benefits of biosolids recycling. Also, provides shortened summary of the biosolids program
EPA/625/R-92/013	Control of pathogens and vector attraction in sewage sludge (U.S. EPA, 1992)	Guidance on pathogen and vector attraction reduction	Enables the regulated community to better understand how to manage pathogens and reduce vector attractiveness
EPA 625/10-84-003	Environmental regulation and technology: use and disposal of municipal wastewater sludge (U.S. EPA, 1989)	Guidance on land application and disposal	Provides perspective
EPA/USDA/NBP 832/D-99-900	A guide to recommended practices for field storage of biosolids and other organic byproducts used in agriculture (U.S. EPA, 1999a)	Draft field storage guidance for comment	Establishes points for critically controlling the field storage process so that it is community friendly. Emphasizes linkage among the generator, transporter, and land applier.
EPA 832-B-92-005	U.S. EPA. 1993. Domestic septage regulatory guidance: a guide to the EPA 503 rule (U.S. EPA, 1993)	Guide for septage management	Provides simplified listing of the Part 503 rules that apply to the use and disposal of septage
EPA-USDA-FDA (1981)	Land application policy for the use of municipal sewage sludge for the production of fruits and vegetables (U.S. EPA, 1981)	Policy as indicated	Positive statement by EPA/the Food and Drug Administration and the U.S. Department of Agriculture on the safety of using biosolids for the production of fruits and vegetables

Table 18.7 EPA Guidance That Can Impact Compost Production and Use (Continued)

Source	Specific Purpose	Contains	Impacts On Composting
49 FR 24358 (6-12-84)	EPA Policy on the beneficial use of biosolids (U.S. EPA, 1984)	Policy as indicated	Statement encouraging beneficial use of biosolids based on research and experience
56 FR 33186 (7-18-91)	InterAgency policy beneficial use of biosolids on federal lands (U.S. EPA, 1991b)	Policy as indicated	Statement encouraging beneficial use of biosolids by USDA, EPA, Interior and other Federal agencies based on research and experience

1. Proactive

One of the most important considerations is to become proactive. That means that each person will actively undertake and participate in those components of this strategy for the agricultural use of organic and inorganic byproducts and wastes for which he or she has any interest or responsibility; think and talk about the role with a colleague; and then begin to carry out this role within 24 hours.

2. Communication

Effective communication is important. An excellent communications strategy was prepared for the Water Environment Federation (WEF) for use by wastewater management professionals regarding the beneficial use of biosolids (NBP, 1999b). The communication objectives included in this strategy are to enhance public perception of biosolids recycling and to gain broad public acceptance of its use; to help advance the goals of EPA and WEF in fostering beneficial use; to support municipal programs that use biosolids; and to improve the environment and protect human health. The strategy stated that the best method of accomplishing this goal is to focus communication efforts on gatekeepers and the process of recycling of biosolids rather than on biosolids themselves. These "gatekeepers" are the individuals and organizations who are asked by the public for their opinion about biosolids utilization because of their expertise, authority, or position. The concern of some members of the public about risks from recycling (Hellstrom and Goran-Dahlberg, 1994) have made undertaking these steps essential, especially with biosolids recycling. Key gatekeeper audiences to educate include (1) academics and agricultural scientists, (2) water quality professionals, (3) farming groups, (4) environmentalists, (5) regulatory officials, and (6) the media.

3. Consistent Technical Information

A sound communication strategy must be supported by sound and consistent technical information. The use of skillfully maintained Internet pages should be fostered. Research must be funded and supported and persons trained so that needed information can be generated. Creativity and teamwork should be fostered. People will need to be found who have the skills and time to provide the needed information, and this information must be validated by a system (such as by peer review) that ensures consistency and technical correctness. These efforts will be aided by incorporating a broad spectrum of stakeholder participation.

4. Stakeholder Partnerships and Teams

EPA has fostered the development of effective biosolids organizations at the local level that can help meet the goals of this strategy. One highly successful regional biosolids organization has been the Northwest Biosolids Management Association (Thompson et al., 1996; Touart, 1997). Another example is the Mountains to Sound Greenway Trust in Seattle, WA, a very successful coalition of public, industry, and municipal authorities. This Trust was formed to use biosolids to maintain a "green corridor" along Interstate 90 from the Cascade Mountains to the Puget Sound (Touart, 1998).

EPA also worked with the WEF, the Association of Metropolitan Sewerage Agencies, the USDA, and other stakeholders to form a National Biosolids Partnership (NBP) (NBP, 1999a). The NBP was formed in 1997 to cooperatively promote environmentally sound and publicly acceptable biosolids management. The NBP will address the needs of its users and the public by addressing scientific, technical, and policy issues; providing assistance to local organizations; fostering professional development and networking; and providing information to all audiences.

5. Oversight and Compliance

With decreased federal and state resources available for the oversight of the utilization of byproducts and wastes like biosolids, stakeholders need to fill the gap. There is an overriding need for stewardship and good practice to minimize nuisances and optimize opportunities for sustained agricultural use. A key component of an implementation strategy is to provide augmentation to state and federal oversight. A high priority of the NBP is to develop a *voluntary* environmental management system (EMS) for biosolids. The EMS (NBP, 1999b) will consist of compliance with all existing regulations as a baseline; a code of good practice; guidance for achieving good practice; a series of uniform steps for developing and measuring the effectiveness of the EMS at critical control/monitoring points; and a system for independent third-party verification of performance.

The "Code of Good Practice" (Table 18.8) is a framework of goals and commitments to guide the production, management, transportation, storage, and use or disposal of biosolids to achieve environmentally-sound, community friendly practice.

6. Relevant Research

EPA has also helped fund a number of research projects about the benefits and risks of using biosolids directly and indirectly through such means as the Water

Table 18.8 Components of the National Biosolids Partnership Environmental Management System (EMS) Code of Good Practice

(1)	To commit to compliance with all applicable federal, state, and local requirements regarding production at the wastewater treatment facility, and management, transportation, storage and use or disposal of biosolids away from the facility
(2)	To provide biosolids that meet the applicable standards for their intended use or disposal
(3)	To develop an environmental management system for biosolids that includes a method of independent third-party verification to ensure effective ongoing biosolids operations
(4)	To enhance the monitoring of biosolids production and management practices
(5)	To require good housekeeping practices for biosolids production, processing, transport, and storage, and during final or use or disposal operations
(6)	To develop response plans for unanticipated events such as inclement weather, spills, and equipment malfunction
(7)	To enhance the environment by committing to sustainable, environmentally acceptable biosolids management practices and operations through an environmental management system
(8)	To prepare and implement a plan for preventative maintenance for equipment used to manage biosolids and wastewater solids
(9)	To seek continual improvement in all aspects of biosolids management
(10)	To provide methods of effective communication with gatekeepers, stakeholders, and interested citizens regarding the key elements of each environmental management system, including information relative to system performance

From National Biosolids Partnership, 1999b. *Improving Biosolids Management Programs and Increasing Public Support through an Environmental Management System.* Water Environment Federation, Alexandria, Virginia. With permission.

Environment Research Foundation. More recent research has shown that the organic and inorganic matrices of the singular and blended waste products can have a major impact on the bioavailability of the pollutants in these matrices (Chaney and Ryan, 1993; Heneghan et al., 1994; Henry and Brown, 1997; Wright et al., 1998). Learning about how these matrices impact pollutant bioavailability favors the development of byproducts management practices that maximize benefit and minimize potential adverse impacts. This knowledge can also guide development of methods to design and tailor-make blends and forms of wastes to enhance their benefits, whether they are used separately or together with conventional fertilizer and soil conditioning materials. Matrix impact research also can help establish appropriate limitations on the use of the compost products (Corey et al., 1987).

7. Promoting Beneficial Use and Educating the Public

The EPA has supported a large number of activities to promote and support the beneficial use of biosolids. Activities have included the aforementioned support of the NBP and an information sharing group (ISG) in New York and New Jersey. This latter group was charged with the task of identifying and then helping overcome factors limiting the beneficial use of biosolids in those two states. As a result of recommendations by the ISG, a large research study was funded in the ecologically sensitive Pinelands of New Jersey; work was undertaken with bankers to help overcome their concern about lending money to farmers who landspread biosolids

(Walker and Allbee, 1994); a series of round table discussions were held in the state of New York to determine a position about the use of biosolids; and Rutgers University and stakeholder efforts were funded in New Jersey to promote sound practice and the beneficial use of biosolids (Rossi, 1994). Another grant was awarded to the National Research Council of the National Academy of Science to determine the appropriateness of the utilization of municipal wastewaters and biosolids (NRC/NAS, 1996). Still another grant went to the WEF to develop the aforementioned communications strategy (Powell Tate, 1993). The WEF was awarded other grants to determine the outlook for the beneficial use of biosolids in each state and to develop state of the knowledge reports on (1) odor as a health issue, (2) appropriate facility design and management techniques for biosolids in wastewater treatment plants to minimize odors when land applying biosolids, and (3) emerging pathogen issues in biosolids, animal manures, and other similar byproducts.

The EPA has sponsored and cosponsored a number of informational workshops to help educate rulemakers, municipalities, their consultants, and the public. The EPA has also produced a series of user-friendly guidances on topics such as the benefits of using biosolids (U.S. EPA, 1994b), the Part 503 Rule (U.S. EPA, 1994a), risk assessment for the Part 503 Rule (U.S. EPA, 1995), and the rules for domestic septage (U.S. EPA, 1991b). The EPA is also supporting excellence via its annual national Beneficial Use of Biosolids Awards Program for operating projects and technology development, research, and public acceptance activities (U.S. EPA, 1999b).

8. Changing the Paradigm

Working together creatively is the essential ingredient for turning environmental problems from inorganic and organic byproducts and wastes into benefits to society. Not only are biosolids, animal manures, food processing wastes, and the recyclable organic fraction of MSW natural fertilizers part of the natural cycle of life, they also are feedstocks for a continuum of uses both agronomic and nonagronomic as discussed by Galloway and Walker (1997). These feedstocks should be and have been used beneficially, but greater attention needs to be given to their properties and the opportunities for altering and tailor making these byproducts as feedstocks for a whole continuum of uses. This greater care and understanding offers great opportunity for their sustainable agricultural use and the potential for enhancing the benefits of other more traditional chemical fertilizer and pesticide products.

The processing of a waste impacts the way a byproduct looks, its odor potential, and its public acceptability. Waste processing methodology also determines end product stability and nutrient content, the extent of reduction of any pathogens, and attractiveness of the byproduct to vectors. The processing also impacts how the end product can be stored and used. Because of all these impacts, the form of processing will lead to varying degrees of restriction on use (Walker et al., 1995).

Knowledge gained from basic and applied research as well as practice is paving the way for creative uses of byproducts and wastes that arise either directly from various municipal and industrial waste treatment processes, or from combinations of special treatments of waste materials to give them desirable characteristics.

REFERENCES

Bastian, R.K. 1997. The biosolids (sludge) treatment, beneficial use, and disposal situation in the USA. *European Water Pollution Control* 7(2):62–79.

Chaney, R. and J. Ryan. 1993. Heavy metals and toxic organic pollutants in MSW-composts: research results on phytoavailability, bioavailability, fate, etc., p. 451–506. In: H.A.J. Hoitink and H.M. Keener (eds.). *Science and Engineering of Composting*. Renaissance Publications, Worthington, Ohio.

Corey, R., L. King, C. Lue-Hing, S. Fanning, J. Street, and J. Walker. 1987. Effects of sludge properties on accumulation of trace elements by crops, p. 25–51. In: A.L. Page, T.G. Logan, and J.A. Ryan (eds.). *Land Application of Sludge: Food Chain Implications*. Lewis Publishers, Chelsea, Michigan.

Galloway, D.F. and J.M. Walker. 1997. An entrepreneurial view of the future use of wastes and byproducts, p. 22–27. In: J.E. Rechcigl and H.C. MacKinnon (eds.). *Agricultural Uses of Byproducts and Wastes*. American Chemical Society, Washington, D.C.

Harrison, E.Z., M.B. McBride, and D.R. Bouldin. 1999. *The Case for Caution: Recommendations for Land Application of Sewage Sludge and an Appraisal of the U.S. EPA's Part 503 Sludge Rules*. Cornell Waste Management Institute, Cornell University, Ithaca, New York.

Haug, R.T. 1990. An essay on the elements of odor management. *Biocycle* 31(10):60–67.

Haug, R.T. 1993. *The Practical Handbook of Compost Engineering*. Technomics Publishing, Lancaster, Pennsylvania.

Hellstrom, T. and A. Goran-Dahlberg. 1994. Swedish experience in gaining acceptance for the use of biosolids in agriculture, p. 9–25 to 9–36. In: Water Environment Federation (WEF). Proceedings of the WEF Specialty Conference on *The Management of Water and Wastewater Solids for the 21st Century: A Global Perspective*, Washington, DC, 19–21 June 1994. WEF, Alexandria, Virginia.

Heneghan, J.B., D.M. Smith, Jr., H.W. Mielke, R.L. Chaney, and J.M. Walker. 1994. *Composts/Biosolids Reduced Soil Lead Bioavailability in Rat Feeding Studies*. The National Conference of the Composting Council, Washington, D.C., Nov. 1994. The Composting Council, Alexandria, Virginia. (Abstract).

Henry, C. and S. Brown. 1997. Restoring a superfund site with biosolids and fly ash. *BioCycle* 38(11):79–83.

Hoitink, H.A.J., M.J. Boehm, and Y. Hadar. 1992. Mechanisms of suppression of soilborne plant pathogens in compost-amended substrates, p. 601–621. In: H.A.J. Hoitink and H.M. Keener (eds.). *Science and Engineering of Composting: Design, Environmental, Microbiological and Utilization Aspects*. Renaissance Publications, Worthington, Ohio.

Logan, T.J. and R.L. Chaney. 1983. Utilization of municipal wastewater and sludges on land-metals, p. 235–326. In: A. Page, T. Gleason, J. Smith, I. Iskander, and L. Sommers (eds.). *Proceedings of a 1983 Workshop on Utilization of Municipal Wastewater and Sludge on Land*, Denver, Colorado, 23–25 Feb. 1983. University of California, Riverside, California.

Logan, T.J., B.J. Lindsey, and S. Titko. 1997. Characteristics and standards for processed biosolids in the manufacture and marketing of horticultural fertilizers and soil blends, p. 63–71. In: J.E. Rechcigl and H.C. MacKinnon (eds.). *Agricultural Uses of Byproducts and Wastes*. American Chemical Society, Washington, D.C.

Millner, P.D., S.A. Olenchock, E. Epstein, R. Rylander, J. Haines, J. Walker, B.L. Ooi, E. Horne, and M. Maritato. 1994. Bioaerosols associated with composting facilities. *Compost Science and Utilization* 2(4):6–57.

National Biosolids Partnership (NBP). 1999a. *NBP 1998–99 Annual Report.* Alexandria, Virginia.

National Biosolids Partnership (NBP). 1999b. *Improving Biosolids Management Programs and Increasing Public Support through an Environmental Management System.* Water Environment Federation, Alexandria, Virginia.

National Research Council/National Academy of Science (NRC/NAS). 1996. *Use of Reclaimed Water and Sludge in Food Crop Production.* National Academy of Science, Washington, D.C.

Powell Tate. 1993. *Communications Plan on Biosolids: Research Findings.* Report prepared by the Powell Tate firm for the Water Environment Federation, Alexandria, Virginia. July, 1993.

Rossi, D. 1994. The New Jersey-New York Information Sharing Group approach to overcoming factors limiting the beneficial use of biosolids, p. 9–37 to 9–43. In: Water Environment Federation (WEF). Proceedings of the WEF Specialty Conference on *The Management of Water and Wastewater Solids for the 21st Century: A Global Perspective,* Washington, D.C., 19–21 June 1994. WEF, Alexandria, Virginia.

Ryan, J. and R. Chaney. 1993. Regulation of municipal sewage sludge under the Clean Water Act Section 503: a model for exposure and risk assessment for MSW-compost, p. 422–450. In: H.A.J. Hoitink and H.M. Keener (eds.). *Science and Engineering of Composting: Design, Environmental, Microbiological and Utilization Aspects.* Renaissance Publications, Worthington, Ohio.

Ryan, J. and R. Chaney. 1995. Issues of risk assessment and its utility in development of soil standards: The 503 methodology: an example. In: *Proceedings of the Third International Symposium on Biogeochemistry of Trace Elements,* Paris, France, 15 May 1995.

Thompson, D., C. Ready, and T. Moll. 1996. From sludge management to biosolids recycling. *Biocycle* 37(2):70–72.

Touart, A.P. 1997. Creating an effective biosolids information network. *BioCycle* 38(8):74–77.

Touart, A.P. 1998. Winning biosolids support. *BioCycle* 39(2):86–90.

U.S. EPA, Food and Drug Administration, and U.S. Department of Agriculture. 1981. *Land Application of Municipal Sewage Sludge for the Production of Fruits and Vegetables: A Statement of Federal Policy and Guidance.* EPA 832-R-81-900; NTIS PB95157483. United States Department of Commerce, National Technical Information Service (NTIS), Springfield, Virginia.

U.S. EPA. 1984. Policy on municipal sewage sludge management. 49 *Federal Register* 24358, 12 June 1984.

U.S. EPA. 1989. *Environmental Regulation and Technology: Use and Disposal of Municipal Wastewater Sludge.* NTIS PB80/200/546. Office of Water, Springfield, Virginia.

U.S. EPA. 1990. National Sewage Sludge Survey; Availabilty of information and data, and anticipated impacts of proposed rule on proposed regulations, proposed rule. Vol. 55, No. 218 *Federal Register* 47210–83, 9 Nov. 1990.

U.S. EPA. 1991a. *Cooperative Testing of Municipal Sewage Sludges by the Toxicity Characteristic Leaching Procedure and Compositional Analysis.* EPA 430/09-91-007. Office of Wastewater Management, Washington, D.C.

U.S. EPA. 1991b. Interagency policy on beneficial use of municipal sewage sludge on Federal Land: Notice. 56 *Federal Register* 33186-88, 18 July 1991.

U.S. EPA. 1992. *Control of Pathogens and Vector Attraction in Sewage Sludge.* EPA/625/R-92/013. Office of Research and Development, Cincinnati, Ohio.

U.S. EPA. 1993. *Domestic Septage Regulatory Guidance: A Guide to the EPA 503 Rule.* EPA 832-B-92-005. Office of Wastewater Management, Washington, D.C.

U.S. EPA. 1994a. *A Plain English Guide to the EPA Part 503 Biosolids Rule*. EPA/832/R-93/003. Office of Wastewater Management, Washington, D.C.

U.S. EPA. 1994b. *Biosolids Recycling: Beneficial Technology of a Better Environment*. EPA 832-R-94-009. Office of Wastewater Management, Washington, D.C.

U.S. EPA. 1995. *A Guide to the Biosolids Risk Assessments for the EPA Part 503 Rule*. EPA/832-B-93-005. Office of Wastewater Management, Washington, D.C.

U.S. EPA. 1999a. *Biosolids Generation, Use and Disposal in the United States*. Office of Solid Waste, Washington, D.C. Final Draft.

U.S. EPA. 1999b. *Nomination Guidance: 1999 Beneficial Use of Biosolids Awards Program: for Operating Projects, Technology Development, Research and Public Acceptance*. EPA 832-B-99-001. Office of Wastewater Management, Washington, D.C.

Walker, J.M. 1991. Fundamentals of odor control. *BioCycle* 32(9):50–55.

Walker, J. M. 1992. Control of composting odors, p. 185–218. In: H.A. J. Hoitink and H.M. Keener (eds.). *Science and Engineering of Composting: Design, Environmental, Microbiological and Utilization Aspects*. Renaissance Publications, Worthington, Ohio.

Walker, J.M. 1994. Production, use, and creative design of sewage sludge biosolids, p. 67–74. In: Soil Science Society of America (SSSA). *Land Utilization and the Environment*. SSSA, Madison, Wisconsin. Miscellaneous publication.

Walker, J.M. 1998. Biosolids management, use and disposal, p. 768–813. In: R.A. Meyers (ed.). *Encyclopedia of Environmental Analysis and Remediation*. John Wiley, New York.

Walker, J.M. and M.J. O'Donnell. 1991. Comparative assessments of MSW compost characteristics. *BioCycle* 32(8):65–69.

Walker, J.M. and R.N. Allbee. 1994. Yes, but what about the liability? *Water Connection*. 11(2):10–11. New England Interstate Water Pollution Control Commission, Wilmington, Massachusetts.

Walker, J.M., R.E. Lee, and R K. Bastian. 1995. The impact of the part 503 and state biosolids rules on distribution and marketing practices and costs. In: *Water Environment Federation (WEF). Proceedings of the 9th Annual Biosolids and Residuals Conference of the WEF*, Kansas City, Missouri. WEF, Alexandria, Virginia.

Walker, J.M., R.M. Southworth, and A. Rubin. 1997. EPA regulations and other stakeholder activities affecting the agricultural uses of byproducts and wastes, p. 72–90. In: J.E. Rechcigl and H.C. Mackinnon (eds.). *Agricultural Uses of Byproducts and Wastes*. American Chemical Society, Washington, D.C.

Wright, R.J., W.D. Kemper, P.D. Millner, J.F. Power, and R.F. Korsack (eds.). 1998. *Agricultural Uses of Municipal, Animal, and Industrial Byproducts*. U.S. Department of Agriculture, Agricultural Research Service, Beltsville, Maryland. Conservation Research Report 44.

Index

A

B

C

D

E

F

G

H

I

J

K

Q

R

S

T

U

V

W

Y

Z

Milton Keynes UK
Ingram Content Group UK Ltd.
UKHW021834071024
449327UK00021B/1493

9 780367 397593